COMPUTATIONAL METHODS FOR LARGE MOLECULES AND LOCALIZED STATES IN SOLIDS

COMPUTATIONAL METHODS FOR LARGE MOLECULES AND LOCALIZED STATES IN SOLIDS

Proceedings of a Symposium, Held May 15-17, 1972,
at the IBM Research Laboratory, San Jose, California

Edited by

F. Herman, A. D. McLean, and R. K. Nesbet

Large-Scale Scientific Computations Department
IBM Research Laboratory, San Jose, California

ℚℙ PLENUM PRESS • NEW YORK-LONDON • 1973

Library of Congress Catalog Card Number 72-93442

ISBN 0-306-30716-2

© 1973 Plenum Press, New York
A Division of Plenum Publishing Corporation
227 West 17th Street, New York, N.Y. 10011

United Kingdom edition published by Plenum Press, London
A Division of Plenum Publishing Company, Ltd.
Davis House (4th Floor), 8 Scrubs Lane, Harlesden, London,
NW10 6SE, England

PREFACE

During the past few years, there has been dramatic progress in
theoretical and computational studies of large molecules and localized states in solids. Various semi-empirical and first-principles
methods well known in quantum chemistry have been applied with
considerable success to ever larger and more complex molecules,
including some of biological importance, as well as to selected
solid state problems involving localized electronic states. Increasingly, solid state physicists are adopting a molecular point
of view in attempting to understand the nature of electronic states
associated with (a) isolated structural and chemical defects in
solids; (b) surfaces and interfaces; and (c) bulk disordered solids,
most notably amorphous semiconductors.

Moreover, many concepts and methods already widely used in
solid state physics are being adapted to molecular problems. These
adaptations include pseudopotentials, statistical exchange approximations, muffin-tin model potentials, and multiple scattering and
cellular methods. In addition, many new approaches are being de-
vised to deal with progressively more complex molecular and localized electronic state problems.

In view of these developments, IBM San Jose Research Laboratory,
in conjunction with IBM World Trade Corporation, organized a research symposium on the application of the computational methods of
quantum chemistry and solid state physics to the study of electronic
properties of complex molecules, localized imperfections in solids,
and disordered condensed systems. Dr. Frank Herman served as
Symposium Chairman. The present volume consists of papers presented at this symposium, which was held on May 15 to 17, 1972 at
the IBM San Jose Research Laboratory.

An important purpose of this symposium was to bring together
active workers in theoretical chemistry and solid state theory for
an exchange of ideas and methodology across the communication
barrier that has traditionally separated these two disciplines.
In both of these fields theoretical and computational methods have

reached a high level of sophistication, and most theoreticians, for natural reasons, normally work within the framework of only one of these disciplines.

The organizers of this symposium believe that such meetings are essential for minimizing provincialism, for making people aware of techniques used in one area which may be ready for exploitation in another, and for exploring in depth the regions that overlap. The subject of provincialism in science is considered further in the Banquet Speech by Professor George S. Hammond. One of the highlights of the symposium was a panel discussion at which representatives of many different schools of thought presented their views and compared notes. This panel discussion is reproduced in the present volume.

The papers included here address the following topics:

(a) scientific goals and challenges for computational studies of complex molecules and localized states in solids;

(b) effectiveness of different approaches to these goals;

(c) discussion of relative accuracy and practicality of traditional methods of quantum chemistry;

(d) potential applications of new computational methods directed specifically to complex molecules and localized states in solids;

(e) discussion of areas of research requiring a synthesis of molecular and solid state concepts, for example, exciton states in molecular crystals, biological macromolecules, polymers, molecules in crystalline environments, localized defects in crystals, surface states, and localized states in disordered solids.

The organizers of this symposium are indebted to many persons who have contributed to its success. We are particularly grateful to Dr. Andrew H. Eschenfelder, IBM San Jose Research Laboratory and Mr. Gary S. Kozak, IBM World Trade Headquarters, New York for their encouragement and financial support; and Mr. Warren C. Edwards, IBM San Jose Research Laboratory, for his expert handling of the local arrangements.

We are also grateful to many of our colleagues within the IBM World Trade Corporation for their cooperation in bringing selected speakers and participants to the symposium. We are particularly grateful to the following: Mr. P. de Blanpre, IBM France; Mr. L. O. Bunka, IBM Canada; Mr. O. Cilius-Nielsen, IBM Denmark; Mr. F. G. Insinger and Mr. J. A. van der Pool, IBM Netherlands; Mr. R. Jacqmin, IBM Belgium; Mr. J. S. Nist, IBM Mexico; Mr. J. J. Peppercorn. Mr.

S. Quigley, and Mr. C. Squire, IBM United Kingdom; Dr. P. Schweitzer, IBM Germany; Mr. A. Serrano, IBM Spain; Dr. J. Vogel, IBM Switzerland; and Mr. G. Wethal, IBM Norway. We are also indebted to Drs. William J. Turner and Bernard J. van der Hoeven, IBM Research, Yorktown Heights for administrative support; and Mr. Robert N. Ubell, President, Plenum Publishing Corporation, for his cooperation.

Finally, we are indebted to many of our colleagues at IBM San Jose Research Laboratory for their important contributions. In particular, thanks are due to Miss Colleen de Long, Mrs. Linda Lopez, Mrs. Jacqueline Mendenhall, and Mrs. Patricia B. Rodgers for extensive typing and editorial assistance; Mrs. Rodgers again for transcribing the tape recordings of the symposium; Mr. Jack De Lany, Mr. Brad Hall, and Mrs. Lorraine Riddle for artistic and graphics support; and Mr. Rick W. Kipp, Mr. Rudy M. Martinez and Mr. Donald E. Schreiber for help with the visual aids.

This effort can be considered a success if the symposium and the present volume help to open up channels of communication between quantum chemists and theoretical solid state physicists.

Frank Herman, A. D. McLean, and R. K. Nesbet
Symposium Proceedings Coeditors

CONTENTS

3. Computational Methods I

4. Computational Methods II

5. Localized States
and Disordered Solids I

6. Localized States
and Disordered Solids II

7. Banquet Speech

8. Symposium Summary

WELCOMING REMARKS

A. H. Eschenfelder, Director

IBM Research Laboratory, San Jose, California 95114

     I just want to take a few minutes to tell you how pleased
I am that you're here for this particular meeting. We have been
in this building for a year now, and one of the advantages of
this new building is that we now have facilities that enable us
to hold conferences on subjects that are dear to our hearts. I
think you all know that scientific computation isn't something
new in San Jose Research Lab; it's been a major theme in this
laboratory for at least a decade. Enrico Clementi, Bob Nesbet,
Doug McLean and others have established a tradition with us of
trying to advance the methods for computing properties of atoms
and molecules. In recent years they have been joined by a number
of other people whom I think you know and who have tried to extend this work to a broader field. In this last five years we
have had a steady increase in the computer power that's available
for these studies. Five years ago we had a 360/50, then we moved
to a /65, a /91, and now we have our 195. During that time we've
also had some very significant collaborations with many of you
outside the Laboratory, and that's been very gratifying to us.

     We continue to be confident that effective use of computation
can have a very dramatic effect on science. And we're determined
to do what we can to play our part in making that come true. I
have a few basic premises regarding work in this area, and I want
to make them known to you. First of all, we do recognize that
bigger and more powerful computers are needed in order to compute
accurately the more complicated molecular structures. But on the
other hand, I'm convinced that more progress is made by the imaginative structuring of a problem and the ways you go about attack-
ing that problem than is made just through the improvements in
raw computer power. I've seen Enrico Clementi get orders of
magnitude reduction in the amount of computer time it took for a

particular problem even on the same computer by changing his
approach to the computation. And so we very much believe that
the emphasis has to be on the imaginative approach to these problems, taking advantage of what increases in computing power come
along, but not waiting for that or relying on it.

Secondly, we understand that a wide variety of computing
approaches is necessary. These range from the very precise ab
initio methods to those which involve a very substantial amount
and different kinds of approximation. Which method we'll use
depends on the purpose of the calculation, the complexity, and
many other factors. Certainly we don't believe that any one
method is the right one; which one is right depends on the particular situation.

We believe that the development of computational methods has
to go hand in hand with advances in theoretical techniques and
also in collaboration with crucial experiments, because these
three factors help to direct and refine each other as they go
forward hand in hand. We think that the computations, theoretical
development and experiments ought to be related to really critical
problems that open up significant new scientific understanding,
and really reveal the value and power of the methods. I don't
know who cares what we do unless we do things that are really
important and uncover knowledge which wouldn't have been possible
without the new techniques.

We also believe that solid state scientists have very much to
learn from molecular scientists in terms of the concepts of the
field and also the methods that are used. And we believe the
reverse is true, that molecular scientists have a lot to learn
from the solid state scientists. We recognize that progress is
being made on all of these fronts, certainly in the development of
the methods and certainly also in the physical and chemical insights, especially in more complex molecular and solid configurations.

So with all these premises and beliefs, we are very happy
to have a meeting like this in our laboratory, and I'm very grateful to Frank Herman for showing the leadership to bring this meeting together, and also to Frank's colleagues who helped and those
of you who've come here to participate. The conference has been
generated in order to bring together a number of people who can
learn from each other. We've been particularly anxious to bring
together the solid state scientists and the molecular scientists;
to bring together proponents and workers in the different computational methods where rapid progress is being made; and certainly
to bring together those people who can stimulate us in our orientation to important problems. I hope this meeting really does
satisfy a real need for all of you, and that it will be a very
productive meeting for you.

1. Scientific Challenges I

INTRODUCTORY REMARKS

Robert S. Mulliken

University of Chicago

Dr. Eschenfelder has so well described the origin and objectives of this symposium that I feel little is left to me at this initial session but to call on the speakers. However, I will take a little time to review some aspects of past history, plus a few pointers toward the future. (In this written account, I am taking the liberty of "extending" the brief remarks which I actually made).

Everybody knows Dirac's statement to the effect that with the advent of quantum mechanics the underlying physical laws necessary for the mathematical theory of a large part of physics and the whole of chemistry are completely known, the difficulty being only that the exact application of these laws leads to equations much too complicated to be soluble. Dirac wasn't then foreseeing what the computer can do. Further, since the time of his statement a number of people have worked very hard toward better and better approximations to the solution of the difficult equations, including the development of numerical methods of dealing with them. We've already come a long way.

Now let me quote from a 1947 paper (1) which I gave at a symposium on molecules late in 1946. We did not on that occasion consider solid state problems. However, Slater had already been dealing with both solid state and molecular problems and the relation between them, as he has continued to do and as a paper tomorrow by Johnson and Connolly will illustrate. Bloch had introduced LCAO molecular orbital theory into the study of solids, and Wigner concerned himself with both molecules and solids, as did Sommerfeld even earlier using the old quantum theory. Seitz's book, "The

Modern Theory of Solids" published in 1940 contains much wisdom
about the use of SCF (self-consistent-field) theory for atoms and
molecules as well as for solids.

This morning's session is devoted mainly to molecules, so I
shall confine my remarks mainly to these. To be sure, Professor
Robinson's talk on electronic excitations in molecular crystals
introduces a theme in which both molecular and solid-state theory
are involved. And later in the symposium, the subject of molecules
as localized spots in solids will come up.

Now I will quote in part from my 1947 paper: "Let us review
in outline the quantum-mechanical methods that may be used in
understanding molecules and their properties.

"As an ideal goal, we might ask for the electronic eigenfunction and energy of every important molecule or class of
molecules in its ground state and its important excited electronic
states, as a function of all possible variations in the nuclear
configuration -- thereby also including all isomers. We should
also like to know how the eigenfunctions and energies vary under
the influence of force fields, particularly those resulting from
the approach of other molecules. From such information, desired
behavior and properties could be deduced by relatively simple
calculations.

"The complete attainment of such a goal would surely take
decades, if not generations. It has taken centuries to apply
Newton's laws to practical situations, and we are not through yet.
In the application of quantum mechanics we have only started.
While working toward our ultimate goal, less complete information
will be very useful at first. This can be obtained by applying
quantum mechanics in a variety of ways differing in their relative
quantitativeness and in their relative content of pure theory and
of empirical data.

"These applications in respect to the proportion of theory
used may be divided roughly into four methods:

(1) The qualitative method: judicious application of systematic qualitative and semiquantitative interpretations
throughout the vast realm of inorganic and organic compounds
in the solid, liquid, and vapor states.

(2) The semiempirical method: systematic theoretical computations of a relatively simple character, based on approximate
wave functions of the molecular orbital or of the atomic
orbital type, but making frequent use of empirical data as a
substitute for, or even an improvement on, certain theoretical integrals. Empirical and semiempirical information

likely to be useful for this purpose includes: frequencies,
intensities, and other data from electronic spectra; excitation
and ionization potentials, probabilities, and processes;
dipole moments; thermal data; electronegativities derived from
thermal data, dipole moments, and other sources; interatomic
distances and bond angles from electron-diffraction, infrared,
and microwave investigations; force constants and other information on bond properties from infrared and Raman spectra,
including information on bond moments and polarizabilities as
a function of internuclear distance, obtained from intensity
measurements and other spectroscopic data.

(3) The approximate theoretical method: purely theoretical
computations, using approximate molecular eigenfunctions of
the molecular orbital or of the atomic orbital type.

(4) The accurate theoretical method: completely theoretical
computations, using really accurate molecular eigenfunctions.

"Computations of all types may be capable of very considerable
extension with the use of the superpowerful computing machines now
being developed.

"Approximate theoretical methods for describing and computing
electronic structures and spectra of molecules include, as is well
known, the AO (atomic orbital) and the MO (molecular orbital)
approximation methods. It is characteristic of the molecular
problem that both of these methods, in different ways, are disappointingly far from exact in quantitative applications. More
exact methods are available, but are far too complicated for application, at least for the present and probably for a long time to
come, to any but very simple molecules".

At that time we were thinking in terms of the simplest approximations to AO's and MO's. I for one had no idea how far and fast
the development of computers would go. For π-electron systems the
Hückel method was beginning to be recognized as very helpful in
organic chemistry (2).

In 1947 I was still favoring semi-empirical methods but was
becoming dissatisfied with the over-simple Hückel theory. Roothaan
took the bull by the horns and developed his SCF-LCAO-MO theory (3).
The great problem then was the evaluation of the interelectronic
repulsion integrals, which in 1950 seemed difficult even for the
two-center cases.

Quoting from a later paper (4): "After the early years of
quantum mechanics, the work of a number of Japanese, English,
American, and other investigators was directed toward breaking the

bottleneck of the difficult integrals, but it was only in the 50's that really substantial progress was made. Among the most active workers were Kotani and his group in Japan, Boys in Cambridge, Coulson and his group at Oxford, Löwdin and his group in Uppsala, Slater's group at M.I.T., and our own group at Chicago".

Scherr at Chicago published his SCF calculations on N_2 in 1955 after two years of work with desk machines. Meantime Boys at Cambridge stimulated Sahni to do an all-electron SCF-LCAO-MO calculation on BH. These were minimal basis ab initio calculations at one internuclear distance. (Who first used the expression ab initio? Was it Robert Parr? (5))

Quoting again: "A major and indeed crucial step beyond the development of formulas for molecular integrals was the programming for large electronic digital computers of the otherwise still excessively time-consuming numerical computation of these integrals, and of their combination to obtain the desired molecular wave functions and related molecular properties. The pioneering work in this field was that of Boys at Cambridge, England". Our paper (4) goes on to summarize early progress on machine computations at Chicago, including that of McLean in the efficient programming of 3-center and 4-center integrals for linear molecules. Since that time, more and more people have got into the ab initio game as computers have become bigger, faster, and better.

Meantime, however, more and more people have been making semi-empirical SCF computations. This started perhaps with the 1938 paper of Goeppert-Mayer and Sklar on benzene, but really got going in a big way with Pariser and Parr, and Pople, on π-electron systems (6). Now we have CNDO/1 and /2, INDO, MINDO, and more besides. It is interesting to note here that digital computers are now being used for the semi-empirical as well as the ab initio calculations.

Professor Pople at a Sanibel meeting once drew a curve whose ordinate indicated the practicable ratio of ab initio to semi-empirical character, while the abscissa represented the number of electrons per molecule. The curve fell rapidly as the number of electrons increased. If Pople's curve had been revised annually, it would have been seen to move rather quickly toward the right, toward ab initio computation of bigger and bigger molecules. So in a million years -- well, no, ten years, five years -- I think we will be doing pretty big molecules -- and that eventually includes solids -- by pretty exact methods. Clementi especially has been pulling us in that direction. Even Pople is already doing some ab initio calculations.

Personally I am happier with these and I might add that there is still much to be done even with small molecules using large computers. Especially is this true as soon as one attempts calculations relevant to the mechanism of chemical reactions. However, for reactions of real chemical interest we are still far to the left on Pople's curve, and Dr. Salem in his talk on Organic Transition States later this morning will doubtless still be in the semi-empirical region. Semi-empirical again, but in a different way, will be Dr. Little's talk on Molecular Modeling by Computer. I understand that he will discuss models of polymeric aggregates, a problem akin to solid state phenomena. Incidentally, he is using minicomputers.

REFERENCES

(1) R. S. Mulliken, Chem. Rev. $\underline{41}$, 201 (1947).

(2) Cf. A. Streitwieser, Jr., "Molecular Orbital Theory for Organic Chemists" (Wiley, 1961).

(3) C. C. J. Roothaan, Rev. Mod. Phys. $\underline{23}$, 69 (1951).

(4) R. S. Mulliken and C. C. J. Roothaan, Proc. Nat. Acad. Sci. $\underline{45}$, 394 (1959).

(5) See the first page of the article by R. G. Parr, D. P. Craig, and I. G. Ross, J. Chem. Phys. $\underline{18}$, 1561 (1950).

(6) R. G. Parr in "The Quantum Theory of Molecular Electronic Structure" (W. A. Benjamin, 1963) gives a good account of developments through 1962, including selected reprints.

QUANTUM CHEMISTRY. THEORY OF GEOMETRIES AND ENERGIES OF SMALL MOLECULES.

John A. Pople

Department of Chemistry, Carnegie-Mellon University,

Pittsburgh, Pennsylvania 15213

INTRODUCTION

In recent years there has been an increasing interest in quantitative attempts to predict molecular structures and stabilities by purely theoretical quantum mechanical techniques. For the most part, such attempts have been based on molecular orbital techniques at various levels of approximation. In this paper, the current status of such theories for geometries and energies of small organic molecules will be reviewed.

It is useful to regard a well-defined computational method as a theoretical model chemistry, if it is such that it can be applied to give a unique wave function and energy for any set of nuclear positions and number of electrons. If such a model is clearly specified and if it is sufficiently simple to apply repeatedly, it can be used to generate molecular potential surfaces, equilibrium geometries, energies and other physical properties. Each such theoretical model can be explored in detail and compared with the real world in considerable detail. If it is found to make predictions in good agreement with experimental observations, its predictions (in areas where data is lacking) clearly gain credibility. Each such level of theory, therefore should be thoroughly tested and characterized before the significance of its predictions is assessed.

This manuscript is mainly concerned with various levels of molecular orbital theory, their merits and their performances in structural studies. Emphasis will be laid on methods developed at Carnegie-Mellon University, although parallel developments have been taking place elsewhere. Some examples of the extent of agreement between theoretical and experimental structures will be given. Finally, the way in which theory can now be applied to molecular energetics will be reviewed.

A SURVEY OF MOLECULAR ORBITAL MODELS

Within the framework of molecular orbital theory, three general types of computational method may be distinguished. The first type is semiempirical in which a number of simplifying approximations are made in the formal analysis and the remaining parameters are chosen partly by appeal to appropriate experimental properties. Such theories are usually easiest to apply because of their simplicity, but, on the other hand, they necessarily involve some subjective judgments in the choice of parameters. The other types of theory are ab initio in the sense that they attempt to avoid any appeal to experiment (except for the values of fundamental constants). They may be subdivided broadly into single-determinant methods with a single molecular orbital configuration and more complicated configuration interaction methods in which the wave function is written as the sum of several determinants.

It is convenient to deal with single determinant molecular orbital methods first. Here it is normal to assign electrons to spatial molecular orbitals ψ_i which are in turn approximated as linear combinations of a set of basis functions ϕ_μ,

$$\psi_i = \sum_\mu c_{\mu i} \phi_\mu \qquad (1)$$

The coefficients $c_{\mu i}$ are then chosen by the variational method (minimizing the calculated total energy). To specify the method, we need to prescribe the type of single determinant and then the basis set. These aspects may be discussed separately.

The most widely used form of single-determinant wave function is the spin-restricted type in which electrons are assigned singly or in pairs (α and β) to real orbitals ψ_i. For the normal diamagnetic closed shell system, all ψ_i are doubly occupied or empty. The corresponding equations for the $c_{\mu i}$ coefficients were given by Roothaan.[1] A second, more general possibility is to use a spin unrestricted type where electrons of

α and β spin may occupy completely different molecular orbitals.
The corresponding equations for the $c_{\mu i}$ can be generalized to
two coupled sets of equations for α and β molecular orbitals.[2]
A third generalization permits the $c_{\mu i}$ to be complex. Even though
it may be shown that the exact solution of the Schrodinger equation may be taken as real, the variationally best single determinant may involve complex molecular orbitals. This happens in
situations where two incompletely occupied orbitals are close in
energy.[3] As a rule, it would seem to be desirable to take the
most general form of determinantal wave function. In practice,
of course, this will often turn out to reduce to the restricted
type.

Turning now to the various semiempirical and ab initio
methods with single determinant wave functions, these may be
arranged in a hierarchy as shown in Table 1. The simplest type

---

TABLE 1. TYPES OF BASIS SET AND MOLECULAR ORBITAL APPROXIMATION

Type of Basis and Approximations	Examples
AB INITIO	
Perfect orbitals	
Split shell + polarization	6-31G + d
Split shell	4-31G, 6-31G
Minimal	STO, STO-3G
SEMI-EMPIRICAL	
Zero differential overlap	CNDO, INDO, MINDO
Independent electrons	EHT

---

of basis set for ab initio work is minimal, that is consists of
just enough atomic-orbital type functions to describe the ground
state of the corresponding atom (1s for hydrogen, 1s, 2s, 2py,
2pz for carbon etc.). An attractive simple minimal set of
analytical functions is that of Slater-type functions or orbitals
(STO).[4]

$$\phi_{1s} = (\zeta^3/\pi)^{1/2} \ \exp \ (-\zeta_1 r)$$

$$\phi_{2s} = (\zeta^5/96\pi)^{1/2} \ r \ \exp \ (-\zeta_2 r)$$

$$\phi_{2px} = (\zeta^5/32\pi)^{1/2} \ x \ \exp \ (-\zeta_2 r) \tag{2}$$

The main difficulty in using such a set is the evaluation of the appropriate integrals. A nearly equivalent procedure is the use of a set of Slater-type functions replaced by a least-squares-fitted combination of N gaussian type functions (STO-NG).[5-8] Here all integrals can be evaluated analytically. The STO-3G basis has now been used extensively in theoretical studies of molecular structure and we shall comment on its performance in the next section.

Below the ab initio minimal basis level, we list the common semiempirical procedures. Almost universally these simulate the single-determinant minimal type of calculation. The crudest level of semiempirical theory is the independent electron model which neglects details of electron repulsion and assumes that all electrons move in the same effective potential field. The most widely used version is the extended Huckel theory (EHT) of Hoffman.[9] The other main type is the zero-differential-overlap methods which are somewhat less approximate and attempt to allow for electron repulsion by appropriate spherical averaging of atomic charge densities. In some versions such as complete neglect of differential overlap[10,11] (CNDO) and intermediate neglect of differential overlap[12] (INDO), parameters are chosen so that the results parallel those of ab initio minimal basis calculations. In other versions such as modified INDO (MINDO) introduced by Dewar,[12] the parameterization is made so that the results fit known experimental data for some molecules. Thus in MINDO, the deficiencies of the single-determinant minimal level are compensated by modifying existing integrals rather than by introducing new features into the theory. Such a model procedure is likely to do quite well for systems close to those used for parameterization, but it remains to be seen how successful it will be in overall prediction.

We now turn to the use of basis sets which are superior to the minimal level. The first improvement is to replace each minimal basis function by a pair of ϕ_μ (inner and outer parts).

Such basis sets are sometimes referred to as 'double-zeta'. At
a slightly simpler level, only the valence shell functions may
be split (split valence basis). Several basis sets of this
sort, both exponential[14] and gaussian[15] have been proposed and
used. We have introduced[16] a contracted gaussian set (4-31G)
which is intended to be a simple split valence basis for extensive organic applications. It has a 4-gaussian inner shell and
valence shell functions with 3-gaussian inner and 1-gaussian
outer parts. A corresponding 6-31G set with an improved inner
shell description is found to give similar results.[17]

Beyond the split shell level, the next step is the addition
of polarization functions with higher orbital angular quantum
number (p-functions on hydrogen, d-functions on carbon, etc.).
Although a number of molecules have been studied with polarization functions, no such single basis has been extensively
applied in a systematic manner. We have recently begun a study
of this sort[18] using a single set of gaussian polarization
functions added to 6-31G.

The final level of basis is a very large one which
approaches completion so that the molecular orbitals ψ_i become
perfect. With a restricted determinantal wave function, this
is the Hartree-Fock limit. In practice, however, this limit
can only be achieved for very small molecules and it will be
exceedingly difficult to implement in an extended manner.
Nevertheless, it remains an important conceptual stage of
theoretical development.

All the methods listed up to this point are based only on
a single configuration of electrons. It is well recognized
that this has considerable deficiencies even at the Hartree-Fock
limit. It is therefore very desirable to develop other systematic models which use configuration interaction (CI). However, although a great many wave functions have been obtained
by CI techniques, there are difficulties in specifying a method
which is genuinely systematic and practicable. The ideal procedure would be to allow for interaction between all configurations that can be constructed from a particular basis set, but
this involves far too much computation except in the very simplest cases. Alternatively, all double excitations from occupied to unoccupied molecular orbitals may be included, but this
still involves a very large number of configurations. Overall,
an adequate model-type theory with CI does not yet seem to
exist. Particular studies with particular configurations are
possible, but there is no general procedure. In the remainder
of this paper CI wave functions will not be used, but we will
rather attempt to assess how well single-determinant theories
work in dealing with equilibrium geometries and energies.

EQUILIBRIUM GEOMETRIES

Having set up a theoretical model chemistry, it should then
be applied extensively to molecules which are well characterized
experimentally so that its performance and limitations are documented. Only then will its predictions in non-experimental
areas become significant. In this section, we will discuss the
results of some such tasks on the geometries and energies of
small organic molecules. Three types of calculation are used--
the semiempirical INDO scheme,[12] the minimal basis STO-3G[8] and
the split valence basis 4-31G.[16] These studies are sufficiently
extensive for some general conclusions to be drawn.

Bond Lengths

Bond lengths to hydrogen have been obtained for a series of
AH_n molecules (A = C, N, O, F), including some excited states.[18-20]
A fairly complete comparison with experiment is possible. For
molecules involving two heavy atoms H_mABH_n (A, B = C, N, O, F)
a very extensive study with STO-3G has recently been completed.[21]
The mean absolute differences between theoretical and experimental values are given in Table 2. Clearly the results improve
steadily with improvement of the theoretical model.

TABLE 2. BOND LENGTH ERRORS ($\overset{\circ}{A}$) IN SIMPLE ORGANIC MOLECULES[*]

	INDO	STO-3G	4-31G
AH Bonds	.046 (22)	.021 (31)	.010 (13)
AB Bonds	.071 (10)	.032 (38)	.011 (7)

[*]Number in parentheses is the number of independent bonds
compared.

Bond Angles

The HAH and HAB bond angles in the same series of AH_n and
H_mABH_n molecules have also been compared with experimental
values. This is displayed in Table 3.

TABLE 3.   BOND ANGLE ERRORS (DEGREES) IN SIMPLE ORGANIC
           MOLECULES

	INDO	STO-3G	4-31G
Mean abs. error	3.1 (27)	3.3 (30)	4.1 (7)

*Number in parentheses is the number of independent angles
 compared.

For the angles, there is not a clear improvement in going to the
extended basis set which still gives rather unsatisfactory
results. The main deficiency at the 4-31G level is a tendency
for bond angles at atoms with lone pair electrons (e.g. H_2O
and NH_3) to be too large. However, there are preliminary indications that these angles are substantially improved when
d-functions are added.[22,23]

Dihedral Angles

Dihedral angles and associated internal rotational potentials are moderately well given by both minimal and split
shell basis sets. However, there are exceptions. Hydrogen
peroxide, for example, has an extremely flat potential curve
with a Slater-type minimal basis[24] and is incorrectly predicted
to be trans if the 4-31G basis is used.[21] On the other hand, it
should be noted that semiempirical schemes such as CNDO have
been found to have some serious faults in dealing with equilibrium values of dihedral angles.[26]

MOLECULAR ENERGIES

It has long been recognized that single-determinant molecular orbital theory cannot lead to a calculated total energy as
low as the experimental value. The difference between the best
molecular orbital energy and the exact answer is often described
as the correlation energy. Its evaluation requires configuration interaction or some other method of improving the wave
function beyond the single-determinant level.

Even though calculation of the total energy may be a very
difficult task, there remains the possibility that single
determinant molecular orbital theory may be adequate for energy
changes in important chemical processes, provided that the same
level of theory is used throughout. We now consider some of
these possibilities.

Dissociation Energies

Some theoretical dissociation energies of bonds to hydrogen (using the 4-31G basis) are listed in Table 4 and compared to experimental values (from heats of formation at 298°K). These experimental values should really be corrected for changes in zero-point vibrational energy, since the theory really applies to stationary (or infinitely heavy) nuclei. Such corrections cannot be made precisely because of the absence of data, but they usually add 5-10 kcal/mole to the values.

TABLE 4.   DISSOCIATION ENERGIES OF BONDS TO HYDROGEN

Reaction	Energy (kcal/mole)	
	Theory (4-31G)	Experimental
$H_2 \rightarrow H + H$	80	104
$CH_2 \rightarrow CH + H$	102	103
$CH_3 \rightarrow CH_2 + H$	85	111
$CH_4 \rightarrow CH_3 + H$	85	103
$C_2H_5 \rightarrow C_2H_4 + H$	41	40
$C_2H_6 \rightarrow C_2H_5 + H$	82	97
$NH_3 \rightarrow NH_2 + H$	83	104
$OH_2 \rightarrow OH + H$	77	119

The results in Table 4 show that the 4-31G basis theory gives poor dissociation energies for the rupture of bonds in which an electron pair is broken leading to separated unpaired electron spins (e.g. $CH_4 \rightarrow CH_3 + H$ giving two doublet radicals). The values for such dissociations are 20-50 kcal/mole too small. Further refinement of the basis set may reduce this error somewhat, but a large discrepancy is to be expected since breaking apart an electron pair will lead to a decrease in correlation energy. On the other hand, it is also clear from Table 4 that dissociation energies which do not lead to an increase in the number of unpaired electrons are given more satisfactorily. Thus $CH_2 \rightarrow CH + H$ is a triplet state going to two doublets and $C_2H_5 \rightarrow C_2H_4 + H$ is a doublet going to a singlet and a doublet.

Reaction Energies Involving Closed Shells

Even though single-determinant molecular orbital theory is unable to give a completely satisfactory treatment of dissociation energies, Snyder[27] suggested some years ago that it might perform better on the energies of reactions where all reactants and products were closed shell molecules. Snyder and Basch[28] tested this hypothesis for a number of such reactions using a double-zeta (split shell) basis. One important set of reactions they considered is hydrogenation as in

$$H_2C=O + 2H_2 \rightarrow CH_4 + H_2O$$

These have also been examined with the basis sets described earlier in this paper. Table 5 gives theoretical and experimental results for the two-heavy-atom molecules for which non-ionic valence structures can be drawn and for which experimental data is available. For this wide range of reactions, the mean absolute deviation between theory and experiment is 6.6 kcal/mole for 4-31G and only 4.1 kcal/mol for the 6-31G set with addition of polarization functions. This suggests that the Snyder hypothesis is very promising for organic molecules if an adequate basis is selected.

Bond Separation Energies

An even more effective application of single-determinant molecular orbital theory is to the energy of reactions in which the number of a given formal type is retained (isodesmic reactions).[30] An example is a bond separation reaction such as

$$H_2C=C=O + CH_4 \rightarrow H_2C=CH_2 + H_2C=O$$

In this reaction the C=C and C=O bonds in ketene are separated into the simplest molecules containing such bonds. Errors inherent in the theoretical description of such bonds will cancel to a large extent and the energy of the reaction really describes the interaction between the bonds.

TABLE 5.   HYDROGENATION ENERGIES (KCAL/MOLE) FOR MOLECULES
           WITH TWO HEAVY ATOMS

Hydrogenation Reaction	Calculated[a]		Expt[b]
	4-31G	6-31G + polarization	
$CH_3CH_3 + H_2 \rightarrow 2CH_4$	-23.5	-21.7	-18.1
$CH_2CH_2 + 2H_2 \rightarrow 2CH_4$	-65.9	-64.7	-57.2
$CHCH + 3H_2 \rightarrow 2CH_4$	-117.8	-117.9	-105.4
$CH_3NH_2 + H_2 \rightarrow CH_4 + NH_3$	-30.9	-27.2	-25.7
$HCN + 3H_2 \rightarrow CH_4 + NH_3$	-83.4	-80.0	-76.8
$CH_3OH + H_2 \rightarrow CH_4 + H_2O$	-32.0	-30.0	-30.3
$CH_2O + 2H_2 \rightarrow CH_4 + H_2O$	-64.3	-59.3	-57.3
$CH_3F + H_2 \rightarrow CH_4 + HF$	-27.4	-26.3	-29.5
$NH_2NH_2 + H_2 \rightarrow 2NH_3$	-50.4	-48.5	-50.0
$N_2 + 3H_2 \rightarrow 2NH_3$	-47.3	-33.7	-37.7
$HNO + 2H_2 \rightarrow NH_3 + H_2O$	-114.2	-107.1	-102.9
$HOOH + H_2 \rightarrow 2H_2O$	-86.3	-93.1	-86.8
$F_2 + H_2 \rightarrow 2HF$	-118.9	-137.0	-133.8

a.   4-31G results from ref. 29. 6-31G + polarization, unpub-
     lished results of P. C. Hariharan.

b.   Corrected to 0°K and adjusted for zero-point vibration
     charges (ref. 29).

    An extensive study of the energies of bond separation for
acyclic molecules with three heavy atoms has been made using
the 4-31G basis.[29] This led to a mean absolute deviation
between theory and experiment of only 3.1 kcal/mole for these
energies, although results with strained cyclic rings are somewhat less successful.[30] It should be emphasized that a satisfactory method of predicting heats of bond separation would be
sufficient to predict the total energy of the large molecule
provided that the energies of the smaller bond separation
products are known, either experimentally or by fuller theoretical studies.

ACKNOWLEDGMENT

This research was supported in part by the National Science Foundation under Grant GP 25617.

REFERENCES

1. C. C. J. Roothaan, Rev.Mod.Phys. 23, 69 (1951).
2. J. A. Pople and R. K. Nesbet, J.Chem.Phys. 22, 571, (1954).
3. J. A. Pople, Intl.J.Quantum Chem. Symposium 5, 175 (1971).
4. J. C. Slater, Phys.Rev. 36, 57 (1930).
5. J. M. Foster and S. F. Boys, Rev.Mod.Phys. 32, 303 (1960).
6. C. M. Reeves and R. Fletcher, J.Chem.Phys. 42, 4073 (1965).
7. K. O-ohata, H. Takata and S. Huzinaga, J.Phys.Soc. (Japan) 21, 2306 (1966).
8. W. J. Hehre, R. F. Stewart and J. A. Pople, J.Chem.Phys. 51, 2657 (1969).
9. R. Hoffmann, J.Chem.Phys. 39, 1397 (1963).
10. J. A. Pople, D. P. Santry and G. A. Segal, J.Chem.Phys. 43, 5129 (1965).
11. J. A. Pople and G. A. Segal, J.Chem.Phys. 44, 3289 (1966).
12. J. A. Pople, D. L. Beveridge and P. A. Dobosh, J.Chem.Phys. 47, 2026 (1967).
13. M. J. S. Dewar and E. Haselbach, J.Amer.Chem.Soc. 92, 590 (1970).
14. E. Clementi, J.Chem.Phys. 39, 1397 (1963).
15. H. Basch, M. B. Robin and N. A. Kuebler, J.Chem.Phys. 47, 201 (1967).
16. R. Ditchfield, W. J. Hehre and J. A. Pople, J.Chem.Phys. 54, 724 (1971).
17. W. J. Hehre, R. Ditchfield and J. A. Pople, J.Chem.Phys. 56, 2257 (1972).
18. P. C. Hariharan and J. A. Pople, to be published.
19. M. S. Gordon and J. A. Pople, J.Chem.Phys. 49, 4643 (1968).
20. W. A. Lathan, W. J. Hehre, L. A. Curtiss and J. A. Pople, J.Amer.Chem.Soc. 93, 6377 (1971).
21. W. A. Lathan, L. A. Curtiss, W. J. Hehre, J. B. Lisle and J. A. Pople, to be published.
22. A. Rauk, L. C. Allen and E. Clementi, J.Chem.Phys. 52, 4133 (1970).
23. D. Neumann and J. W. Moscowitz, J.Chem.Phys. 49, 2056 (1968).
24. R. M. Stevens, J.Chem.Phys. 52, 1397 (1970).
25. A. Veillard, Theor.Chim.Acta 18, 21 (1970).
26. O. Gropen and H. M. Seip, Chem.Phys.Letters 11, 445 (1971).

27. L. C. Snyder, J.Chem.Phys. <u>46</u>, 3602 (1967).
28. L. C. Snyder and H. Basch, J.Amer.Chem.Soc. <u>91</u>, 2189 (1969).
29. L. Radom, W. J. Hehre and J. A. Pople, J.Amer.Chem.Soc.
     <u>93</u>, 289 (1971).
30. W. J. Hehre, L. Radom and J. A. Pople, J.Amer.Chem.Soc.
     <u>92</u>, 4796 (1970).

ORGANIC TRANSITION STATES

Lionel Salem*

Department of Chemistry, Harvard University

Cambridge, Mass.   02138

The theoretical basis of Organic Chemistry lies in the understanding of organic reaction mechanisms. The reaction mechanism is
generally a cursory description of the pathway followed by the
different atoms in the molecule (or molecules) during the reaction.
One important feature of the pathway is the actual geometry of the
col, or potential barrier: the so-called transition state. Transition states are not amenable to direct experimental observation;
only indirect gross information is available via experimental activation energies, entropies of activation, etc. Computation
therefore seems an extremely appropriate tool for elucidating the
structure of transition states. Of course the lack of available
experimental data will be a drawback for any direct comparison;
computation of the potential surface, to which we will restrict
ourselves, would have to be followed by computation of dynamical
trajectories before any meaningful comparison of rates, for instance,
could be made. However the calculated transition state, and its
energy relative to other competing points, can give information on
the likely products to be obtained in the reaction.

To give you an example of the variety of transition states
which are "reasonable" candidates in very simple reactions, and
which can be distinguished by appropriate labeling, consider the
methylene--cyclopropane rearrangement (1):

There are at least <u>eight</u> transition states which would have to be considered in a serious calculation, and which in a fully labeled molecule would lead to different sets of products.

These transition states are
    1)   Woodward-Hoffmann allowed

    2)   Woodward-Hoffmann forbidden (but subjacent allowed) (2)

    3)   Diradical:

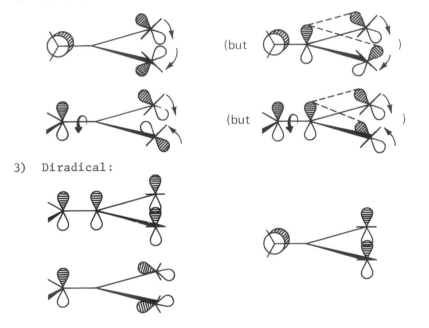

(but   )

(but   )

    4)   Via Mobius "twist" trimethylene methane (3)

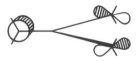

Let me now turn to several points which can be important in a
thorough calculation of a transition state. First of all it is
essential, at least in reactions which can possibly involve both
concerted "closed-shell" and non-concerted "diradical" pathways, to
include--in a single-determinant type SCFMO calculation--configura-
tion interaction. The type of configuration interaction required
might be called zero-order configuration interaction (that which is
necessary, for instance, to distinguish the 3 UV bands of benzene).
Here configuration interaction between the three configurations is

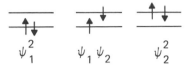

required for a restricted-type calculation; in an unrestricted
type calculation four configurations would be involved. Only in
this manner can one obtain the correct (valence-bond + λ ionic)
type wave function for diradicals with two very weakly interacting
electrons. One can avoid any discontinuities in the energy (4)
and also avoid "loading" the calculation in favor of a "closed-
shell" pathway or of an "open-shell" pathway by doing two separate
CI calculations. In one of these ψ_1^2 is the basic configuration,
with CI to the virtual configurations $\psi_1\psi_2$ and ψ_2^2. In the other,
the starting configuration is $\psi_1\psi_2$, with CI to the virtual configurations ψ_1^2 and ψ_2^2. The dramatic difference between the behavior
of single configurations and that of states (6) is illustrated in
Fig. 1 for internal rotation in the trimethylene diradical.

A second point is that unfortunately there is no truly ration-
al method for searching transition states. All methods used up till
now--even the interesting attempt (7) by McIver and Komornicki to
search for all points where all energy gradients are zero by minimizing the sum of squares of these gradients, and then to look at
the behavior of the energy along all normal coordinates--are vastly
simplified if some "chemically reasonable" assumptions are made on
the paths along which to concentrate the search. One mildly useful
property of reaction paths exists in narcissistic reactions (8),
which are equivalent to pure reflexion. In these reactions one can
distinguish between type I pathways, with symmetric midpoints, and
type II pathways, which are non-synchronous and do not go through
any symmetric situation.

To conclude, we wish to show the transition state which we
calculated for the geometrical isomerization of cyclopropane (9).
This transition state appears clearly on a two dimensional surface
where energy is plotted against rotation of the terminal methylene
groups. The valence angles and bond lengths all have values typical of ground state geometries. At least using the internal
coordinates, the curvature of the energy is negative for only one

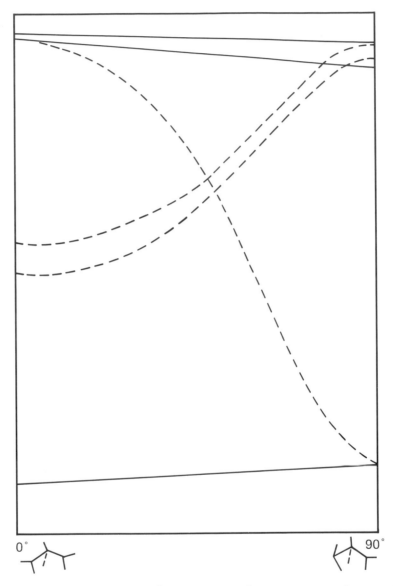

Figure 1   Configurations (dotted lines) and states (solid lines)
            in the internal rotation of trimethylene (6).

coordinate, positive for all twenty others. The surprising concerted motion of the terminal methylene groups is illustrated in
a 5-minute movie.

* Laboratoire de Chimie Théorique (490), Université de Paris-Sud,
  Centre D'Orsay, 91 Orsay, France. (Permanent address)

REFERENCES

(1) W. Von E. Doering and L. Birladeanu, Tetrahedron, submitted
     for publication (and references therein).

(2) J. A. Berson and L. Salem, J. Am. Chem. Soc., 94 (1972)
     (in press).

(3) H. E. Zimmermann, Acc. Chem. Res. 4, 272 (1971).

(4) A. R. Gregory and M. N. Paddon-Row, Chem. Phys. Letters 12,
     552 (1972).

(5) For typical use of dual CI calculations, see a) J. S. Wright
     and L. Salem, J. Am. Chem. Soc., 94, 322 (1972); b)
     M. D. Newton and J. M. Schulman, ibid., submitted for
     publication.

(6) L. Salem and C. Rowland, Angew. Chemie Intern. Ed. 11,
     92 (1972)

(7) J. W. McIver and A. Komornicki, J. Am. Chem. Soc., 94,
     2625 (1972).

(8) L. Salem, Acc. Chem. Res. 4, 322 (1971).

(9) a) Y. Jean, L. Salem, J. S. Wright, J. A. Horsley, C. Moser
        and R. M. Stevens, Pure Appl. Chem. Suppl. (23rd Congress),
        1, 197 (1971).
     b) J. A. Horsley, Y. Jean, C. Moser, L. Salem, R. M. Stevens
        and J. S. Wright, J. Am. Chem. Soc., 94, 289 (1972).

SOLID STATE CONCEPTS IN RADIATION CHEMISTRY AND BIOLOGY

G. Wilse Robinson*

California Institute of Technology,[†] Pasadena,
California 91109; and University of Canterbury,
Christchurch, New Zealand

ABSTRACT

The processes taking place between the initial deposition of
ionizing energy and the chemical utilization of it in radiation
chemistry and radiation biology are discussed from a solid state
point of view. This realm, often referred to as the physicochemical realm of radiation chemistry, occurs in the time range
$10^{-16} < t < 10^{-13}$ sec, and its understanding is of paramount
importance in the interpretation and prediction of channels for
energy dissipation in radiation chemistry and biology. A recently
discovered exciton fission process is suggested as the important
electronic relaxation mechanism in the physicochemical realm. The
better known events of "internal conversion" and "intersystem
crossing" follow the higher energy relaxation mechanisms and help
to localize the excitations in low-lying neutral or ionic states
of the aggregate so that chemistry can eventually occur.

ENERGY STATES OF AGGREGATE SYSTEMS

The intuition that one has about radiation chemistry in
aggregate systems, such as polymers, solids, liquids, and solutions,
is based almost entirely on what has been learned about excitation

---

*Guggenheim Foundation Fellow, 1971-72; Firth Visiting Professor,
Department of Chemistry, University of Sheffield, England, 1971;
Erskine Fellow, Department of Chemistry, University of Canterbury,
Christchurch, New Zealand, 1971.
†Arthur Amos Noyes Laboratory of Chemical Physics, Contribution
No. 4545.

29

and ionization mechanisms in the gaseous phase. In the gas-phase
view, each subunit of the aggregate is envisioned as a separate
entity with similar if not identical electronic properties as the
gaseous molecule. However, a correct discussion of the interaction
of an electromagnetic field with polymers or other aggregate
systems cannot strictly be based upon the quantum energy levels of
the subunits of which they are composed. This is because the
electrons and nuclei of a subunit interact coulombically with the
electrons and nuclei belonging to other subunits and molecules in
the "solvent environment."

   The interaction matrix elements, which are of the excitation
transfer type* $A^{\ddagger}B^{\circ} \leftrightarrow A^{\circ}B^{\ddagger}$, depend upon the nature of the subunit
and its environment in their excited states and the spatial
proximity and relative orientations of subunit molecules with each
other and with environmental molecules. Such intermolecular interactions are responsible for the quantum mechanical mixing together
of electronic characteristics among the subunits and their environmental neighbors. It is therefore not correct to consider the
deposition of energy in a radiolysis experiment as taking place
locally in a single subunit. The inadequacies of the independent
molecule picture are particularly severe when the subunit interaction energies are very large or when the electronic levels are
closely bunched. Both these conditions are frequently met in the
highly excited electronic states of aggregate systems reached in
radiation chemistry and radiation biology experiments. For such
states there may be little resemblance between the electronic
properties of the isolated subunits and those of the aggregate that
they compose. Such points have been stressed in the past by Fano[1]
and others but have not been wholly embraced by radiation chemists.
See, however, Burton et al.[2] where a beginning has been made. The
present paper qualitatively extends and makes explicit some of
these previous ideas.

Crystal States

   Let us now consider the electronic states of aggregate systems.[3]
Because of the high symmetry of a perfect crystal, a detailed
theoretical description of the states for this type of aggregate is
possible. A brief qualitative discussion will be presented here
before proceeding to more relevant systems.

---

*The symbol ($\ddagger$) refers to electronic excitation while ($^{\circ}$) refers to
the electronic ground state of subunit or environmental molecules A
and B. The interaction matrix elements $\langle A^{\ddagger}B^{\circ}|V|A^{\circ}B^{\ddagger}\rangle$ are often
called transfer integrals. V here is the sum over intermolecular
coulombic interactions. Parts of the Hamiltonian involving spin
interactions will not be considered.

The excited electronic energy states of ordered, crystalline solids fall into four broad categories: Frenkel exciton states, ion-pair states, Mott-Wannier exciton states, and conduction band states. Figure 1 schematically depicts orbitals in a "one-electron approximation" for the excited electron in each of these states. For simplicity, the figure has been drawn in one dimension and possible nodes of the orbitals have been omitted. In the formation of these orbitals through a one-electron Hamiltonian, the excited electron can be considered to be under the influence of an array of potential minima, one corresponding to the "positive hole" created on the molecule from which the electron is excited, the other minima being caused by the interaction of the excited electron with neighboring molecules. The situation is fully analogous to the one in semiconductor physics where an electron interacts with an impurity center, except in the present case the "impurity" is the molecular core* itself. Two extreme cases are depicted in Figs. 2a and 2b, respectively--the first where the excited electron is only weakly bound to environmental molecules as in the case where these molecules have but a small electron affinity--and the other case where there is rather strong interaction between the excited electron and the environmental molecules. The latter case would be represented, for example, by a "hydrated electron" strongly self-trapped in the polarization field of its environment.[†]

An important distinction between two types of localization may be made by referring back to Fig. 1. In a low-lying excited electronic state of a free, gas phase molecule, the excited orbital is localized about a "positive hole" (the molecular core). For a Frenkel exciton, sometimes referred to as the tight-binding limit, the electron is likewise localized about the molecular core. Looking at the potential energy diagram in Fig. 2a, for instance, it is obvious that the lower excited orbitals of such a Hamiltonian are localized and quite molecular-like. In theoretical treatments of the tight-binding limit, the excited states in the molecular crystal are taken to be identical, aside from translational symmetry, with those in the free molecule. In practice, this is almost true for molecular vibrational excitations and the lowest-lying singlet and triplet electronic excitations, all of which form good examples of Frenkel exciton states in molecular crystals.[4]

Not apparent from Fig. 1, however, is an important distinction between the free molecule excited state and the Frenkel exciton state. The former is localized in a second sense: the excitation

---

* The core consists of the molecule's nuclei in addition to the N-1 remaining electrons.

[†] The points along the one-dimensional crystal in Figs. 2 do not necessarily represent atomic or molecular centers.

will remain on the molecule during the entire lifetime of the
excitation. In the crystal, on the other hand, excitation
localized on one molecule will tend not to remain there but,
because of the intermolecular interactions, will "spread" to
neighboring molecules. In Fig. 2 this type of delocalization would
correspond to the concomitant displacement of the electronic orbital
and its associated hole to other crystal lattice positions. For
Frenkel excitons, however, one should not visualize this spreading
process as a form of orbital delocalization, since, unlike the
case of the Mott-Wannier exciton, no electron density builds up
between the molecules. The spreading in the case of a Frenkel
excitation simply reflects the fact that in a perfect crystal,
given a long enough time, there is equal probability that each
molecule in the crystal carries the excitation. This latter type
of delocalization we shall call exchange delocalization and is
somewhat analogous to kinetic translational displacements in a free
molecule (vide infra). The excited electron in a Rydberg state of
an isolated molecule at rest would provide an example of orbital
delocalization without exchange delocalization.

Referring again to Fig. 1, it is seen that the Mott-Wannier
exciton orbital, like a Rydberg orbital in a free molecule, is
orbitally delocalized.* In fact, the Mott-Wannier exciton state
is just the crystal counterpart of the Rydberg state in a free
molecule. As seen in Figs. 2a and 2b, one-electron orbitals have-
ing energy farther up on the potential energy diagram, where the
electron interacts weakly with its positive hole, would be expected
to be delocalized into the environment. This would be all the more
true if the affinity of the electron for the environment were large.
This affinity causes the Mott-Wannier orbitals to be orbitally
delocalized to a much greater extent even than the analogous
Rydberg orbitals of a free molecule; thus the Mott-Wannier exciton
extends over many molecules in the crystal. The simplest theoretical
description of a Mott-Wannier excitation is given in terms of an
electron with an effective mass in a hydrogenic orbital embedded in
a uniform dielectric medium, but this is rather a too simplistic
model for most of the relevant systems in radiation chemistry (see
Appendix). In addition to being orbitally delocalized, the Mott-
Wannier exciton is also exchange delocalized--in other words, it
moves.†

---

* The schematic representations in Knox's book[3] also present useful
ways of looking at these orbitals, but it should be stressed that
the parentage of the Wannier orbitals need not be orbitally
localized atomic orbitals distributed over the lattice sites as
depicted in Fig. 10 of Knox's book.

† The distinction between orbital and exchange delocalization is of
course not sharp. This distinction really only depends on the
magnitude of nearest neighbor intermolecular interaction terms and

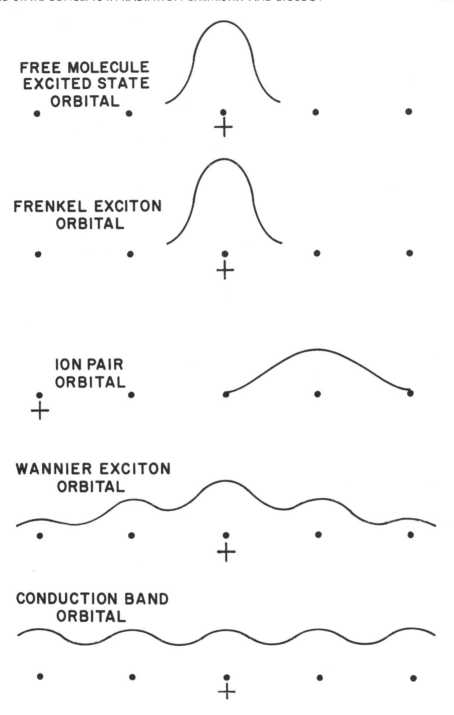

Figure 1. Schematic one–electron orbitals.

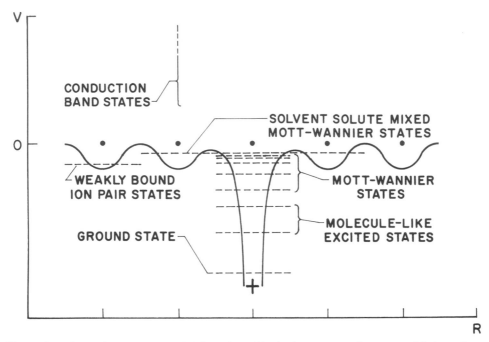

Figure 2a. One–electron potential function. Weak electron–environmental interaction.

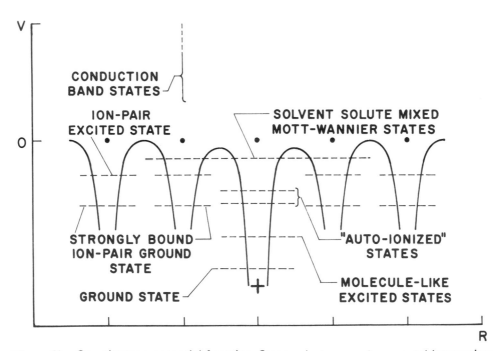

Figure 2b. One-electron potential function. Strong electron-environmental interaction.

An orbital for a conduction band electron (see Appendix for clarity) in a molecular crystal is delocalized over a very large region of the crystal. The process of producing a conduction band electron in the crystal is analogous to ionization of a free atom or molecule in the gas phase. Just as an energy level diagram of an atom or molecule shows the Rydberg states converging to an ionization potential, the Mott-Wannier states of the crystal converge to one of the conduction bands of the crystal. These orbitals lie in the unbound region of the potential energy diagrams of Fig. 2.

Ion-pair states of a crystal are formed when the electronic orbital is not concentric with the positive hole. In fact, in the most general sense the sibling hole or electron may not be present in the system at all. The electronic orbital for an ion-pair state may be Mott-Wannier-like or Frenkel-like. A Frenkel-like orbital for an ion-pair state might simply correspond to the ground state orbital of a molecular negative ion. Mott-Wannier-like orbitals of the ion-pair correspond to orbitally delocalized excited states of the negative ion. Exchange delocalization is also possible for ion-pair states in the crystal since both the hole and the electron can move. The polaron and the hydrated electron are just special cases of ion-pair states where the energy of the state is at least partly determined by electron-"lattice" interactions. Figure 2b, where the electron affinity of the environment is quite large, is an example of a one-electron potential energy diagram that is apt to lead to long-lived metastable ion-pair states.

Throughout the paper we shall use the term _excitation_ to describe the creation of all these excited states. We shall usually not require the more specific term _ionization_ when referring to aggregates. This is not only for semantic simplicity but also because we wish to keep reminding the reader of the distinction between ionization of a free molecule on the one hand and the

---

the resulting time evolution of the crystal wavefunction after a sharp "wave packet" of excitation is produced at a molecular position in the crystal. For a Frenkel exciton, the time evolution shows a rather slow ($>10^{-13}$ sec) buildup of free-molecule-like excitations on neighboring molecules either by a coherent or hopping process, while in the case of a Mott-Wannier exciton, the Mott-Wannier orbital itself first forms, sometimes rapidly ($\sim 10^{-14}$-10^{-15} sec), with electron density building up on but also between neighboring molecules. The time evolution of the crystal wavefunction corresponding to migration (exchange delocalization) of the Mott-Wannier exciton is often a much slower process, its rate depending not so much on nearest neighbor intermolecular interactions, but rather on the more general interactions responsible for simultaneous creation and annihilation of the exciton around positive holes in different parts of the crystal.

creation of conduction electrons or ion-pair states of the
aggregate on the other.

Disordered Aggregates

Biologically important polymers such as DNA, or aggregates
such as membranes or even whole cells are like crystals in one
sense: the various molecular subunits are closely packed and
therefore are capable of inter-subunit interaction. Even the
fluid portion of cells has this property. However, but for a few
exceptions, disorder disallows a rigorous parallel to be drawn
between the energy states of a pure crystal and those of a general
biological system. Disorder in chemical or biological aggregates
occurs as substituent disorder, orientation disorder, and distance
disorder. Substituent disorder is akin to the problem of chemically
or isotopically mixed crystals, while orientation and distance
disorder are met as problems of physical defects in ordinary
crystals.

Because of the inherent difficulty of the subject, the theory
of excitations in disordered crystals is still incompletely understood. However, much progress has been made in recent years, and
it is clear that two limits exist.[5,6] Different excited states of
the same disordered aggregate may lie near either limit or somewhere
in between these limits. One limit, called persistence-type in
mixed crystal theory, is that where the excitation transfer energy
is small compared with the energy of disorder. The energy of
disorder is just the range of appropriate excitation energies that
the aggregate would have if there were no subunit interaction.
For example, in the lowest excited singlet state of DNA, the energy
of disorder would be the energy range spanned by the lowest singlet
excited states of A, T, G, C in the hypothetical case* where all
excitation transfer interactions

$$A^{\ddagger}T^{\circ} \leftrightarrow A^{\circ}T^{\ddagger} \, , \qquad A^{\ddagger}G^{\circ} \leftrightarrow A^{\circ}G^{\ddagger} \, , \qquad T^{\ddagger}G^{\circ} \leftrightarrow T^{\circ}G^{\ddagger} \, , \text{ etc.,}$$

are ignored. Note that the energy of disorder includes van der Waals
energy shifts caused by the varying local environments of the bases
along the DNA molecule, but not the shift[†] caused by the excitation
transfer interaction. Its biggest contribution, however, arises
from the fact that the lowest singlet states of the bases simply
occur at different energies. An estimate of the energy span may
therefore be obtained from absorption spectra of the monomeric
subunits.

---

* The quantum mechanical zero-order representation.

† The "D term" in molecular crystal theory.[4]

In the persistence-type limit, surprisingly enough, relatively well isolated exciton band-like states for the separate subunits persist throughout the entire 0-100% concentration range, with the widths of the bands depending upon concentration. In the case of a dilute solution or a dilute mixed aggregate, the "exciton band" of the solute becomes very narrow and we look upon such states as isolated states of the guest molecule. In the persistence-type limit, excitation can be trapped in these isolated exciton bands of dilute or concentrated mixed crystals for times of chemical importance ($\gtrsim 10^{-13}$ sec).

The second limit, called <u>amalgamation-type</u>, is that where the inter-subunit excitation exchange interaction energy is much larger than the energy of disorder. This is a limit where, if excitation could be localized in some way, it would quickly spread, wave-packet style, throughout a large part of the aggregate: this, in spite of the fact that the aggregate is a disordered one. Subunit or impurity states are virtually "amalgamated" into the overall exciton band of the mixed crystal and no trapping is possible at any concentration. Different states of the same aggregate may lie in different limits. For example, the lowest singlet excited state of DNA probably lies near the persistence-type limit while the second excited singlet state is expected to lie close to the amalgamation-type limit. Excited states of the disordered aggregate in the amalgamation-type limit are very much like those of pure crystals, having Mott-Wannier-like and conduction-band-like states. Even Frenkel-like exciton states of a disordered aggregate can exist if the energy of disorder is small enough.[7]

Electronic states of gaseous molecules lying in the 5-20 eV range are almost always bunched up. In the aggregate, the inter-subunit interactions become large as the subunit wavefunctions encroach upon the intermolecular space. Thus, for such excited electronic states of disordered aggregates, excitation transfer interactions are usually large compared with the energy of disorder, and the amalgamation-type limit is appropriate. It is these states of disordered aggregates and biological systems that are likely participants in some of the early events in radiation chemistry and radiation biology.

TIME SCALES

It seems reasonable that excitation energy must be orbitally localized for a relatively long period of time ($>10^{-13}$ sec) in order for chemistry to take place. In fact, chemistry itself can be thought of as a radiationless transition between orbitally localized states of the system in competition with other radiative and nonradiative processes that can occur. It should not be

forgotton that for chemistry to happen, nuclei must move,* and
this takes time.

    Let us now follow the path taken by the excitation energy
when a disordered aggregate system, such as a biological system,
absorbs ionizing radiation, paying particular attention to the time
scales involved. To do this we hypothetically create a sharp pulse
of relatively monochromatic X-radiation, let's say, of 0.15 nm
wavelength ($\sim$Cu Kα). Because of the fundamental relationship,

$$\Delta t \cdot \Delta \nu \gtrsim 1$$

between the temporal width and the frequency width of a pulse of
electromagnetic radiation, the pulse width can be no sharper than
10^{-16} sec if a 0.5% monochromaticity ($\Delta \nu / \nu$) is to be retained.
Note that during the subjection of the system to this short a pulse
the atomic nuclei in the aggregate move only a negligible distance,
certainly less than a picometer. This can be seen when it is
realized that the root-mean-square velocity of a nucleus carrying
out molecular vibrations is no larger than a few hundred meters/sec,
while the motion of nuclei on the much flatter, more anharmonic
potential surfaces associated with chemical transformations is
probably at least ten times slower than this.[†]

                            Primary Processes

    Subjection of an aggregate system to electromagnetic radiation
of wavelength λ and energy E results in resonance absorption or
scattering (E $\lesssim$ 50 eV), photoelectric absorption (250 KeV $\lesssim$ E $\lesssim$ 40 KeV),
Compton scattering of the primary beam (40 KeV $\lesssim$ E $\lesssim$ 10 MeV) or pair
production (E > 1.02 MeV).[8] For 8.26 KeV (λ = 0.15 nm) radiation,
by far the most important process for biological aggregates is the
photoelectric effect of hydrogen and the first- or second-row
atoms (C, N, O, P, S). In this mechanism a K-electron is ejected
from the atom with kinetic energy $h\nu$ - I, I being the ionization
potential for the K-electron in the absorbing atom. In this way
electrons having from $\sim$6-8 KeV of kinetic energy are produced in
the aggregate by the incident 8.26 KeV X-rays. Electrons of this
energy are supposed to have a range of about a micron in a

---

*Except, of course, for electron-transfer reactions.

[†]For example, instantaneous creation of an electronic charge 0.3 nm
from a bare proton initially at rest would cause the proton to move
only 0.008 pm in 10^{-16} sec. The root-mean-square velocity of a
hydrogen atom in a large molecule, oscillating with zero-point
motion, is about 400 m/sec yielding a displacement of 0.04 pm in
10^{-16} sec. Heavier nuclei would move more slowly.

biological system.[9] The hole in the K-shell is filled by capture
of an electron in an outer orbital, and either a soft x-ray or a
second electron is emitted. For atoms of biological importance, it
is more probable[8,9] that a second electron is emitted (Auger process)
with a maximum kinetic energy equal to the difference in energy
between the K-orbital and the outer orbital from which electron
capture occurs. The kinetic energy of the Auger electron is thus
quite low ($\sim$500 eV) compared with that of the primary K-electron
and has a correspondingly shorter range ($\sim$10 nm).

The K-electron is ejected in a time well within the width of
our hypothetical radiation pulse (10^{-16} sec). The time required to
produce the Auger electron depends on the Coulombic interaction
between electrons in the K-orbital and the outer orbital as first
shown by Wentzel,[10] and is of the order of 10^{-15} sec. Thus the
stage is set for radiation chemistry or biology wherein an X-ray
photon is absorbed, two electrons are produced, one of relatively
low kinetic energy, and a doubly-charged positive hole is formed;
all of this takes place so rapidly ($\lesssim 10^{-15}$ sec) that nuclei are
incapable of moving more than a few picometers during these initial
events. Thus no chemistry is possible during the setting of this
stage. Much the same initial scene would arise if other types of
ionizing radiations were employed, the dramatis personae again
being electrons typically in the 0.5 KeV-1 MeV range.

Secondary Excitations

The discussion in the last section does not deviate from the
standard discussions of radiation effects in chemistry or biology.
It is the description of the excitations produced by the fast
electrons that does begin to deviate.

A gas phase molecule perturbed by a highly energetic primary
electron may be excited, superexcited (autoionized), or ionized.
Such resonance excitations or ionizations cause the energy of the
primary to be reduced by an equivalent amount. There is a close
relationship between the energy levels and the optical spectrum of
the molecule and its ability to absorb a part of the energy of the
primary electron. The transfer of energy to the molecule from the
primary electron is generally quite a small fraction of the total
energy of the electron, many transfers being in the $\sim$20 eV range, as
evidenced from typical loss spectra. Some ionizations lead to
secondary electrons (delta rays) that are themselves sufficiently
energetic to create excitations or ionizations of their own. Again,
most of these subsidiary processes produce excitations or ionizations
of the molecule in the 10-100 eV range. A quantitative discussion
depends upon a knowledge of the energy of the primary electron and

the states of electronic excitation and ionization of the gaseous
molecule. The process has been discussed in great detail in the
literature.[11]

    At this point all that need be said about the radiation
chemistry of aggregate systems that is different from the usual
gas phase or subunit description is that--<u>it is the resonance states
of the aggregate, rather than those of the subunit or gaseous
molecule states, that are the ones available for excitation by the
field of the primary electrons and delta rays</u>.* The well-documented
study of the band structure of solids by x-ray spectroscopy[12]
provides ample evidence that it is the aggregate states of the
system that must be considered in these processes. As mentioned
earlier, these states may have little resemblance to the states of
the subunits, especially in the energy range 5-20 eV. Thus conduction band states, Mott-Wannier and Frenkel exciton states, as well
as ground or excited ion-pair states of ordered or disordered
aggregates are eventually formed along the track of the primary
electron. The "spurs" and "blobs" of radiation chemistry[13] are
produced by delta-ray excitations of the aggregate in the vicinity
of excitation by the primary. It is the evolution of these states
with time, their subsequent relaxation and eventual chemical fate
that pose the central questions of radiation biology or chemistry
in the solid or liquid state.

                              Delocalized States

    Excitation of the aggregate states by primary electrons and
delta rays occurs sufficiently rapidly ($<10^{-15}$ sec) that a fairly
precise start-time may be defined for the beginning of chemical
transformation processes. As mentioned earlier, the states of the
disordered aggregate reached in the primary excitation events are
very likely in the <u>amalgamation-type</u> <u>limit</u>, and consequently these
states are not only <u>orbitally</u> <u>delocalized</u> but are also <u>exchange
delocalized</u>. The fact that these states are in the amalgamation-
type limit means that defect or trap states simply do not exist--
the states of the system are mixed in all their subunit components
and properties, and <u>trapping</u> <u>cannot</u> <u>localize</u> <u>these</u> <u>excitations</u>.
There is simply no distinction between impurities and the aggregate,
solvent and the solute, base pairs and the phosphate linkages, or
chlorophyll and the lipoprotein complex. The fact that the states
are orbitally delocalized means that no chemistry is possible, since

---

*
Primary electrons and delta rays in the aggregate system are more
accurately thought of as conduction band electrons, but the distinction is academic when the total energy of the electron far exceeds
the potential energy terms in the aggregate Hamiltonian.

the influence of the excited electron on any one chemical bond is
minimal. The fact that the states are exchange-delocalized means
that the excitation can move away from its point of origin to use
or lose its energy elsewhere.

Thermalization of Delocalized Orbitals

In order for chemical transformations to take place, the energy
must become orbitally localized in the vicinity of sensitive bonds
in the aggregate and remain there sufficiently long for the nuclei
to move into a new conformation. This process can be envisioned
as a type of rapid electronic relaxation through a series of
electronic states by which the orbitals shrink down towards
orbital localization, while the nuclei are merely spectators to
the rapidly changing fields about them, wanting to respond but
being unable to because of their sluggishness. It is thus likely
that these early processes cannot involve much, if any, phonon
participation.[14] They must therefore be purely electronic in nature.
The exciton fission process of organic crystals[15] seems an excellent
candidate for this type of electronic radiationless transition. The
interaction matrix element in this case is the interelectronic
interaction.

The electronic relaxation process wherein electronic energy is
converted to molecular vibrational energy and heat has already been
extensively dealt with in the literature in discussions of internal
conversion and intersystem crossing among low-lying electronic states
of molecules. Theoretically, relaxation among lower electronic
states of complex molecules is very much like predissociation, the
dense vibrational manifold in the final electronic state furnishing
the "continuum" for the process. It is known that, like predissociation, this relaxation mechanism depends primarily on two factors,
a Franck-Condon factor and a coupling mechanism between the
electronic states involved. The reader is referred to the literature[16,17] for a more detailed account.

SUMMARY

In closing, it seems worthwhile to recapitulate the points that
have been made.

1. The physics of the interaction of a high-energy electron (or
    photon) with a disordered aggregate system is not much differ-
    ent from the dense-gas picture and has been adequately
    described in the literature. These events make up the physical
    realm of radiation chemistry. The sequence of events leading
    to chemical transformations begins with the deposit of energy

in ∿10–100 eV packets along the track of a fast primary
electron.

2. The events that occur during these excitations and the orbital
    localization that takes place afterwards are largely unchar-
    acterized for aggregate systems. This realm, which lies
    between the physical and the chemical realms, has been called
    the physico-chemical realm in radiation chemistry. The point
    of departure between earlier models of radiation chemistry
    and the present model lies in this physico-chemical realm.
    This realm can be divided into an excitation part followed by
    an electronic relaxation part.

3. Excitation of the aggregate system should be thought of in
    terms of the quantum states of the aggregate rather than the
    states of the subunits of which the aggregate is composed.
    The aggregate states are qualitatively different from the
    subunit states, the most important difference being one of
    localization.

4. One should distinguish between two kinds of delocalization in
    aggregate systems: orbital delocalization where there is
    considerable electron density in the intermolecular regions and
    where the orbitals of the aggregate are quite unlike the
    molecular orbitals of the subunits; and exchange delocalization
    where the orbitals of the aggregate are equivalently defined in
    terms of different coordinate origins in the aggregate. A
    Rydberg orbital of an isolated free molecule at rest is
    orbitally delocalized but not exchange delocalized; a Frenkel
    exciton orbital of a molecular crystal is exchange delocalized
    but not orbitally delocalized; a Mott-Wannier orbital of a
    crystal is both orbitally and exchange delocalized.

5. Heavily-doped mixed crystal theory suggests that for disordered
    aggregates, such as biological systems, the states reached in
    the post-initial excitation processes are probably very much
    like ordered crystal states in that there is orbital as well as
    exchange delocalization.

6. The transition from orbitally delocalized states to more
    localized, thus chemically active ones, is nothing more than
    part of the generalized electronic relaxation process. Little
    is known as yet experimentally or theoretically about the
    overall process except that it is probably very fast. The
    recent discovery of fast exciton fission processes in organic
    crystal exciton phenomena probably has bearing on the early
    parts of this generalized relaxation, while the later parts are

undoubtedly the "intersystem crossing" and "internal conversion" processes of organic photochemistry.

7. The decision by the aggregate system as to whether it wants to relax by way of ion-pair states or "neutral" excited electronic states (Fig. 2) would seem to depend entirely on how strongly the excited electron interacts with environmental molecules as compared to the interaction with its own positive core.

8. The chemical realm of radiation chemistry includes some of the later relaxation processes discussed above. The diffusion and chemical reaction of the various ions, free radical, and metastable excited states produced by the earlier fast processes make up the remainder of the chemical realm.

All the speculations in this paper are useless without experimental data. While radiation chemistry data proliferate the literature, what is really needed does not exist. This is work at liquid helium temperatures on well-defined, chemically pure single crystals with the fewest possible physical defects, and on which optical experiments have been extensively performed and interpreted. After this, a systematic branching out into more complicated systems of radiation chemistry could profitably occur.

APPENDIX

MOTT–WANNIER EXCITON STATES

Mott-Wannier exciton orbitals $\psi_{n\ell m, K}$ in a rigid crystal are typically described by hydrogenic-like functions $F_{n\ell m}(r)$ suitably modified by the transformation $Z \rightarrow \varepsilon^{-1}$, where ε is a dielectric constant; and multiplied by a function $\exp i[K \cdot \rho]$ to take account of the translational symmetry of the crystal. Here K is the exciton wave vector, ρ is the position vector defining the electron-hole center of mass, and $n\ell m$ are the atomic quantum numbers. The reduced mass μ appearing in the resulting energy expression,

$$E_n(K) = - \frac{\mu e^4}{2\hbar^2 \varepsilon^2 n^2} + \frac{\hbar^2 K^2}{2(m_e^* + m_h^*)} \tag{A1}$$

for the system is defined in terms of the effective electron and hole masses, m_e^* and m_h^*, respectively. K is a vector in the reciprocal lattice, whose values range over the first Brillouin zone, $|K|_{max}$

being of the order of π/ℓ and $|K|_{min} = 0$, where ℓ is a unit cell dimension.

The correct choice of the dielectric constant ε has been a matter of some discussion. For exciton orbitals of low principle quantum number n, the electron kinetic energy is high, and the other valence electrons are not able to follow the rapid motion. In this case the high frequency dielectric constant (ε_∞) is used. For large orbitals, however, even whole atoms are able to follow the electronic motion and the static dielectric constant (ε_0) should be used. The distinction is a bit academic because of the approximate nature of the model.

Equation (A1) is approximate in three important ways: a) it ignores the dynamic interaction of the excited electron with the lattice vibrations, b) it reflects neither the molecular intricacies of the electron-hole interaction for small orbitals nor the complexities of the interactions of the electron with the static environment for large orbitals (see Fig. 2), and c) it does not have the proper K-dependence. For large Wannier orbitals, the electron kinetic energy is so small, particularly for small $\underset{\sim}{K}$, that the Born-Oppenheimer approximation cannot be used. The electron is scattered off the vibrating environmental molecules many times during an orbital excursion and the very concept of a well-defined electronic state breaks down. In addition, the potential of inter-action between the electron and the neighboring molecules is expected to modify considerably the spherically symmetric charge cloud of the Mott-Wannier picture, particularly when that interaction is large.

The K^2 dependence given by Eq. (A1) can be shown from general considerations (see, for example, Ref. 3) to arise from the assumption of "parabolic energies,"

$$W(k) = const \pm \frac{\hbar^2(\underset{\sim}{k} + \underset{\sim}{G})^2}{2m*} \qquad (A2)$$

of the electron (+ sign) and hole (- sign) states, respectively. Here $\hbar k$ is the electron momentum associated with the valence or the conduction band and $m*$ is the appropriate effective mass, $m*_e$ or $m*_h$. $\underset{\sim}{G}$ is a vector in the reciprocal lattice defining the locations of the various Brillouin zones. The considerations leading to the K^2 dependence $[\underset{\sim}{K} = \underset{\sim}{k}_e - \underset{\sim}{k}_h]$ may be summarized as follows: the k^2 energy dependencies in Eq. (A2) arise from a rigid lattice model where there is a single electron (or hole) in a lattice with translational symmetry but where the periodic electron (and hole)-lattice potential and the electron-hole inter-action both vanish. Introduction of a small electron (or hole)-lattice potential gives rise to mixing between levels in adjacent

Brillouin zones near the zone boundaries, while introduction of a large potential of this type, as could occur for many systems of interest in radiation chemistry, would strongly mix the k_e (or k_h) states throughout the various zones. Physically, all this means is that the concept of a free particle in such a lattice has completely broken down and the energy levels would not be expected to be even vaguely similar to the k^2 dependent ones of a free particle. Note however that in the absence of electron-hole interactions k_e and k_h are quantum-mechanical constants of the motion even when the electron (or hole)-lattice interaction is big. Thus, only levels with wave vectors related to each other by symmetry translations in the reciprocal lattice can mix--e.g., a level k_h can mix with a level $k_h + G$. When the kinetic energy of the quasiparticle (electron or hole) increases sufficiently compared with the periodic potential of interaction with the lattice then the free particle description again becomes a good approximation. However, it would be presumed that this latter limit is hardly ever reached in problems of interest in the physicochemical realm of radiation chemistry since repulsive and attractive interactions must be considered, and one or the other is often quite large.

Introduction of the electron-hole interaction effectively destroys the periodicity of the lattice as far as the electron or hole separately are concerned, and neither k_e nor k_h are any longer constants of the motion. In other words, the electron-hole interaction mixes states of different k_e and k_h. As in molecular physics, Coulomb and exchange interactions involving the excited electron and unexcited electrons may enter into the energy expressions and the k_e, k_h dependence may be very complicated. However, even though k_e and k_h are not good "quantum numbers," the "exciton momentum" $\hbar K = \hbar(k_e - k_h)$ is a constant of motion for the problem. The K-states derived from the inclusion of the electron-hole interaction in the Hamiltonian contain the exciton states. Such states $\Phi_q(K, \beta)$ can be expanded in terms of zero-order states $\Phi_{mn}(k_h, k_e)$ (electron excited from m^{th} valence band to n^{th} conduction band),

$$\Phi_q(K, \beta) = N^{-1/2} \sum_{mn} \sum_{k_e} A_{q;mn} e^{-i\beta \cdot k_e} \Phi_{mn}(k_h, k_e) \qquad (A3)$$

where $k_h = k_e - K$. Here β is a vector measuring the electron-hole separation in units of the lattice constants. In the case where the electron-lattice and hole-lattice interaction both vanish (but not the electron-hole interaction) one expects the electron-hole pair to behave like a free quasiparticle in a lattice. The existence of the K^2 energy dependence depends only on the proper choice of coordinate system, namely the center of mass coordinate system of the pair. Again, in the presence of a potential energy interaction of the electron-hole pair with the lattice, $\underset{\wedge}{K} + \underset{\wedge}{G}$ levels of the various zones are mixed and the simple energy expression is not expected.

In summary then it is seen then that the electron-hole pair behaves as a quasiparticle in the crystal. Two separate approximations are common: Ignoring the electron-hole interaction causes the electron and hole separately to behave as independent quasiparticles. Ignoring the periodic potential of interaction between these quasiparticles and the static lattice causes the particles to behave like free particles with kinetic energy only, giving rise to the familiar $\hbar^2 k^2 / 2m^*$ energy expression. Putting in a nonvanishing periodic potential mixes states of the same $\underset{\wedge}{k}$ in the various zones, and the simple free particle description is destroyed. Putting in the electron-hole interaction also destroys the free particle description of the electron and hole separately but the pair may behave as a free particle providing there is no interaction of the "particle" with the static lattice. If the electron-hole interaction is strong enough, bound states as well as continuous states of the electron-hole pair can occur. The bound states are the Frenkel and Wannier excitons while the continuous states, while not strictly conduction band states of a free electron, become more like them when the total energy is large compared with the electron-hole potential energy. Introduction of the static lattice potential, as before, mixes states of the same $\underset{\wedge}{K}$ in the different zones causing a complicated $\underset{\wedge}{K}$ dependence in $E(\underset{\wedge}{K})$.

ACKNOWLEDGEMENTS

The John Simon Guggenheim Memorial Foundation is gratefully acknowledged for support that allowed the author to spend a part of the academic year away from his usual duties so he could more effectively carry out research in photo- and radiation biology. The hospitality of the Chemistry Departments at both Sheffield and Canterbury Universities is sincerely acknowledged.

REFERENCES

1.   U. Fano, in <u>Radiation Research</u>, Proceedings of the Third International Congress of Radiation Research, Cortina d'Ampezzo, Italy, 1966, G. Silini, Ed. (North-Holland Publishing Co., Amsterdam, 1967). The following statements are very pertinent.

In this connection (that of condensed compared with gaseous matter), the following problem appears particularly bothersome at the present time. As the passage of a charged particle tends to produce activations by disturbing the electrons of different molecules, the responses of electrons of adjacent molecules in a condensed material are not necessarily independent. The most likely responses to radiation by adjacent molecules--or (adjacent) molecular groups-- may interact with one another so strongly as to modify the absorption spectrum substantially (i.e., the responses may interact). .... (We usually regard molecules as independent constituents of matter because they interact through weak van der Waals or polar forces when they are in their ground states, but related forces become much stronger in the presence of excitation.) .... Conceivably, it might not even be possible to localize certain activations initially within a volume smaller than hundreds of atoms across. .... unless localization occurs sufficiently rapidly, its development might afford an opportunity for the energy to be funneled preferentially towards particular atoms or groups of atoms.

2.   M. Burton, K. Funabushi, R. R. Hentz, P. K. Ludwig, J. L. Magee, and A. Mozumder, in Transfer and Storage of Energy by Molecules. Electronic Energy, G. M. Burnett and A. M. North, Eds. (Academic Press, New York, 1969), Vol. I, pp. 161-217.

3.   R. S. Knox, Theory of Excitons (Academic Press, New York, 1963); J. C. Phillips, Solid State Physics 18, 55 (1966).

4.   G. W. Robinson, Ann. Rev. Phys. Chem. 21, 429 (1970).

5.   Y. Onodera and Y. Toyozawa, J. Phys. Soc. Japan 24, 341 (1968).

6.   F. Yonezawa and T. Matsubara, Progr. Theoret. Phys. (Kyoto) 35, 357, 759 (1966); ibid. 37, 1346 (1967).

7.   H.-K. Hong and G. W. Robinson, J. Chem. Phys. 52, 825 (1970); ibid. 54, 1369 (1971).

8.   A. H. Compton and S. K. Allison, X-Rays in Theory and Experiment (D. Van Nostrand and Co., Inc., New York, 1935).

9.   D. E. Lea, Actions of Radiations on Living Cells (Cambridge University Press, 1947).

10.   G. Wentzel, Phys. Zeits. 29, 333 (1928).

11.   R. L. Platzman, in Radiation Research, G. Silini, Ed. (North-Holland Publishing Co., Amsterdam, 1967), pp. 20-53; A. Mozumder and J. L. Magee, J. Chem. Phys. 45, 3332 (1966).

12. N. F. Mott and R. W. Gurney, Electronic Processes in Ionic
     Crystals (Oxford University Press, 1940).

13. A. Mozumder and J. L. Magee, Radiat. Res. 28, 203 (1966).

14. M. Lax, J. Phys. Chem. Solids 8, 66 (1959).

15. M. Pope, N. E. Geacintov, and F. Vogel, Mol. Cryst. 6, 83
     (1969); R. P. Groff, R. E. Merrifield, and P. Avakian, Chem.
     Phys. Letters 5, 168 (1970).

16. J. Jortner, S. A. Rice, and R. M. Hochstrasser, Advan.
     Photochem. 7, 149 (1969).

17. G. W. Robinson, "Molecular Electronic Radiationless Transitions,"
     in Advances in Electronic Excitation and Relaxation. I.
     Radiationless Transitions: The Past, the Present, and the
     Future (to be published by Academic Press, Inc., New York,
     1972), Vol. I.

MOLECULAR MODELING BY COMPUTER

W. A. Little

Department of Physics, Stanford University

Stanford, California 94306

The history and development of the computer has long been intertwined with the development of calculations of the energies, wave functions and charge distributions of atoms and molecules. This type of calculation is characterized by the need for massive "number crunching" techniques and because of this has contributed significantly to the evolution of large machines well matched to this type of problem. More recent developments in computers have led to more subtle uses where the interaction with the environment or with the operator plays a greater role than mere computational power of the computer alone. In this situation the computer acts not as a slave but more as an aid or intelligence amplifier to the operator, expanding his view and aiding his visualization of complex phenomena. This can be accomplished by various forms of interactive graphic displays which do not need the huge computational power of the large machines but which can be operated with the new generation of minicomputers. I will discuss this aspect of computation as it applies to the design and visualization of large molecules and molecular complexes. It is a more subtle problem than numerical calculation and as such is less amenable to an analytical solution.

We became interested in molecular modeling through our interest in superconductivity. For the past decade we have been studying the possibility of synthesizing a polymeric compound which we hope and expect will be superconducting at relatively high temperatures. This has required the synthesis of a linear polymeric conductive spine along which electrons can move with relative freedom; and, to which must be attached an array of polarizable substituents as shown in Fig. 1. The purpose of the polarizable groups is to modify the effective dielectric constant between the electrons on

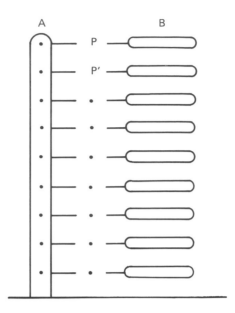

Figure 1. Proposed superconducting structure of conductive spine, A and array of polarizable substituents, B.

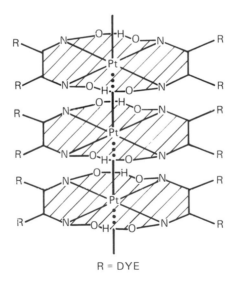

R = DYE

Figure 2. Square planar Pt–complex forming filamentary conductive spine. R–group is methyl but can be substituted by a polarizable dye–like molecule.

the spine and under certain conditions this can lead to an
attraction which then may lead to superconductivity. While this
may appear to be a very special problem, it is really an extreme
example of a fairly common one. In the vicinity of any large
molecule containing polarizable substituents the Coulomb field
will be perturbed by the polarizability of the environment. For
example at the active site of an enzyme the presence of nearby
polarizable bonds will modify the effective Coulomb field and thus
reduce its symmetry below that found in free space. This may well
contribute to the specific catalytic property of the enzyme. In
the problem of superconductivity we have reached the point now
where we have synthesized materials which satisfy our criteria
for the conductive spine. These are linear chains of transition
metals held by organic ligands, such as shown in Fig. 2. We have
successfully prepared other ligands which carry both the transition
element which acts as the nucleus of the spine, and the polarizable
substituents. However, in attempting to form the conductive spine
using these ligands we have failed. The compounds invariably
choose a different molecular conformation thereby frustrating our
efforts to produce our model compound. Presumably, these other
conformations of the compound have a lower free energy than the
one which we seek.

Our problem then is how do we modify the design of our
monomeric unit so that our choice of the ultimate three dimensional
tertiary structure will result rather than some other possibility.
In this it is related to the problem of calculating the tertiary
structure of an enzyme given the sequence of its amino acid residues.
We wish to know the probable outcome of our design prior to synthesis.
Virtually the only way to tackle this type of problem is by explor-
atory modeling. We considered several possibilities using conventional plastic molecular models. We were appalled at the time and
physical effort needed to construct moderately complicated structures.
Moreover the Corey-Pauling-Koltun space-filling models were designed
specifically for biochemical systems, while our problem places much
more emphasis on metal-organic systems and for this the existing
models were found to be inadequate. For this and other reasons we
turned to computer simulation using a graphic display system.

The system we have developed allows us to construct molecular
models on the screen of a color graphic display. The models can
be viewed from all directions and may be assembled into larger
units. As such it is simply a model building kit, however, it
offers a number of advantages over the conventional plastic models.
These are:

   (a)   The models are weightless so that the construction of
         large structures such as DNA is not limited in accuracy
         by the weight of the structure.

(b)   Any kind of atom or group of atoms can be used as
       building blocks by merely adding their properties to
       the appropriate tables.

(c)   Variable bond lengths can be used and even bond lengths
       which depend upon some external parameter such as bond
       order or the oxidation state of the atom.

(d)   These models can be made very quickly and can be
       duplicated virtually instantaneously.

(e)   The same system can display ball and stick models,
       space-filling or crystal models with equal ease.

(f)   A strain relief program can be made part of the system
       and thus arbitrary forms of force constants for the
       bond lengths and bond angles can be used.

(g)   Not only the Van der Waals' interaction but all forms
       of interaction-ionic; hydrogen bonding, hydrophobic
       and hydrophilic forces can be incorporated into the
       modeling system as a guide to the more likely conforma-
       tions.

(h)   One can readily vary the scale of the molecule to see
       the overall features of a complex or to zoom in and
       examine certain specific details.

(i)   The atomic coordinates are available at all times so any
       subsequent calculations can be done with these as input
       data.

(j)   Finally we have found it to be a useful teaching tool
       and a form of "interdisciplinary cement" bringing
       together as it does the physicist and the chemist - to
       the same console to discuss the same problem from
       different points of view.

Indeed, it appears to have all the desirable features suggested
twelve years ago by John Platt (1) in his plea in "Science" for
Better Macromolecular Models.

I would like to give credit to Carl Berney and Frank DeFrancesco
of Aware Systems who have worked with us for the past two years on
this project and who were responsible for the design of both the
hardware and the software of the system.

A film of the modeling system was then shown.

System

The system consists of a Varian 620f computer, 16K core of 16 bit words, a color display using a specially designed deflection yoke and deflection amplifiers, a 500K removable platter single disc memory, data tablet and joystick. As many as 350 atoms can be displayed at one time and the structures rotated or translated in real time. Molecular fragments may be duplicated or may be assembled into larger structures using a simple strain-relief join routine. The atoms and bonds are color-coded and space-filling models are displayed by generating circles at the atomic positions with the Van der Waals radius. The display generates perspective views of the structures which serves to enhance the illusion of depth. Ball and stick, space-filling or crystal models may be displayed by appropriate choice of commands.

1. Platt, J. R., Science 131, 1309 (1960).

2. Scientific Challenges II

INTRODUCTORY REMARKS

John A. Barker

IBM Research Laboratory

San Jose, California 95114

The papers in this session are devoted to physical chemistry
and biological chemistry. Physical chemistry and the physical
part of biological chemistry are areas in which computational
quantum chemistry has so far had relatively little impact. This
is certainly not because there are not important questions to be
answered. The area of interactions between molecules or parts of
the same molecule, and rotational barriers in polymer molecules,
could in principle benefit from quantum mechanical calculations.
The problem is rather that the interactions of interest in physical chemistry are relatively weak, ranging for the most part from
.002 k cal/mole (the interaction of two helium atoms at the van
der Waals minimum) to about 10 k cal/mole (a strong hydrogen
bond, or a rotational barrier in a polymer molecule).

It is only in the upper part of this range that present
methods permit quantum mechanical calculations of useful accuracy
-- naturally the physical chemist would like to know the interactions to an accuracy of one percent or so. In this situation,
physical chemists have been obliged to develop their own experimental and empirical methods for determining energy levels and
interactions, and in many cases these are highly sophisticated.
Some examples of this are given in the papers of this session.
Nevertheless, not all problems can be solved in this way. Thus
the challenge from physical chemistry to computational quantum
chemistry is a call for calculations of sufficient accuracy to
determine interactions between molecules or parts of molecules to
within a few percent.

On the other hand, one must not underestimate the amount of qualitative insight which has already been gained and will be gained through computation.

The first of these two papers discusses the chemistry of the physiological action of hemoglobin with particular emphasis on the cooperativity of the oxygenation, and draws some analogies with the behavior of enzymes. The second is devoted to a survey of problems arising in connection with the nucleation of a new phase.

STRUCTURAL CHARACTERISTICS AND ELECTRONIC STATES OF HEMOGLOBIN

Mitchel Weissbluth

Department of Applied Physics

Stanford University, Stanford, California 94305

OXYGENATION PROCESS

The main physiological function of hemoglobin is to pick up molecular oxygen in the lungs, transport it to the tissues and deposit it there. On the face of it, this is not especially remarkable; it is only when the process is examined in some detail that a number of unusual features emerge.

Hemoglobin is constructed of four polypeptide chains (subunits) which are of two varieties, labeled α and β. Each chain contains a heme group consisting of an iron protoporphyrin IX with two axial ligands one of which is an attachment of the heme to the α- or β- chain. In oxyhemoglobin, the bound oxygen is the second axial ligand; when the oxygen is detached (deoxyhemoglobin) the position originally occupied by the oxygen molecule is left vacant.

The oxygenation process may be described as follows: In the lungs, the partial pressure of oxygen is sufficiently high so that hemoglobin becomes almost totally saturated with oxygen; in the tissues, the partial pressure of oxygen is lower and the saturation of hemoglobin falls to a low value, that is, a large fraction of the oxygen originally bound to hemoglobin has been detached. When investigated quantitatively it is found that the curve which relates saturation to partial pressure (also known as an equilibrium curve) has a sigmoidal shape. The conclusion drawn from this character- istic, as well as from other lines of evidence, is that the oxygena- tion process in hemoglobin is cooperative. This means that when an oxygen molecule becomes attached to a particular subunit, the probability of attachment to another subunit is increased; the effect is cumulative so that the probability of attachment to the last subunit is many times greater than to the first subunit. The same thing happens in the reverse direction, upon deoxygenation: When one subunit releases its oxygen, the chances are improved for oxygen to be released from another subunit.

Cooperativity in hemoglobin is not only an interesting phenomenon from the standpoint of understanding the workings of a complicated molecule; it is also of vital importance physiologically. Moreover it turns out that the behavior of certain enzymes contain features which bear close resemblance to cooperativity in hemoglobin. It is therefore quite likely that we are dealing with a phenomenon which is not restricted to hemoglobin alone, but is fundamental for an important class of biological molecules.

What exactly is the problem in understanding cooperativity in hemoglobin? X-ray diffraction studies have disclosed that the four subunits clump together to form an approximate sphere of 55-60 Å in diameter and that the binding sites, that is, the positions of the iron atoms, are separated by distances ranging from 25 to 40 Å. These distances are far too great for any known electromagnetic interactions to be of importance. How, then, does one iron atom "communicate" to another iron atom the fact that an oxygen molecule is or is not attached to it?

Attempts to answer the foregoing question have been based on the notion that in hemoglobin, and in other biological molecules as well, conformational changes may occur and that they provide the mechanism whereby subunits can interact with each other. To apply this approach it is obviously necessary to know the detailed structure of the molecule in all of its important configurations. Fortunately, such information exists for hemoglobin mainly as a result of many years of painstaking work on the part of Max Perutz and his collaborators at the Laboratory of Molecular Biology in Cambridge, England. In a paper entitled "Stereo-chemistry of Cooperative Effects in Haemoglobin," Perutz (1) traces in great detail the sequence of conformational changes associated with the oxygenation process and shows how various chemical and biochemical observations, seemingly unrelated, are encompassed within a single unified framework.

Briefly, it goes like this: Each subunit may be found in either of two possible conformations (also known as tertiary structures) denoted by oxy and deoxy. In the oxy conformation the subunit has a high affinity for oxygen and in the deoxy, a low affinity. It is assumed that a subunit in the oxy conformation always contains a bound oxygen molecule while a subunit in the deoxy conformation has no bound oxygen. We let α^O and α^D stand for oxy and deoxy conformations, respectively, of α-chains; β^O and β^D are the

corresponding conformations for β-chains. The four chains com-
prising the whole molecule are held together in a certain spatial
arrangement. This is the quaternary conformation (or structure)
and again there are two types--oxy and deoxy. A completely
deoxygenated molecule could be symbolically described by
$(\alpha_1{}^D\alpha_2{}^D\beta_1{}^D\beta_2{}^D)^D$ which would be interpreted to mean that all four
subunits are in the deoxy conformation and the quaternary conforma-
tion is also deoxy. Suppose now that α_1 becomes oxygenated. We
then have $(\alpha_1{}^O\alpha_2{}^D\beta_1{}^D\beta_2{}^D)^D$; Perutz (1) shows that in this condition
certain interchain linkages have been broken so that it is now
easier for another subunit, say α_2, to make the transition
$\alpha_2{}^D \rightarrow \alpha_2{}^O$. When this occurs additional interchain linkages are
broken. In this way the constraints imposed by the deoxy quaternary
conformation to maintain the tertiary conformations in the deoxy
form are gradually weakened until at some point, after several
subunits have become oxygenated, the quaternary conformation goes
over into the oxy form. The remaining subunits which have so far
not been oxygenated are now urged by the oxy quaternary conformation
to make the transition from deoxy to oxy.

We see then that this scheme depends on the interplay between
tertiary and quaternary structures with the latter imposing constraints, yet influenced by the former. There are no complicated
interactions connecting one iron atom with another. The so-called
communication between iron atoms is accomplished through the spatial
relations among subunits so that the probability of a deoxy-oxy
transition depends on whether other subunits have or have not
already made such a transition. Hence the cooperative effect.
Professor Martin Karplus (in a private communication) has analyzed
the Perutz model and has demonstrated how cooperativity is quantitatively deduced from the assumptions of the model.

ELECTRONIC PROPERTIES

We turn now to a consideration of the electronic properties of
iron in hemoglobin. Ideally we should be able to make contact with
the Perutz theory; our objective, then, is to examine how this may
be accomplished. First we summarize a few facts:

(1). Iron in hemoglobin exists either as Fe^{2+} or Fe^{3+}.

(2). Physiologically active hemoglobin, that is, hemoglobin
that is capable of binding oxygen reversibly, contains Fe^{2+}
(ferrohemoglobin). The iron remains in the ferrous state whether
or not oxygen is bound to it. However, in oxyhemoglobin, the six
3d-electrons are spin-paired yielding $S = 0$ while in deoxyhemoglobin $S = 2$. Another important distinction is that the iron atom
in oxyhemoglobin is located in the porphyrin plane unlike

deoxyhemoglobin in which the iron atom is displaced from the
porphyrin plane by about 0.8 Å in the direction of the polypeptide
chain.

(3). In the hemoglobin containing Fe^{3+} (ferrihemoglobin), the
five 3d-electrons produce either $S = 1/2$ or $S = 5/2$ depending on
the nature of the axial ligand, e.g.,
$CN^-(S = 1/2)$, $N_3^-(S = 1/2)$, $F^-(S = 5/2)$, $H_2O(S = 5/2)$.

It is ironic that we know more about the electronic properties
of ferrihemoglobin which does not bind oxygen than of ferrohemoglobin which does. The reason is that one of the most revealing
methods in the study of electronic states is electron spin resonance
(ESR). Oxyhemoglobin with $S = 0$ is, of course, incapable of producing ESR signals but neither have they been observed in deoxyhemoglobin despite the fact that in the latter $S = 2$. Nevertheless
the information obtained on ferrihemoglobin by ESR methods is not
without interest for ferrohemoglobin while other methods like
magnetic susceptibility (2), Mossbauer resonance (3) and others are
applicable to ferrohemoglobin and have contributed useful
information.

We now examine the electronic states of Fe^{3+} in ferrihemoglobin.
In the free ion the d-orbitals are 5-fold degenerate (without spin);
in the heme group Fe^{3+} is 6-coordinated and the dominant symmetry is
cubic. The d-orbitals are therefore split into a 3-fold degenerate
set (t_2) and a 2-fold degenerate set (e) and the e-levels are raised
in energy with respect to the t_2-levels by an amount Δ.

The manner in which the five d-electrons of Fe^{3+} will distribute
themselves in the t_2- and e-orbitals will depend on the magnitude
of Δ relative to Coulombic and exchange energies. When Δ is
small, high-spin states are favored so that the electronic configuration of the ground state is expected to be $t_2^3 e^2$ and the
resulting term is $^6A_1$; at the other extreme--for large Δ--the
configuration is t_2^5 and the term is $^2T_2$. For some value of Δ,
say Δ_c, the energy of $t_2^3 e^2$ $^6A_1$ is equal to that of t_2^5 $^2T_2$.
There is good reason to believe (2) that Fe^{3+} in hemoglobin is characterized by a value of Δ approximately equal to Δ_c, that is, the
high and low spin states are energetically close to one another and
both varieties can coexist if the temperature is not too low. In
Kotani's (2) terminology such an electronic structure is said to be
"soft." This appears to be a unique property of hemoglobin and may,
according to Kotani, have biological significance. We shall return
to this point later.

ESR experiments provide additional details. In low-spin
ferrihemoglobin three distinct g-values, all lying within the range
1-3, are observed. Analysis indicates that the ligand field has a

rhombic component and that the t_2-orbitals interact through spin-orbit coupling. The total effect is to remove the 3-fold degeneracy of the t_2-orbitals so that the separation between adjacent orbitals is of the order of 1000 cm^{-1}.

Similar experiments in high-spin ferrihemoglobin yield ESR spectra with two g-values, 2 and 6, which are quite accurately described by a spin Hamiltonian of the form DS_z^2. Two effects contribute to the form of the spin Hamiltonian: (a) spin-orbit coupling, in second order, between the ground state $t_2^3e^2 \, {}^6A_1$ and an excited state $t_2^4e \, {}^4T_1$ which comes close to ${}^6A_1$ when $\Delta \approx \Delta_c$ and (b) an axial field which splits ${}^4T_1$ into two components but leaves ${}^6A_1$ intact. Spin-orbit coupling alone, without the axial field, or vice versa would be completely ineffective in splitting ${}^6A_1$. The value of D in the spin Hamiltonian varies from one ferrihemoglobin derivative to another but, typically, is in the range of 5-10 cm^{-1}. This is an unusually high value compared to other ferric complexes in which D is a small fraction of a wave number. The explanation again resides in the fact that $\Delta \approx \Delta_c$ which permits ${}^4T_1$ to come close to and thereby interact strongly with ${}^6A_1$.

Turning next to the electronic states of Fe^{2+} in ferrohemoglobin, we find certain similarities with those of Fe^{3+} in ferrihemoglobin. Most importantly, ferrohemoglobin is also a soft structure in Kotani's sense so that again $\Delta \approx \Delta_c$. From this it follows that low-spin (S = 0) oxyhemoglobin must be, energetically, comparable to high-spin (S = 2) deoxyhemoglobin. With six 3d-electrons in Fe^{2+} the ground state of oxyhemoglobin is $t_2^6 \, {}^1A_1$ and that of deoxyhemoglobin is $t_2^4e^2 \, {}^5T_2$.

The electronic state of iron in oxyhemoglobin deserves additional comment. There is evidence, e.g., from Mossbauer resonance (3) and x-ray fluorescence (4) experiments that electronic charge is transferred from iron to oxygen so that a more accurate description would be to regard the iron as Fe^{3+} with S = 1/2. This would then leave the iron in a ${}^2T_2$ state. Though this point is not entirely clear theoretically, it does appear to be consistent with existing molecular orbital calculations (5).

THE JAHN-TELLER HYPOTHESIS

To this stage two aspects of hemoglobin have been discussed: the stereochemical theory of cooperativity and the electronic states of iron. An obvious and central question would then be--what is the connection, if any, between these two aspects? We shall attempt to give an answer to this question but it should be stated

at the outset that what follows is speculative and must therefore
be viewed with caution.

The basic assumption is to invoke the Jahn-Teller effect as a
triggering mechanism for conformational changes (6). In hemoglobin
the argument might go as follows: Suppose that initially hemoglobin
is oxygenated; as we have seen, the Fe^{2+} ion is then in the low-spin
state $^1A_1$ and is located in the center of the porphyrin plane.
Because of the proximity of the high-spin state $^5T_2$, it is assumed
that there is a nonvanishing probability for thermal excitation
from $^1A_1$ to $^5T_2$. Once this occurs, the system is subject to a
Jahn-Teller effect which will tend to distort the molecular framework so as to remove the 3-fold degeneracy of the T_2-state. The
basic theorem limits the distortions to those which obey certain
symmetry requirements. In the present case only two kinds of distortions are possible and one of them, which also appears to be the
most effective one, is a tetragonal distortion along any one of
the three coordinate axes. We now suppose that (a) Jahn-Teller
distortions within the heme plane are less likely than one perpen-
dicular to the plane, (b) the heme group, which sits in a pocket
formed by the polypeptide chain, has constraints imposed upon it
by the local geometry and (c) the combination of (a) and (b) makes
it energetically more favorable for the iron atom, once it gets
into the $^5T_2$-state, to move out of the heme plane. It is seen that
Kotani's conjecture that the energetic proximity of high- and low-
spin states may have biological significance is now embodied in
the Jahn-Teller hypothesis.

The view presented here is that the Jahn-Teller effect provides
a driving force which results in a local distortion which in turn
acts as a trigger for the complex series of conformational changes
that follow. The contact between the quantum-mechanical description
of electronic states and the conformational changes which are
responsible for cooperativity is therefore presumed to reside in
the triggering mechanism mediated by the Jahn-Teller effect.

We conclude these speculations with a few comments regarding
enzymes and the manner in which the foregoing ideas may be applicable there. The central problem concerning enzymes is to understand the basis for their enormous catalytic power. Estimates
indicate that we need factors of 10^9 to 10^{12} to explain any
enzymatic reaction (7). One approach to this problem is the
induced-fit theory proposed by Koshland (7) who postulated that
"(a) small molecules could induce conformational changes on binding
to enzymes; (b) a precise orientation of catalytic groups was
needed to cause reaction; and (c) substrates induced the proper
alignment of residues whereas nonsubstrates did not." Here too, we
have a situation where a small molecule, which binds in some
localized region of the enzyme, is capable of triggering a series

of complex conformational changes. In its general characteristics
this particular feature bears a striking resemblance to hemoglobin
so that it is not unreasonable to suppose that if the Jahn-Teller
effect has anything to contribute toward understanding the behavior
of hemoglobin, it would probably also be helpful in elucidating the
early stages of enzymatic activity.

It is evident that many uncertainties remain regarding the
Jahn-Teller hypothesis but certainly some of them could be cleared
up by a good calculation on a heme group.

REFERENCES

1.   M. Perutz, Nature 228, 726 (1970).

2.   M. Kotani, Adv. Quant. Chem. 4, 227 (1968).

3.   J. E. Maling and M. Weissbluth in Solid State Biophysics (ed.
     J. Wyard), McGraw-Hill (1969).

4.   A. S. Koster, J. Chem. Phys. 56, 3161 (1972).

5.   M. Zerner, M. Gouterman and H. Kobayashi, Theoret. Chem. Acta
     6, 363 (1966).

6.   M. Weissbluth, J. Theoret. Biol. 35, 597 (1972).

7.   D. E. Koshland, Jr. and K. E. Neet, Ann. Rev. Biochem. 37,
     359 (1968).

This work was supported by NIH grant GM16690.

MOLECULAR THEORY OF NUCLEATION*

J. W. Corbett, H. L. Frisch, D. Peak and M. St. Peters

State University of New York at Albany

Albany, New York 12222

ABSTRACT

The science of nucleation is briefly surveyed and many attendant challenges are noted. The emerging molecular-level theory
of nucleation is discussed as is the importance of a diffusion-
controlled reaction kinetic treatment in that theory. Experimental study of nucleation at the molecular level is discussed.

- - - - -

A thermodynamic phase change will occur whenever conditions
of either instability or metastability are established. In the
case of instability there is no energy barrier to be surmounted
for the phase change, while in the metastable case there is such
a barrier; as we will discuss more fully, nucleation relates to
passage over that barrier.

There are many common instances which involve nucleation,
e.g., the various forms of precipitation (rain, snow and hail),
fog, clouds, smoke, the various facets of pollution and smog,
etc. These, and others we will consider later, frequently have
substantial economic impact. Further nucleation has many
scientific challenges[1] as we shall discuss; these challenges involve classical physics, statistical mechanics and quantum

---

*Research sponsored in part by the Metallurgy and Materials
Program of the Division of Research, U. S. Atomic Energy
Commission, under Contract No. AT-(11-1)-3478.

mechanics; they are theoretical and experimental and at every stage require substantial computer usage. Thus nucleation is an eminently suitable topic for this conference.

Most of the nucleation which occurs in nature is hetero-geneous, i.e., the droplet (or the volume of whatever is the stable phase) is formed on a "foreign" substance, be it a sur-face, particle or even an ion. The obverse of heterogeneous nucleation is homogeneous nucleation, i.e., the formation of a droplet without a foreign substance. We will confine our dis-cussion mainly to homogeneous nucleation which will present ample challenges, and simply note that a complete understanding and explication of heterogeneous nucleation is even more remote.

There is a classical theory of homogeneous nucleation which is based on the work of Volmer and Weber[2] as elaborated by Becker and Döring[3] and by Zeldovich[4]. In this theory the forma-tion, say, of a liquid droplet in a supersaturated vapor would be divided into two stages: First, the formation of a quasi-equilibrium distribution of various size clusters of molecules (these clusters being termed embryos), culminating in a steady-state concentration of so-called critical nuclei; Second, the growth of these critical nuclei to form macro-drops. The free energy of a cluster is expressed as a sum of a positive surface energy ($\propto R^2$) and a negative bulk energy ($\propto R^3$) as shown in Fig. 1;

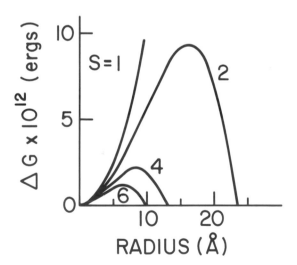

Fig. 1. Free energy of droplet formation versus droplet radius, for various saturation ratios, S. (After E. A. Boucher in Ref. 1, p. 527).

thus versus increasing cluster size there is a maximum in free energy. It is this maximum that characterizes the metastability, i.e., a fluctuation must carry at least a portion of the system, e.g., a proto-droplet, over this maximum for the phase change to occur. The maximum in energy is equivalent to an energy barrier in a reaction kinetic equation with one notable difference; the bulk free energy is readily related in, say, the vapor-liquid case to the equilibrium vapor pressure; the difference then in free energy between the vapor and liquid - the driving force for the phase change, is related to the "excess" vapor pressure or the supersaturation, S; since the supersaturation will vary, the bulk free energy and hence the energy barrier will vary.

The classical theory uses this free energy and calculates the equilibrium concentration of clusters of various size. It is then assumed that the non-equilibrium concentrations involved in the phase change do not differ significantly from the values obtained from equilibrium fluctuation theory up to the critical nucleus size. The steady-state rate of formation of critical nuclei is calculated from the concentrations. The rate of macro-droplet formation is obtained by including the diffusion-con-trolled growth of these critical nuclei.

Szilard in unpublished work quite early recognized that nucleation was in fact a succession of molecular reactions:

$$\alpha_1 + \alpha_1 \rightleftarrows \alpha_2$$

$$\alpha_2 + \alpha_1 \rightleftarrows \alpha_3$$

$$\alpha_3 + \alpha_1 \rightleftarrows \alpha_4 \qquad\qquad (1)$$

$$\vdots$$

$$\alpha_{N-1} + \alpha_1 \rightleftarrows \alpha_N \; - \; \text{the Critical Nucleus}$$

$$\alpha_N + \alpha_1 \rightleftarrows \alpha_{N+1} \; - \; \text{Nucleation}[5]$$

$$\left.\begin{array}{l} \alpha_{N+1} + \alpha_1 \rightleftarrows \alpha_{N+2} \\[4pt] \alpha_{N+2} + \alpha_1 \rightleftarrows \alpha_{N+3} \\[4pt] \vdots \end{array}\right\} \; \text{Growth}[5]$$

and Farkas[6] following this suggestion discussed this formulation explicitly. It was not until the advent of the modern computer that the hierarchy of equations in (1) could begin to be treated.

For example, Courtney[7], using simple (i.e., non-diffusion-
controlled) kinetic rate equations with the growth rates and
evaporation rates in Eqn. 1 characterized by macroscopic thermodynamic variables, solved 100 or so simultaneous differential
equations in treating non-steady-state nucleation; he concluded
that steady-state concentrations were a good approximation to the
non-steady-state ones, except for a usually insignificant time
lag. (We will return to these kinetic equations shortly.) While
the calculation of equilibrium concentrations may seem straightforward, we must emphasize that there is a major, statistical
mechanical, controversy on this very point[8]. Frenkel[9], Kuhrt[10]
and Lothe and Pound[11] recognized that the classical theory ignored the Brownian motion of the clusters, i.e., their translational and rotational degrees of freedom. There have been various prescriptions[12] for including these degrees of freedom but
no unanimity as to how the equations expressed in macroscopic
thermodynamic parameters are to be modified. The discrepancy
in some cases between the classical theory and, say, the Lothe-
Pound prescription[13] is ca. 10^{17}! A controversy involving such
large numbers would at first glance seem readily resolvable,
either experimentally or theoretically; there are a number of
reasons why this is not so. In principle the statistical mechanics of the heirarchy of clusters could be solved directly,
but this is not at present feasible, consequently physical argu-
mentation and approximation must be invoked in prescriptions to
treat the additional degrees of freedom. Some of these prescriptions[10,12,14,15] give numbers approximately the same as the
classical result and others are intermediate between it and the
extremal Lothe-Pound values. Nor is the matter readily solved
experimentally. A factor of 10^{17} certainly does predict a
difference in the nucleation rate, but parameterizing the experiment is not trivial; in terms of the law of mass action the
concentration of the critical nuclei (C_N) is proportional to the
N-th power of the monomer concentration: $C_N \propto (C_1)^N$, i.e., extraordinarily sensitive to the supersaturation. Further the
properties of small droplets, e.g., surface tension, are not
known and appropriate adjustment of these parameters will give
agreement between any of these theories and data for any particular fluid. Even so comparisons between theory and experiments
using the flat-film surface tension has been carried out and it
was found that ammonia[16], benzene[17], chloroform[17], and tri-
chlorofluoromethane[17] agreed with the Lothe-Pound theory but
water[16], many derivatives of ammonia and of benzene and many
other organic compounds[18] agreed with classical theory. Thus the
question of what is the correct formulation remains to be settled.
In an attempt to resolve these problems a number of programs have
arisen using the computer to follow directly the cluster dynamics
including one effort[19] which has followed the physical clustering
through the phase transformation for a highly idealized model

system. These projects are quite promising.

In addition to the challenges discussed thus far we note that there are important physical processes in nucleation which remain to be fully described; when a monomer encounters a cluster it has a kinetic energy and releases a substantial binding energy; if, as is readily evident in a collision between two atoms, that energy is not dissipated[20], the monomer will re-emerge from an elastic collision; this physics gives rise to the inter-related concepts of the heat- and mass-accommodation coefficients. While there has been some work[21,22,23,24] on these concepts, we are a long way from a satisfactory theory (much less any experimental information), particularly of the small clusters.

The theory discussed thus far is based on macroscopic variables, even though it is explicitly recognized that the critical nucleus may involve a very few molecules, e.g., ten to a hundred. It is the thesis of this paper that science is currently witnessing the coalescence, nucleation and growth of a molecular-level theory of nucleation and growth, in fact one which treats nucleation and growth in a unified fashion with diffusion-controlled reaction kinetics throughout. This theory is not completely formed but, we hold, all the necessary elements exist and it is emerging. Indeed the theory is being demanded by experiment, as we will discuss shortly. The molecular-level theory requires a knowledge of the total energies of the various clusters; in the crudest form these are available from valence-bond theory, but, as Professor Pople made clear[25], each variant of theoretical chemistry predicts its own "chemistry", which can, with substantial success, treat such reaction energetics. As we have mentioned the requisite hierarchy of simple reaction kinetic equations has been employed using energies scaled from macroscopic parameters; such equations could be used with molecular energies to treat non-steady state nucleation at the molecular level. Schematically the hierarchy can be expressed in terms of the concentration of the gth cluster in the following equation

$$\partial_t \, c_g = G_{g-1} \, c_{g-1} + E_{g+1} \, c_{g+1}$$

$$- G_g \, c_g - E_g \, c_g \tag{2}$$

with G_g the coefficient characterizing the growth of the gth cluster by addition[26] of a monomer yielding a $(g+1)$ cluster, E_g the coefficient for evaporation of a monomer from the gth cluster, and the coupled equations are implied. But it has been known since Smoluchowski[27] that such equations do not completely describe diffusion-controlled reaction kinetics. The requisite mathematics _is_ now at hand however. We can illustrate the physics by considering diffusion to a sink of a fixed radius, R_o. In Fig. 2a we show the concentration of diffusing species as a function of space and time for the case of initially a uniform distribution and for the boundary condition at the sink, $c(R_o, t > 0) = 0$ – the so-called Smoluchowski boundary condition (SBC) since this is precisely the problem he treated. In Fig. 2b we show the same problem with the more general[28] radiation-boundary condition (RBC), $\partial_R \, c(R_o, t > 0) = \beta \, c(R_o, t)$; this problem has been discussed by several authors[29,30,31]. The radiation boundary condition permits a partial reflection of the diffusing species at the sink boundary and is the diffusion-controlled analog of the mass accommodation coefficient. The mathematics summarized in Figs. 2a and 2b can apply directly, but non-trivially, to the growth terms in Eqn. 2, since the capture radius of a cluster of size g is constant and concentration profiles shown can just as well describe the ensemble pair distribution of g-mers and monomers. The figures illustrate the famous "fast" kinetics which occurs due to g-mers which have close-lying diffusing molecules. We simply note that for large g-mers nucleation emphasizes the fast kinetics. Figs. 2a and 2b do not apply to the evaporation terms since evaporation is an event independent of accretion and is <u>not</u> described by the reflection at the boundary in the radiation boundary condition. The mathematics required for the evaporation term also exists, however. It is shown schematically in Fig. 2c where the evaporation process, for purposes of illustration, is characterized as yielding a delta-function distribution at $\langle R \rangle$ which then diffuses both towards the sink and away from it; this problem was solved for the SBC by Waite[31] and Peak <u>et al</u>[32,33] for the RBC. They showed that the probability (ϕ) of recapture of an evaporated molecule by the sink is given by

$$\phi_{RBC} = \frac{\beta R_o^2}{\beta R_o + 1} \frac{1}{\langle R \rangle} \quad \xrightarrow[\beta \to \infty]{} \quad \phi_{SBC} = \frac{R_o}{\langle R \rangle}$$

Applying this result to the Eqn. 2 we observe that as g increases the corresponding R_o increases[34], but $\langle R \rangle$ is a fixed increment larger than R_o. Hence $R_o / \langle R \rangle$ approaches unity with increasing g. That is, a direct consequence of diffusion-controlled kinetics is that the effective escape of an evaporated molecule decreases for larger g-mers. This is, of course, the same physics that leads to Ostwald ripening (the growth of large droplets at the expense of small ones) at a late stage in the growth process and emphasizes the importance of including diffusion-controlled kinetics throughout.

In summary the requisite mathematics for the diffusion-controlled kinetics exists. Exploiting it, however, will require

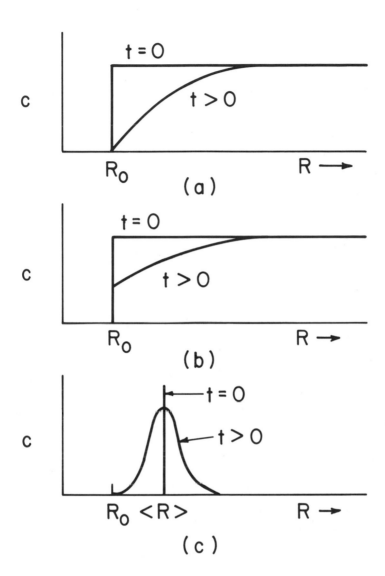

Fig. 2. Diffusion-controlled concentration profile as a function of time. a) Uniform initial distribution beyond sink radius R_0, Smoluchowski Boundary Condition; b) Same as (a) but Radiation Boundary Condition; c) Initial delta function distribution at $<R>$.

substantial effort, computer use, a knowledge of the g-mer energies and such currently unsolved questions as the proper value
of β (and whether β should be a function of g).

Why bother?

In part the answer lies in the hope and reasonable expectation that it will help clarify the existing controversies in
nucleation, but, more simply, experiments are directly requiring it.
Time and time again experiments at the "molecular" level encounter the hierarchy of reactions in Eqn. 1, and can, and must,
follow the reactions quite far to begin to unravel the physics
at the monomer level. For those interested in the monomer
physics, the hierarchy is a nuisance, but "sweet are the uses of
adversity" and that nuisance is precisely the foundation of the
experimental side of molecular-level nucleation.

Perhaps the most advanced line of experiments in this regard
is the study of quenched-in vacancies in metals[35], where a hierarchy of (non-diffusion-controlled) reaction kinetic equations,
complete with energies, has been employed in an effort to unravel
the reactions at the vacancy-level (the single vacancy being the
pertinent monomer). Further, experiments with the field-ion mi-
croscope[36] have been extensively employed on this problem; and
they can directly observe each monomer and g-mer; a g-mer here is
a collection of vacancies - a void in the metal. Far from being
an oddity of only academic interest such voids, enhanced by elec-
tromigration of the vacancies[38], are a major source of failure in
integrated circuits[39]. Voids are also created in irradiation
damage experiments and are also a source of considerable financial
penalty; in such experiments they occur in alkali halides and
other ionic crystals (e.g., UO_2), as well as semiconductors and
metals; these voids cause a swelling in cladding materials in fast-
flux reactors - Huebotter and Bump[40] estimate that achievement
of a reduction of a factor of three in that swelling would save ca.
two billion dollars in the cost of the fast-flux reactors planned
for construction by the year 2000 for the U. S. alone.

Of course, void formation in irradiation damage experiments
involves the competing annihilation reaction between interstitials
and vacancies. The formation of the voids in metals apparently[37]
results from the slight preference in interstitial reaction at,
say, dislocations over that at vacancies and voids[41].

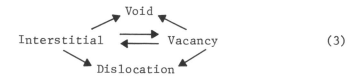

$$\tag{3}$$

The interstitial reactions cause the critical nucleus size to be much larger than it would be for the vacancy reactions alone[42]. We note, however, that in non-metals charge-state effects such as shown in Eqn. 4,

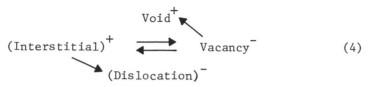

$$(Interstitial)^+ \rightleftharpoons Vacancy^- \qquad (4)$$

as might occur in an ionizing field, (and for the void and dislocation be re-established after the reaction), would have the character of uncoupling the interstitial from the void and lead to a smaller critical nucleus size. Similar charge state, and photochemical effects are thought to be important in pollution problems.

Returning to the molecular-level, as Vook and Brower[43] make clear, substantial progress has been made in identifying vacancy g-mers in irradiated silicon using EPR, setting the stage for following the study of void growth there. Such resonance techniques (EPR, NMR, ENDOR and the Mössbauer Effect) are directly applicable to a variety of precipitation studies in solids. Ordinary spectroscopy encounters such effects as well in solids, and absorption and Raman spectroscopy hold promise for direct measurements on g-mers in a gas in view of the work of Pimentel and others[44,45] in identifying the differences in absorption between monomers, dimers and trimers.

In summary nucleation presents many challenges both theoretical and experimental. And coping with any of them will require the heavy use of computers!

REFERENCES

1. For recent surveys see "Nucleation", Edited by A. C. Zettlemoyer (Marcel Dekker, N. Y., 1969).

2. M. Volmer and A. Weber, Z. Phys. Chem. (Leipzig) $\underline{119}$, 277 (1926).

3. R. Becker and W. Döring, Ann. Physik $\underline{24}$, 719 (1935).

4. J. B. Zeldovich, Acta Physicochimica, USSR $\underline{18}$, 1 (1943).

5. Of course the assumption of the classical theory is that there is no dissociation for the nucleation and growth reactions.

6. L. Farkas, Z. Phys. Chem. 125, 236 (1927).

7. W. G. Courtney, J. Chem. Phys. 36, 2009, 2018 (1962).

8. For a clear review of these matters see P. P. Wegener and J.-Y. Parlange, Naturwissen. 57, 525 (1970).

9. J. Frenkel, J. Chem. Phys. 7, 200 (1939) and in "Kinetic Theory of Liquids" (Oxford Univ. Press, Clarendon, 1946), p. 381.

10. F. Kuhrt, Z. Phys. 131, 185, 205 (1952).

11. J. Lothe and G. M. Pound, J. Chem. Phys. 36, 2080 (1962).

12. See W. J. Dunning in Ref. 1, p. 1.

13. See J. Lothe and G. M. Pound in Ref. 1, p. 109.

14. H. Reiss and J. L. Katz, J. Chem. Phys. 46, 2496 (1967).

15. H. Reiss, J. L. Katz and E. R. Cohen, J. Chem. Phys. 48, 5553 (1968).

16. H. L. Jaeger, E. J. Willson, P. G. Hill and K. C. Russell, J. Chem. Phys. 51, 5380 (1969).

17. D. B. Dawson, E. J. Willson, P. G. Hill and K. C. Russell, J. Chem. Phys. 51, 5389 (1969).

18. J. L. Katz, Proceedings of the Clark Univ. Conf. on Nucleation, May 1972, to be published.

19. C. Carlier and H. L. Frisch, Phys. Rev., to be published.

20. In fact the release of this heat in hypersonic nozzle experiments causes an increase in pressure which is what is actually measured.

21. J. P. Hirth and G. M. Pound, Prog. Mater. Sci. 11, 35 (1963).

22. M. Okuyama and J. T. Zung, J. Chem. Phys. 46, 1580 (1967).

23. G. O. Goodman, J. Chem. Phys. 53, 2281 (1970).

24. F. O. Goodman and J. D. Gillerlain, J. Chem. Phys. 54, 3077 (1971).

25. J. A. Pople, this conference.

26. We ignore here the growth via coalescence of clusters.

27. M. V. Smoluchowski, Z. Phys. Chem. $\underline{92}$, 129 (1917).

28. The Smoluchowski boundary condition obtains for $\beta \to \infty$.

29. F. C. Collins and G. E. Kimball, J. Coll. Sci. $\underline{4}$, 425 (1949).

30. H. L. Frisch and F. C. Collins, J. Chem. Phys. $\underline{20}$, 1797 (1952).

31. T. R. Waite, Phys. Rev. $\underline{107}$, 463, 471 (1957); J. Chem. Phys. $\underline{28}$, 103 (1958).

32. D. Peak and J. W. Corbett, Phys. Rev. $\underline{B5}$, 1226 (1972).

33. D. Peak, H. L. Frisch and J. W. Corbett, Rad. Eff. $\underline{11}$, 149 (1971).

34. Roughly as $g^{1/3}$.

35. For a recent summary see "Vacancies and Interstitials in Metals", Edited by A. Seeger, D. Schumacher, W. Schilling and J. Siehl (North-Holland Press, Amsterdam, 1970).

36. See, for example, E. W. Müller in Ref. 35, p. 557 and R. W. Balluffi and D. N. Seidman in Ref. 37, p. 563.

37. "Radiation Induced Voids in Metals", Edited by J. W. Corbett and L. C. Ianniello, (U.S.A.E.C., Washington, 1972).

38. H. P. Huntington in "Encyclopedia of Chemical Technology" Suppl. Volume 2nd Edition (John Wiley & Sons, N.Y., 1971) p. 278.

39. See, for example, I. A. Blech and E. S. Meieran, Appl. Phys. Letters $\underline{11}$, 263 (1967).

40. P. R. Huebotter and T. R. Bump in Ref. $\underline{37}$, p. 84.

41. It is also felt that some gas, e.g., He, in the void is required to stabilize the void morphology; therefore this is heterogeneous nucleation.

42. R. W. Powell and K. C. Russell, Rad. Eff. $\underline{12}$, 127 (1972).

43. F. L. Vook and K. L. Brower, this proceedings.

44. E. Whittle, D. A. Dows and G. C. Pimentel, J. Chem. Phys. $\underline{22}$,

1943 (1954).

45. See B. Meyer, "Low Temperature Spectroscopy" (Amer. Elsevier Publ., N. Y., 1971).

3. Computational Methods I

APPROXIMATIONS FOR LARGE-MOLECULE CALCULATIONS

Frank E. Harris

Department of Physics, University of Utah

Salt Lake City, Utah 84112

The papers in this section deal with the actual computational methods used in studies of the electronic structures of large molecules and localized states in solids. Such studies differ from those applicable to small molecules through the necessity of choosing methods which do not outstrip our computing capacities, and differ from conventional infinite-crystal studies through a lack of exploitable translational symmetry. The regime under consideration is thereby characterized by a maximum in difficulty of precise calculation, but this difficulty is to some extent compensated by the fact that relatively crude results can often be highly useful.

A main thrust in current work in this field is the study and identification of approximations which are satisfactory for various purposes. The bulk of the effort, and all the contributions in this section, deal with methods within the framework of conventional self-consistent-field (SCF) theory and the use of superpositions of orbital configurations (configuration interaction, CI). However, there has been much recent work on applications of perturbation theory to molecular systems (1), and several workers have introduced approaches based on Green's-function or second-quantization formalisms (2). An important aspect of such formalisms is that they can lead naturally to approximations which have no simple direct expression in configuration space, and some of the approximations thereby obtainable are comparable in operational simplicity with well-known semi-empirical procedures such as the Hückel molecular orbital theory. It is probably still too early to assess critically the future of these methods, but they certainly should not be dismissed from further consideration.

Within the more conventional calculational schemes, we may distinguish between non-empirical (ab initio) methods and semi-empirical methods in which experimentally-identified parameters appear. The non-empirical methods have both the advantages and disadvantages of the absence of adjustable parameters, plus the feature of containing clearly identifiable approximations and limitations. On the other hand, the parameters of semi-empirical formulations can include effects apparently beyond the scope of the formalism; for example the independent-electron potential described parametrically in the Hückel method can be viewed as an approximation to a true many-body optical potential rather than as an approximate self-consistent-field potential. A similar situation applies to the parametric approximate characterizations of exchange energy as used in the SCF Xα method (3). These observations imply that parameters in semi-empirical SCF-like methods can be set to produce results in maximum correspondence either with accurate SCF studies or directly with experiment. It is probably easier to characterize semi-empirical methods on the former basis, but the parametrization to experiment can lead to highly accurate descriptions.

Concentrating on the non-empirical approach, let us review the steps involved in conventional molecular calculations, with particular attention to the numerical difficulties involved and the means presently available for their circumvention. We start by identifying the types of calculation presently possible. Most studies have been carried out at the restricted Hartree-Fock (RHF) level or at approximations thereto. Such studies are generally adequate when processes involving rearrangements or changes in the numbers of electron pairs are not involved. For large RHF studies, the rate-determining steps usually include the calculation of integrals and the construction therefrom of the Fock matrix, and the obtaining of convergence in the iterative process by which the Fock equations are solved. A properly designed iterative process can eliminate much difficulty in the latter step, but well-engineered processes are presently the exception rather than the rule. The matrix diagonalizations entering RHF calculations are no longer a significant numerical problem. The main opportunity for labor-saving in RHF studies is therefore in the elimination of insignificant integrals; this saves not only their calculation time, but also that involved in subsequent data handling.

When bond-breaking and accompanying electronic rearrangements are to be studied, the RHF approach is inadequate, and either a spin-extended SCF procedure (4) or a CI must be employed. Such procedures are also needed to discuss spin densities and other properties not given qualitatively correctly by RHF wavefunctions. These procedures involve the same integrals as RHF calculations,

but the difficulty of obtaining convergence has caused extended
Hartree-Fock methods thus far to have seen application only to very
small systems. Thus, CI procedures have become the main approach
to structures of larger molecules not adequately handled by the RHF
method.

The main problem with CI methods is the choice of configurations, as even relatively modest basis sets provide the possibility
for astronomical numbers of configurations. Consequences of basis
set and configuration choice are reviewed in a paper to follow
(Bagus et al). Once a limited set of configurations is chosen, the
construction of matrix elements (5) and diagonalization of the
resulting secular matrix (6) become routine steps, and one way to
proceed is simply to take as many configurations as circumstances
permit. Better procedures are to choose configurations according
to criteria derived from perturbation theory or from prior calculations; another possibility is to transform the orbital basis set
so as to make a limited set of configurations as effective as
possible. When carried out exhaustively, this is the multiconfig-
uration SCF (MCSCF) procedure (7); it has been successfully applied
only to small systems, and with of the order of 10 configurations.
Part of the advantage of the MCSCF procedure can be realized by
obtaining natural orbitals from a CI calculation and then using
these approximate natural orbitals as the basis for a new CI. This,
the iterative natural orbital (INO) procedure (8), does not necessarily converge, and successive iterations do not always produce
improved wavefunctions and energies. Nevertheless, the first few
iterations often yield a striking improvement in a molecular wavefunction. Another approach toward obtaining effective orbitals
for CI studies is to transform an orbital set to a maximally
localized form; this topic is the subject of one of the presenta-
tions to follow (Ruedenberg). It is also possible to use CI methods
to evaluate electron-pair interaction energies which may then be
combined to yield a (non-variational) estimate of the pair contributions to the correlation energy (9).

All the above discussion ignores an area of possible approximation which may be very useful for polymers or for localized
states in solids, namely the replacement of the infinite system by
a finite system, possibly with periodic boundary conditions. It
then becomes possible to investigate the results of calculations
as a function of the size of the cluster of atoms taken. An example
of this approach follows (Watkins and Messmer).

Turning now to items of more detail, the first practical
question is the choice of a basis set. The size of the basis determines the number of integrals entering the computations as well as
the degree to which an accurate wavefunction can be produced. By
using Gaussian-type orbitals (GTO's) the time of integral evaluation

can be decreased over that required for the same number of Slater-
type orbitals (STO's), but a GTO basis must be larger than an STO
basis giving results of comparable accuracy. Frequently larger
GTO bases are "contracted" into orbitals consisting of given combinations of GTO's with fixed coefficients; contracted GTO's are
comparable to STO's in efficiency of wavefunction representation.
Completely numerical bases have been used for atoms and small
molecules, but there appear to be great practical difficulties in
application to larger systems. If a basis set has n members, the
number of integrals required for a complete energy calculation will
be approximately $n^4/8$, and even for a minimum basis this number
becomes inconveniently large for molecules of interest. However,
irrespective of this difficulty it is necessary to use a basis
sufficient to describe the effects being studied. One possibility
for approximation, however, should be noted. Inner shells of
electrons may be viewed as fixed charge distributions and their
effect completely described by a local potential, which may include
a consideration of exchange as well as classical coulombic contributions. Orbitals for these electrons are then no longer needed.
This type of approximation has most frequently been used in the
crudest possible form: inner-shell electrons are combined with the
nuclei to yield lowered effective nuclear charge, and orbitals
corresponding to inner shells are omitted from the basis set.

Next we consider the question of integral evaluation. For
GTO's there are no problems except those attendant upon the large
numbers of integrals involved. Much data handling must accompany
the construction of integrals for contracted GTO's from basic GTO
integrals. For STO's, analytical (10) and accurate numerical
methods (11) exist but they are presently too slow for use in large
molecules. However, fairly accurate evaluations result if an STO
is approximated by a linear combination of GTO's (12); some work
has also been done on representing two-center STO products by
GTO's so as to reduce the number of basis GTO integrals finally
needed (13). Cruder evaluations result from Mulliken-type approximations in which two-center charge distributions are replaced by
single-center functions; for consistent results it is necessary to
go beyond Mulliken's original suggestion when the two-center overlap integral vanishes (14). This level of approximation is well
represented by the LEDO method introduced by Bloor and collaborators (15).

It is this author's opinion that, whether GTO's or STO's are
used, practical considerations will for some time dictate the use
of approximations in which large numbers of integrals are not calculated at all, i.e. they are assumed to vanish. This need not be
as crude as the vanishing of individual integrals seems to imply,
because electron-nuclear and electron-electron interactions can be
grouped into largely cancelling pairs which a consistent approxi-

mation scheme can take into account. We have obtained encouraging results for small systems based on the concept of balancing largely compensating errors (16). Large-scale neglect of integrals with two-center charge distributions corresponds to the assumption of zero differential overlap (ZDO) for these charge distributions, and many calculations have been made in this approximation (CNDO/2 and related methods)(17). The status of ZDO and other approximations is reviewed in a paper to follow (Zerner).

In summary, we note that much progress has been made in computational methods applicable to large molecules. Looking at the survey presented herein, we see that matrix diagonalization has ceased to be a key bottleneck, and that integral evaluation is rapidly becoming less troublesome. For SCF calculations, our main remaining problems center around the handling of large data sets. On the other hand, CI methods face problems whose solutions are not yet fully evident. New insight and techniques will be needed for the selection of orbitals and configurations yielding satisfactory convergence, or for the identification of alternate methods avoiding these problems. It seems reasonable to expect considerable progress in this area in the near future.

We close with a note of caution: by and large, insufficient attention has been paid to implementing calculations by programs which are modular, largely machine-independent, and well-documented. The result is that most calculations fail to take advantage of much that is known, and large amounts of tedious work are repeated unnecessarily. As large-molecule calculations become more sophisticated and complex, it will become increasingly important to achieve a reasonably effective level of communication among the contributing investigators.

ACKNOWLEDGMENT

This work has been supported in part by National Science Foundation Grant GP-31373X.

REFERENCES

1. See, for example, P.O. Löwdin and O. Goscinski, Int. J. Quantum Chem. 85, 685 (1971); P.R. Certain and J.O. Hirschfelder, J. Chem. Phys. 52, 5977 (1970); B. Kirtman, Chem. Phys. Lett. 1, 631 (1968).

2. See, for example, T.H. Dunning and V. McKoy, J. Chem. Phys. 47, 1735 (1967); J. Linderberg and Y. Ohrn, J. Chem. Phys. 49, 716 (1968); B. Schneider, H.S. Taylor and R. Yaris, Phys. Rev. A 1, 855 (1970); J. Paldus, J. Cizek, and I. Shavitt, Phys. Rev. A 5, 50 (1972).

3. J.W.D. Connolly and K. H. Johnson, Chem. Phys. Lett. $\underline{10}$, 616
    (1971); see also their contribution in this volume.

4. P.O. Löwdin, Phys. Rev. $\underline{97}$, 1509 (1955); see also U. Kaldor
    and F. E. Harris, Phys. Rev. $\underline{183}$, 1 (1969); W.A. Goddard, Phys.
    Rev. $\underline{157}$, 73 (1967).

5. K. Ruedenberg, Phys. Rev. Lett. $\underline{27}$, 1105 (1971); F.E. Harris,
    J. Chem. Phys. $\underline{46}$, 2769 (1967).

6. I. Shavitt, J. Comput. Phys. $\underline{6}$, 124 (1970), and unpublished
    work.

7. See, for example, N. Sabelli and J. Hinze, J. Chem. Phys.
    $\underline{50}$, 684 (1969); P.S. Bagus, N. Bessis and C.M. Moser, Phys.
    Rev. $\underline{179}$, 39 (1969).

8. C.F. Bender and E.R. Davidson, J. Phys. Chem. $\underline{70}$, 2675 (1966).

9. R.K. Nesbet, Phys. Rev. $\underline{155}$, 51 (1967).

10. K.G. Kay and H.J. Silverstone, J. Chem. Phys. $\underline{51}$, 956, 4287
    (1969); $\underline{53}$, 4269 (1970).

11. See, for example, F. E. Harris and H. H. Michels, Adv. Chem.
    Phys. $\underline{13}$, 205 (1967).

12. A. M. Lesk, Int. J. Quantum Chem. $\underline{3}$, 289 (1969); R.F. Stewart,
    J. Chem. Phys. $\underline{52}$, 431 (1970).

13. H. J. Monkhorst and F.E. Harris, Chem. Phys. Lett. $\underline{3}$, 537
    (1969).

14. R. Rein and F. E. Harris, Theor. Chim. Acta $\underline{6}$, 73 (1966).

15. F. P. Billingsley and J. E. Bloor, J. Chem. Phys. $\underline{55}$, 5178
    (1971).

16. J. M. Herbelin and F. E. Harris, J. Am. Chem. Soc. $\underline{93}$, 2565
    (1971).

17. See, for example, J. A. Pople and D. L. Beveridge, <u>Approximate
    Molecular Orbital Theory</u>, McGraw-Hill, New York, 1970.

AB INITIO COMPUTATION OF MOLECULAR STRUCTURES THROUGH CONFIGURATION

INTERACTION*

P. S. Bagus, B. Liu, A. D. McLean, and M. Yoshimine

IBM Research Laboratory

San Jose, California   95114

INTRODUCTION

From its very beginning with the calculation on the hydrogen molecule by Heitler and London,[1] quantum chemistry has been faced with severe computational difficulties in the application of its theories and models. This is particularly true because it is concerned with interactions of atoms and molecules which by their very nature cannot be described concisely in a single coordinate system. Yet, these computational hurdles must be overcome if the subject is to advance to the level of providing tools for quantitative prediction, as well as evaluating approximations used in less rigorous applications in the vast range of chemical systems for which the quantum mechanical equations of motion must be severely approximated.

The first demonstration of the power of computing solutions of the Schrodinger equation to provide a detailed knowledge of the electronic structure of a molecule, and to reproduce experimentally determined properties, was a calculation by James and Coolidge,[2] again on the ground state of the hydrogen molecule. This important calculation has an ambivalent role in the history of quantum chemistry. On the one hand it showed convincingly that quantum mechanics does provide a theoretical basis for all of chemistry. Yet on the other hand, the complexity of the calculation and the difficulty of interpreting the resulting wave functions in more

*This research was supported, in part, by the Advanced Research Projects Agency of the U.S. Department of Defense and was monitored by U.S. Army Research Office-Durham, Box CM, Duke Station, Durham, North Carolina 27706, under Contract No. DAHC04-69C-0080.

familiar terms, bred a wide body of opinion that this calculation
did not open up profitable areas of research, but rather was the
end of the road. One factor contributing to this opinion was the
specialized methods employed by James and Coolidge, who worked in
a coordinate system useful only for light homopolar diatomic
molecules, and who explicitly made the wave function dependent on
the interelectronic distance, r_{12}, which leads to problems so far
insuperable in systems with more electrons.

The exciting possibilities opened by the James and Coolidge
calculation have, however, spurred two generations of research
workers in so called ab initio molecular quantum mechanics to find
more generally applicable ways of solving Schrodingers equation, so
that detailed information can be obtained for atomic and molecular
properties and processes which are not accessible experimentally.
In fact, even for H_2 the refined calculations of Kolos and
Wolniewicz[3] uncovered an error of ~ 8 cm^{-1} in the very carefully
measured value of the dissociation energy, as well as giving
valuable information on the coupling of electronic and nuclear
motions. These efforts, with the essential assist of large scale
computing machinery, have now reached the point where reliable
information on the properties of ground and excited states of small
molecules can be determined. Work in this area has been critically
reviewed recently by Schaefer.[4]

The problem to be addressed is that of finding solutions of the
Schrodinger equation with an accuracy superior to that of the
independent particle model Hartree-Fock (HF) approximation. Computational and numerical problems have been thoroughly investigated
for the HF procedure, to the point that computational tools exist[5]
which make it possible for research workers, not specialists in
computation, to determine HF wave functions. This is not at all the
case for ab initio methods which strive to resolve the correlation
errors[6] inherent in the HF model. The general method of determining
these correlation effects that has attracted the most attention,
and with which the largest number of calculations have been made,
is the method of configuration interaction (CI). A Hartree-Fock
wave function may be considered as a special case of a CI wave
function whose expansion in terms of configuration has been
restricted to one term. This paper discusses our current knowledge
of how to quantitatively apply the method of configuration interaction, and attempts to establish guidelines to ranges of application.

Unfortunately, there still exists a sentiment that large CI
expansions do not offer useful descriptions of molecular structure,
capable of interpretation through basic concepts. It is held that
the most complicated mathematical description capable of being so
interpreted is the Hartree-Fock wave function. What is overlooked

is that CI expansions, no matter how long, can bè cast in a form in which a small number of leading terms dominate. In fact, in cases where a Hartree-Fock description is reasonably good, the lead term of a CI expansion written in terms of its natural orbitals[6] will be very close to the Hartree-Fock wave function. These lead terms are amenable to the same type of interpretation as Hartree-Fock functions; the added terms bridge the gap from semiquantitative to quantitative description through inclusion of the many body effects neglected in the simpler description.

AB INITIO COMPUTATION

In this paper the discussion of ab initio computation is restricted to the determination of the eigenfunctions and eigen-values of the n-electron clamped nuclei Hamiltonian

$$H = -\frac{1}{2} \sum_i \nabla_i^2 - \sum_{A,i} \frac{Z_A}{r_{Ai}} + \sum_{i<j} \frac{1}{r_{ij}} + \sum_{A<B} \frac{Z_A Z_B}{r_{AB}} \tag{1}$$

In Eq. (1) the terms, in order, correspond to kinetic energy, nuclear attraction, electronic repulsion and nuclear repulsion with i,j running over electrons, A,B over nuclei. Terms responsible for spectral fine structure and hyperfine structure, such as magnetic interactions, as well as those which couple electronic and nuclear motions are not included. These effects are normally small and can be accurately treated through perturbation theory.[4,7]

Approximations to the true eigenstates of Eq. (1) are deter-mined by diagonalizing the Hamiltonian in an n-particle basis set of configuration state functions (CSFs) constructed from a 1-particle basis set of orbitals. Calculations within the con-straints of these two basis sets are made without approximation; it is this fact that qualifies the computation as ab initio. In the limit of complete 1-particle and n-particle basis sets, diagonalization of the Hamiltonian yields the exact spectrum of Eq. (1). The practical task discussed in this paper is the determination of basis sets which yield solutions with predictable accuracy. A series of properly designed calculations can investi-gate convergence of predictions with respect to improvement in both 1-particle and n-particle bases. Through extrapolation, the com-plete basis set limits can be estimated[8] and the sequence of calcu-lations can establish its own error bounds. This latter point is, of course, of importance for useful prediction of properties where experimental checks are not available. The following discussion will focus on the requirements for accurate ab initio computation, where a molecular property is to be determined within a satisfactory error range through the stability of the calculation itself with respect to change in the two basis sets. We emphasize that different properties will have different requirements.

We can think of approaching an exact eigenstate of Eq. (1) by following a sequence of 1-particle basis limits, where each limit represents the results of diagonalizing H in the complete n-particle space which can be constructed from the given 1-particle space. In these terms the questions to be asked are: (1) For a given 1-particle basis how closely do we approach the 1-particle basis limit, when we are using less than the full set of possible n-particle functions?, and (2) How closely does the 1-particle basis limit approach the exact eigenstate to be determined? It is apparent that there must be some appropriate balance between the two basis sets. The 1-particle basis limit should be close to the exact result, and a sufficiently large n-particle basis set must be used so that this limit may be approached. An unbalanced calculation, for example one with an n-particle basis set sufficiently large to approach an inadequate 1-particle basis limit, can be justified in terms of insight gained into the nature of CI expansions, but not in terms of the value or accuracy of predicted properties.

ELEMENTARY FUNCTIONS AND THE 1-PARTICLE BASIS

Spanning the space of the 1-particle basis is a set of elementary functions, $\{\chi\}$, chosen to provide a flexible starting point for expanding orthogonal orbital functions, $\{\phi\}$, from which the n-particle basis functions are to be constructed.[9] The orthogonal orbital functions are the 1-particle basis; the expansion from $\{\chi\}$ is according to

$$\phi_{i\alpha} = \sum_{p=1}^{N} c_{i\alpha,p}\chi_p \qquad (2)$$

where in Eq. (2), i is an orbital serial number, α counts over functions forming a basis for an irreducible representation of the ith orbital, p is a serial number of an elementary function and $c_{i\alpha,p}$ is a linear expansion coefficient.

The elementary functions are normally a non-orthogonal set of functions centered on the various nuclei in the molecule, chosen largely for their ability to describe the electronic structure of the separated atoms in the states to which the desired molecular state or states dissociate. In addition to these functions, which can be determined in simpler atomic calculations the elementary functions must include functions which efficiently describe electronic polarization effects in the molecular field, functions spanning the space of molecular Rydberg states which may correlate with separated atom occupied orbitals, and functions required to describe many-body molecular correlation effects, particularly those that are geometry dependent. Different regions of the space spanned by the elementary functions may get different treatments, depending on the information required from the calculation. For example, where the space of the valence shells is of greatest importance a much more approximate description of inner shells may be satisfactory. One set of properties where particular care does have to be taken with inner shells is the electric field and its gradients at the molecular nuclei. For these properties, polarization effects involving shell interactions, of the anti-shielding type for example, though the entire shell structure can be important. Thus, appropriate 1-particle basis functions for polarizing all shells will be necessary in computing these properties.

It is a common practice to choose the elementary functions by supplementing well optimized atomic basis sets with polarization functions, chosen by symmetry type, with exponents based on principles of maximum overlap or from previous experience.[10] These elementary functions, particularly the polarization ones, can be optimized in the system under study,[11] but the process is expensive and time consuming and does not permit sufficient experimentation. We propose here a method for choosing optimum polarization functions by carrying out modified atomic calculations. Atomic basis sets supplemented by polarization functions are optimized to reproduce atomic properties in external electric fields. We stress that these atomic calculations may be of HF or CI type. If the elementary functions used in such atomic calculations are capable of describing the distortions in the electric field, as indicated by the accuracy of computed atomic polarizabilities and atomic shielding factors, then they must be appropriate for describing the atomic distortions in a molecule. In fact, functions required to reproduce atomic polarizabilities should be suitable for describing molecular valence shell polarization effects; functions required to reproduce atomic shielding factors should be suitable for describing polarization phenomena contributing to electronic properties at the atomic nuclei. A systematic study of atoms in external electric fields along these lines could be a great value in providing a starting point for molecular calculations. The method has been applied with considerable success to the determination of the quadrupole coupling constant in LiH.[12]

Functional forms of the elementary functions most used are

(i) Slater-type, $r^{n-1}e^{-\zeta r}Y_{\ell,m}(\theta,\phi)$, with $n > \ell \geq |m|$ integers, and ζ referred to as the elementary function exponent. These have proven superior for use in diatomic and linear molecular systems. They have achieved more limited use for systems of more complicated geometry in which matrix element evaluation of the electronic repulsions has proved a severe obstacle.

(ii) Gaussian-type in either spherical polar coordinates $r^{n-1}e^{-\alpha r^2}Y_{\ell,m}(\theta\phi)$ or more commonly in cartesians $x^{n_1}y^{n_2}z^{n_3}r^{n_4}e^{-\alpha r^2}$ with certain constraints on the integer powers of coordinates, related to the methods of matrix element evaluation.

(iii) Numerical-type $R(r)Y_{\ell,m}(\theta,\phi)$ where the radial dependence of the single-center function has been determined numerically in some previous calculation. As a starting point for molecular calculations care must be taken to see that elementary functions of this type are sufficiently flexible to allow expansion of near optimal molecular orbitals. This could be done using numerically tabulated atomic orbitals, and their derivatives, in the molecular basis set.

As a guide for the number of elementary functions needed to determine a 1-particle basis adequate for accurate ab initio work Table 1 shows the requirement in terms of Slater-type functions with carefully chosen orbital exponents, for the lighter atoms. The requirements for individual applications may diverge somewhat, but the table should provide a useful starting point. These recommended minimal sets largely correspond to "double-zeta + polarization (DZ + P)"[10c] sets, the double-zeta referring to two Slater-type function exponents, reflecting the fact that the minimal require-ment for an accurate description of an atomic shell is two well chosen elementary functions.

The use of a d-type elementary function for the lightest atoms H → Be, which in their ground states do not even have an occupied p shell, deserves comment. The electric field in the vicinity of the light atom in a molecule can have an important quadrupolar component. With orbitals constructed from elementary functions of various symmetry types, on a single atomic center, Table 2 shows the symmetry of charge distribution components relative to the atomic center. Since a quadrupolar field is D in character, the interaction of the charge distribution with the field requires D character in the charge distribution. This can only be achieved by elementary function products of sd type. (Those of pp type, which do contain a D component, cannot describe the quadrupolar interaction because the amount of p character in the orbitals is dictated by the response to the dipolar field.) The importance of d functions in the very light atoms has been demonstrated in many calculations.[4] Use of a 3dπ function on Li in LiF and a 3dπ function on H in HF_2^- single configuration wave functions are two examples.[10c] This latter result should be generally true in studies of hydrogen bonding.

Table 1.   Elementary Slater-type functions for accurate descriptions
of atoms in molecular calculations.

Atom	Elementary Function Count[a]
H,He	$2(s) + 1(p) + 1(d)$
Li,Be	$4(s) + 2(p) + 1(d)$ $[3(s) + 2(p) + 1(d)]^b$
B→Ne	$4(s) + 3(p) + 1(d)$
Na,Mg	$6(s) + 4(p) + 1(d)$
Aℓ→Ar	$6(s) + 5(p) + 1(d)$ $[6(s) + 5(p) + 2(d)]^c$
K,Ca	$8(s) + 6(p) + 2(d)^c$
Sc→Zn	$8(s) + 6(p) + 3(d) + 1(f)$
Ga→Kr	$8(s) + 7(p) + 3(d) + 1(f)$

[a]In molecular calculations the degeneracy of these functions given, in the Table, in terms of atomic symmetry must be taken into account.   Thus, in a molecule of arbitrary geometry p represents three functions p_x, p_y, p_z.

[b]This illustrates the possibility of using only 1 s function for the K-shell, although care must be exercised in doing atomic calculations to choose the three s functions.   This is necessary to avoid a description in which two functions describe the K shell and only one the L shell; the situation favored in energy optimization[13] of the basis functions.   To avoid this problem, we suggest that the atomic calculation should be done optimizing four s exponents, then freeze the two lower ones which essentially describe the L-shell, and then optimize a third exponent in a subsequent calculation with three s functions.

[c]Even though no atomic d shell is occupied in the ground state atoms, the importance in molecular calculations may justify two functions, particularly for Ca.

Table 2.   Symmetry of charge distribution components arising
           from elementary function products.

Elementary Function Product	Symmetry of Charge Distribution
ss	S
sp	P
pp	S + D
sd	D
pd	P + F
dd	S + D + G

In terms of other types of elementary functions the numbers
required must be the equivalent of Table 1. Using Gaussians, for
example, it is common to use fixed linear combinations of elementary Gaussian functions as the equivalent of a Slater-type
function describing an occupied atomic shell. Several tabulations
of atomic elementary functions which normally provide the molecular
starting point are available.[10,14]

So far the discussion has been on the elementary functions
rather than the 1-particle basis. This emphasis is not misplaced
since the results of molecular computation are ultimately dependent
on these functions; no inadequacy in the elementary functions can
be made up by further computation. A complete linear transformation
on the elementary functions, Eq. (2), leads to a set of orbitals of
the same dimension and spanning the same space as the starting set
of elementary functions. If it was computationally feasible to
diagonalize H in the complete n-particle space, the same results
would be obtained whether the n-particle basis functions were
constructed from orbitals or directly from the elementary functions
(or orthogonalized elementary functions). This however is never
possible, at least in systems with more than three electrons, and
orbitals (the 1-particle basis) must therefore be near optimally
chosen in order to achieve the best possible results from a severely
truncated n-particle basis. Determination of orbitals is strongly
coupled to the choice of n-particle basis and will therefore be
deferred until after the description of the n-particle basis in
the next section, to where we describe various computational
procedures.

n–PARTICLE BASIS

The simplest n–particle functions which can be constructed
from the 1–particle basis, and satisfy the physical requirements
of antisymmetry with respect to electron exchange, are Slater
determinants. In a Slater determinant, electrons occupy a subset
of the spin orbitals, $\psi_j = \phi_{i\alpha} \begin{Bmatrix} \alpha \\ \beta \end{Bmatrix}$, which can be constructed from
an orbital member of the 1–particle basis multiplied by either spin
α or spin β. Each Slater determinant

$$\frac{1}{\sqrt{n!}} \begin{vmatrix} \psi_1(1)\psi_2(1) & \cdots & \psi_n(1) \\ \psi_1(2)\psi_2(2) & \cdots & \psi_n(2) \\ \vdots & & \vdots \\ \psi_1(n)\psi_2(n) & \cdots & \psi_n(n) \end{vmatrix}$$

is normalized, and the set of all Slater determinants is orthonormal,
the orthogonality following from spin and orbital orthogonality.
Slater determinants are, in general, not eigenfunctions of the
symmetry operators for the system. A preferable n–particle basis
is formed by taking linear combinations of Slater determinants which
are symmetry eigenfunctions, configuration state functions (CSFs),
and in particular we will be interested in CSFs that belong to
certain n–particle subpaces which we classify below.

We define an _electronic configuration_, C, as an assignment of
electrons to orbitals, thus

$$C = C(n_1, n_2, n_3, \cdots n_{n_I}, n_{n_I+1}, \cdots n_{n_I+n_E}) \tag{3}$$

where the occupation number n_i is associated with orbital ϕ_i of
Eq. (2). The notation of Eq. (3) shows a division in the orbital
set between ϕ_{n_I} and ϕ_{n_I+1}. The first n_I orbitals are called
internal orbitals, and the remaining n_E, spanning the remainder of
the 1–particle space, are called _external orbitals_. The dividing
point between the internal and external set is not unique, and
depends to some extent on the desired accuracy in the calculation,
but the internal orbital set must contain all orbitals which are
used in the construction of what will turn out to be the leading

terms in the final CI expansions. The number of Slater determinants spanning the n-particle space of configuration C, corresponding to all possible spin orbital assignments consistent with
the occupation numbers n_i, is

$$N_C = \prod_{i=1}^{n_I+n_E} \frac{(2\lambda_i)!}{n_i!(2\lambda_i - n_i)!} \tag{4}$$

where λ_i is the dimension of the irreducible representation of the
ith orbital. This set can be transformed to form a basis of
irreducible representations of the symmetry group for the n-electron
system called <u>configuration state functions (CSFs)</u>. There are
standard techniques for deriving the CSFs, and the matrix elements
of various operators between them, which will not be discussed
here.[15] In certain open shell cases, the same irreducible representation may be spanned several times by the CSFs of a configuration. Thus, in the case of the axially symmetric configuration
$\sigma'\pi'\sigma\pi^2$ in which two σ electrons are distributed between orbitals
σ and σ', and three π electrons are distributed one to orbital π',
two to orbital π, there are 96 CSFs spanning $7\times^2\Pi$, $5\times^4\Pi$, $1\times^6\Pi$,
$2\times^2\Phi$, $1\times^4\Phi$, irreducible representations. The space of CSFs of a
given configuration spanning a given symmetry irreducible representation can be partitioned into subspaces in a way that can be
of practical importance in truncating the length of CI expansions.
This partitioning is done on the basis of Rayleigh-Schrodinger
perturbation theory using, as zeroth order wave function, a function
expanded in a zeroth order set of "reference" CSFs denoted $\{0,0;0\}$.
The CSFs in $\{0,0;0\}$ are, by definition, constructed from internal
orbitals and must be the dominant CSFs of the final CI wave function,
truly serving the role of a good zeroth order approximation. CSFs
in $\{0,0;0\}$ may be associated with several different configurations
$C_q(n_1^{(q)}, n_2^{(q)}, \ldots n_{n_I}^{(q)}, 0, \ldots 0)$. CSF's associated with

any configuration $C_a(n_1^{(a)}, n_2^{(a)}, \ldots n_{n_I}^{(a)}, n_{n_I+1}^{(a)} \ldots n_{n_I+n_E}^{(a)})$
belong to sets $\{n,m;i\}$ with

$$n = \left(\frac{1}{2} \sum_{i=1}^{n_I+n_E} \left| n_i^{(a)} - n_i^{(q)} \right| - m \right)_{min} \tag{5}$$

$$m = \sum_{i=n_I+1}^{n_I+n_E} n_i^{(a)} \tag{6}$$

where min refers to the smallest value that n can take after
C_a is compared with all C_q: All CSFs associated with a single
C_a belong to {n,m;i} with common values of n,m, but possibly
several values of i which denotes the order of the perturbative
correction to the zeroth order wave function to which a CSF first
contributes. The number n is the minimum number of electrons
excited into internal orbitals in C_a compared with all the C_q,
and m is the number of electrons excited into external orbitals.
It is now apparent that our notation {0,0;0} for the zeroth order
CSFs is consistent with the definition of {n,m;i}.

     Our partitioning task, mentioned in the previous paragraph,
in the case that CSFs of C_a form several independent bases for
generating an irreducible representation of a given symmetry, is
to determine linear combinations of CSFs that belong to a uniquely
defined value of i. In other words, members of {n,m;i} span the
space of those, and only those, functions which for a given type of
orbital excitation, described by n,m, contribute first to the ith
order perturbation correction to a wave function expanded in
{0,0;0}. Thus, functions assigned to the ith order sets have
non-zero matrix elements through the Hamiltonian with at least one
member of the (i - 1) order sets. A simple algorithm for determining basis functions spanning the space {n,m;i} has recently been
given.[16]

     The space spanned by {n,m;i} which includes all functions of
that class that can be constructed out of the 1-particle basis,
·is invariant to a linear transformation (rotation) of the external
orbitals. Thus, in particular, the first order subspace is invariant
to a unitary transformation on the external orbitals. A CI wave
function expanded in the zeroth and first order subspaces will be
invariant under such a transformation since it is variationally
determined in an unchanged space. However, the Rayleigh-Schrodinger
first order perturbed wave function in this space is not invariant
to rotation on the external orbitals since it is not variationally
determined.

CONFIGURATION INTERACTION CALCULATIONS

     A configuration interaction calculation involves diagonalizing
the Hamiltonian, Eq. (1), in a selected n-particle basis. It seeks
roots of

$$(\underline{H} - E\underline{S})\underline{C} = \underline{0} \qquad (7)$$

in which $\underline{H}$ is the Hamiltonian matrix in the selected basis, and $\underline{S}$
is the unit matrix in the case under consideration of an orthonormal

n-particle basis. E is an eigenvalue of H in this basis, with
$\underline{C}$ the associated eigenvector. If the n-particle basis functions
are Φ_K, then

$$H_{KL} = \int dx_1 - dx_n \, \Phi_K^*(x_1 - x_n) H \Phi_L(x_1 - x_n) \tag{8}$$

$$S_{KL} = \int dx_1 - dx_n \, \Phi_K^*(x_1 - x_n) \Phi_L(x_1 - x_n) = \delta_{KL} \tag{9}$$

where, in Eqs. (8) and (9), integration variables x include space
and spin coordinates of the suffixed electron. An eigenvector of
Eq. (7) is an expansion in Φ_K

$$\Psi = \sum_K C_K \Phi_K \tag{10}$$

If the Φ_K include all n-particle CSFs that can be constructed
from the 1-particle basis for the problem, then the eigenvalues
of Eq. (7) and the wave function Ψ of Eq. (10) are invariant under
a linear transformation of the 1-particle basis. This would be a
complete CI, yielding eigenvalues and eigenvectors which are at the
1-particle basis set limit. However, as pointed out in a previous
section, such calculations are impracticable and in fact are
unjustified if there exists a method of approaching the 1-particle
basis set limit using a severely truncated n-particle basis.
Current research in configuration interaction is focussed on
seeking and testing possible methods that achieve this result; we
proceed to discuss some aspects of the problem.

    If the natural orbitals of a complete CI wave function are
known, then the expansion of this same wave function in CSFs constructed from the natural orbitals has certain optimal convergence
properties.[6,17] We argue that these same convergence properties
should approximately apply for CI expansions in certain truncated
n-particle spaces, therefore a desirable 1-particle basis for
constructing CSFs is the set of natural orbitals spanning the same
space as the given 1-particle basis. However desirable, they are
not available since they can only be obtained from the complete CI
expansion of some eigenstate which we have supposed to be
inaccessible. Our approach is to design a sequence of practicable
calculations which yield sufficiently good approximations to the
natural orbitals of an eigenstate to permit a severely truncated
n-particle basis to have a large overlap with the complete
n-particle basis. Expansion of the eigenstate in this basis should
closely approach the complete CI limit, and a well designed sequence
of calculations will hopefully permit extrapolation to this limit.

Computational methods for achieving severe truncation of the
n-particle space through use of approximate natural orbitals as a
1-particle basis and perturbation theory for CSF selection are
described in the following sections.   They have evolved through
experience, both our own research group's and others, but have still
not been adequately tested and must therefore be regarded as
tentative.   They do outline directions which we plan on thoroughly
investigating in our future research on molecular systems.

   We proceed to define the natural orbitals which we use, constructed subject to symmetry and equivalence constraints, namely
that orbitals span the basis of irreducible representations of the
spatial symmetry group and that the same orbitals are used for both
α and β spin.

   Consider the expectation value of a totally symmetric spinless
one electron operator from the wave function of Eq. (10).   This
wave function is constructed from CSFs with common space and spin
quantum numbers.   The expectation value is

$$\langle \Psi | \mathrm{Op} | \Psi \rangle = \int dx_1 - dx_n \Psi^*(x_1 - x_n) \left(\sum_{i=1}^{n} \mathrm{Op}_i \right) \Psi(x_1 - x_n)$$

$$= \sum_{K,L} C_K C_L \langle \Phi_K | \mathrm{Op} | \Phi_L \rangle \tag{11}$$

with

$$\langle \Phi_K | \mathrm{Op} | \Phi_L \rangle = \int dx_1 - dx_n \Phi_K^*(x_1 - x_n) \left(\sum_{i=1}^{n} \mathrm{Op}_i \right) \Phi_L(x_1 - x_n)$$

$$= \sum_{i,j} c_{ij}^{KL}(i|\mathrm{Op}|j) \tag{12}$$

In Eq. (12)

$$(i|\mathrm{Op}|j) = \int dr_1 \phi_{i\alpha}^*(r_1) \mathrm{Op}_1 \phi_{j\alpha}(r_1) , \tag{13}$$

the C_{KL} are numerical coefficients and   i and j   run over the full
1-particle basis for a fixed (and arbitrary) value of α.   Clearly
c_{ij}^{KL} is $\neq 0$ only if the occupation number $n_i \neq 0$ (c.f. Eq. (3))
for Φ_K, $n_j \neq 0$ for Φ_L, and both $\phi_{i\alpha}$ and $\phi_{j\alpha}$ belong to the
same irreducible representation.   This result ensues because the
integral of Eq. (13) exists for a totally symmetric operator only
if the two orbital functions belong to the same species

(irreducible representation) and subspecies; further, the value of the integral is independent of subspecies. Combining Eqs. (11) and (12) we have

$$\langle \Psi | Op | \Psi \rangle = \sum_{i,j} \sum_{K,L} C_K C_L C_{ij}^{KL} (i | Op | j)$$

$$= \sum_{i,j} \gamma_{ij} (i | Op | j)$$

$$= \sum_{i,j} \int dr \; Op(r) \gamma_\alpha(r | r') \Big|_{r=r'} \tag{14}$$

where

$$\gamma_\alpha(r | r') = \phi_\alpha \gamma \phi_\alpha^\dagger \tag{15}$$

In Eq. (14) the final integrand is obtained by allowing $Op(r)$ to operate on $\gamma_\alpha(r | r')$ and then removing the primes from the resulting expression. In Eq. (15), ϕ_α is a row vector containing a single function from each degenerate orbital set, with subspecies label α. γ is the matrix of γ_{ij}, defined in Eq. (14). The function $\gamma_\alpha(r | r')$ and the matrix γ are referred to as symmetry constrained density matrices. We note that functions $\gamma_\alpha(r | r')$, $\gamma_\beta(r | r')$, derived from ϕ_α, ϕ_β respectively in which the two row vectors contain functions of a different subspecies, are different. However, both are associated with the same matrix γ, a result which is obvious from the properties of Eq. (13). A totally symmetric $\gamma(r | r')$ can be constructed from the $\gamma_\alpha(r | r')$ combining different α, and it too is associated with the same matrix γ. Starting with a different wave function Ψ, degenerate with the original, will again yield this same matrix γ.

Diagonalization of the symmetric matrix γ can be achieved with the unitary matrix U,

$$U^\dagger \gamma U = d \tag{16}$$

In the case of a CI wave function invariant to the orbital rotation

$$\phi_\alpha' = \phi_\alpha U \tag{17}$$

the symmetry constrained reduced density matrix $\gamma_\alpha(r|r')$ must be unchanged so that

$$\gamma_\alpha'(r|r') = \gamma_\alpha(r|r') = \underset{\sim}{\phi}_\alpha \gamma \underset{\sim}{\phi}_\alpha{}^\dagger = \underset{\sim}{\phi}_\alpha U^\dagger \gamma U \underset{\sim}{\phi}_\alpha{}'{}^\dagger$$

$$= \underset{\sim}{\phi}_\alpha' d \underset{\sim}{\phi}_\alpha' = \sum_{i=1}^{n} d_i \phi_{i\alpha}'^* \phi_{i\alpha}' \qquad (18)$$

showing that in the rotated orbital basis the symmetry constrained density matrix is diagonal.[18] The complete CI wave function is invariant under any orbital rotation. The orbital set ϕ_α' of Eq. (17) for a complete CI wave function are the <u>symmetry constrained natural orbitals (SCNOs)</u> of the wave function. Orbital sets obtained from Eq. (17) for a truncated CI expansion will be <u>approximate symmetry constrained natural orbitals (ASCNOs)</u>. There are certain important types of truncated CI expansions which are invariant to ASCNO orbital rotations which do not mix the internal and external orbitals.[19] Under such rotations the density matrix of the truncated wave function, expanded in CSFs built from the ASCNOs, will be diagonal. In general, however, a truncated CI expansion built from CSFs constructed from ASCNOs will not have a diagonal density matrix.

The diagonal element γ_{ii} of $\underset{\sim}{\gamma}$, the occupation number of the ith orbital in the CI wave function, is given by

$$\gamma_{ii} = \sum_K n_{iK} |C_K|^2 \qquad (19)$$

where the sum is over all CSFs, indexed by K, and n_{iK} is the occupation number of the ith orbital in the Kth CSF (Eq. (3)). For any wave function invariant to an orbital transformation which diagonalizes the symmetry constrained density matrix we can re-expand Eq. (10) in CSFs constructed from the rotated orbitals

$$\Psi = \sum_K C_K \Phi_K(\phi) = \sum_K A_K \Phi_K(\phi') \qquad (20)$$

In the ϕ' representation the diagonal density matrix elements are the eigenvalues of $\underset{\sim}{\gamma}$, equal to

$$d_i = \sum_K n_{iK} |A_K|^2 \qquad (21)$$

Suppose that orbitals i, both in the transformed basis ϕ' with occupation number d_i, and the untransformed basis ϕ with occupation numbers γ_{ii} are ordered to decreasing occupation number. Then the optimal convergence property of CI expansions, in CSFs built from orbitals which yield a diagonal density matrix, comes from the theorem that the sum of the r highest eigenvalues of a symmetric matrix is greater than or equal to the sum of any r diagonal elements

$$\sum_{i=1}^{r} \sum_{K} n_{iK}|A_K|^2 \geq \sum_{i=1}^{r} \sum_{K} n_{iK}|C_K|^2 \qquad (23)$$

and the corollary, if the dimension of the basis is N,

$$\sum_{i=N-r+1}^{N} \sum_{K} n_{iK}|A_K|^2 \leq \sum_{i=N-r+1}^{N} \sum_{K} n_{iK}|C_K|^2 \qquad (24)$$

If the sum of Eq. (23) is over all N, the equality holds and each expression is equal to the number of electrons, reflecting the fact that the trace of the density matrix is invariant under the orbital transformation.

If an occupation number is zero, from Eq. (21) all CSFs in the CI expansion must enter with zero mixing coefficient A_K, and might just as well have been dropped from the n-particle basis in the first instance. We argue that this is approximately true when the occupation number is approximately zero, and use Eq. (24) to state that the n-particle space is better truncated by omitting CSFs built from ϕ' rather than from ϕ. Whether any truncation of CSFs built from ϕ' is permissible depends on both occupation number and the class $\{n,m;i\}$, defined in the previous section, to which the CSF belongs. Our hope is that for an adequate $\{0,0;0\}$ the ϕ' needed for constructing CSFs belonging to $\{n,m;i\}$ with $i \geq 2$ can be restricted to just the first few external orbitals.

With this background we can now offer an adequate description of a path that we believe holds promise of developing accurate CI expansions with practicable effort and with a minimal ingredient of arbitrariness in the selection of the n-particle basis. Three steps are involved: (i) the determination of ASCNOs for the internal orbitals and the associated choice of $\{0,0;0\}$, (ii) the determination of ASCNOs for the external orbitals, and (iii) construction of a final CI expansion in CSFs built from orbitals obtained in the previous two steps, building up the n-particle space by inclusion of functions in $\{n,m;i\}$ in an order that permits

establishing various convergence limits which will help determine
the reliability of the final wave function.

DETERMINATION OF INTERNAL ORBITAL ASCNOs

A prerequisite to internal orbital determination is selection
of the members of $\{0,0;0\}$. This set of zeroth order CSFs must
include all n-particle functions which play an important role,
as measured by their expansion coefficients, in the CI expansions
of the states being studied. This appears to be another situation
where the answer is needed before the calculation begins, and to
some extent this is true. However, for ground states and some
excited states, the most important members can usually be written
down from considerations of elementary molecular orbital theory,
combined with considerations of separated, and possibly united, atom
limits, depending on the range of nuclear geometry considered in
the study. There will undoubtedly be cases where the selection of
zeroth order CSFs is less obvious, particularly for excited states.
In these cases, the required CSFs can be selected in the course of
carrying out the calculations. An initial trial set will be chosen,
and various CI calculations set up using them as $\{0,0;0\}$. These
calculations will be aimed simultaneously at satisfactory orbital
determination and confirmation of the chosen $\{0,0;0\}$. The results
can be used to modify $\{0,0;0\}$ in the event that the initial choice
is not confirmed, both by taking out unimportant ones and by adding
functions from higher order spaces which appeared with large
expansion coefficients. In subsequent CI calculations with the
modified $\{0,0;0\}$, results should be consistent with this new choice.

The aspect of using a preliminary set of calculations to drop
terms from an initially chosen $\{0,0;0\}$ is anticipated to be of
particular importance in the study of excited states of systems
containing many atoms with open shell ground states. In the absence
of any superior knowledge, a desirable initial calculation would be
a complete CI in a 1-particle basis which spans the space of
occupied separated atom orbitals. This is followed by an orbital
transformation based on the diagonalization of the first order
density matrix (formed from a superposition of eigenstates under
study), and repetition of the complete CI in the new orbital basis.
Analysis of this latter CI calculation should permit a sensible
choice for the number of internal orbitals, and initial approximations for them, as well as the composition of $\{0,0;0\}$. This initial
calculation will be recognized as the equivalent of a valence bond
treatment of the system.

The importance of members of $\{0,0;0\}$ is dependent on the
orbitals from which they are constructed, thus calculations must be
aimed at internal orbital determination and a consistency check on

$\{0,0;0\}$. From the discussion of natural orbitals it is clear that
it is desirable to find internal orbitals which, in the final
wave functions of the study, will have the largest possible
occupation numbers under a linear transformation on the full
orbital set. Two methods of orbital determination which produce
orbitals that <u>approximately</u> have this property are (i) to find the
solutions of Hartree-Fock-like equations and (ii) to follow an
iterative natural orbital procedure with a particular type of
n-particle basis.

In the framework of the expansion methods discussed in this
paper, the Hartree-Fock-like problem is to find orbital expansions
in a given 1-particle basis which minimize the energy expectation
value of a wave function set up as an expansion in $\{0,0;0\}$. Both
n-particle expansion coefficients and 1-particle expansion
coefficients are variationally determined. With one member in
$\{0,0;0\}$ this is the traditional Hartree-Fock (HF) calculation;[13]
with more than one member it is referred to as a multi-configuration
Hartree-Fock (MCHF) calculation.[20] The true HF and MCHF wave
functions are obtained in the limit where any increase in the
1-particle basis does not change the wave functions. The equations
to be solved are derived through the calculus of variations; they
are complicated and are normally solved by iterative techniques
which have been plagued by problems of non-convergence, although
recent developments[20] show promise of advancing them to a routine
level. When further development of the desired wave function
involves natural orbital transformations involving the internal
orbitals, it can happen that changes in MCHF internal orbitals are
big enough to cause a redistribution of CI expansion coefficients,
in the new orbital basis, to the point that the members of
$\{0,0;0\}$ must be changed, thus destroying the consistency of the
calculation. This annoying feature can be avoided by not allowing
the internal orbitals to change in later calculation; a satisfactory
procedure as long as all important CSFs are represented in the MCHF
wave function. Alternatively, and in a way that checks the important CSFs, the MCHF orbitals can be used to construct a CI in the
complete n-particle space that can be derived by distribution of
valence electrons through the internal orbitals. The orbitals are
now transformed by the matrix which diagonalizes the first order
density matrix constructed from the appropriate root of this CI, and
the wave function is expanded in the transformed orbitals. The
internal orbitals for subsequent computation are the transformed
orbitals and the members of $\{0,0;0\}$ are taken from the transformed
wave function. Subsequent improvement in the wave function, by
mixing in CSFs involving excitation into external orbitals, and
transformations based on diagonalizing a first order density matrix
will make only small perturbations on these internal orbitals.

An important second type of approach, which uses iterative natural orbital procedures,[21] also offers a way of confirming a choice of $\{0,0;0\}$ and determining internal orbitals which are satisfactory approximations to SCNOs. Using a trial set of orbitals, both internal and external, spanning the full 1-particle space, a CI calculation is done in a subspace of the n-particle functions in $\{0,0;0\}$, $\{a,0;i\}$, $\{b,1,1\}$. Members of the first two sets span the space of a complete CI in the internal orbitals, while members of the third set include CSFs in which one electron has been excited into an external orbital and which have an interaction through the Hamiltonian with $\{0,0;0\}$. A density matrix constructed from the eigenstates of interest in this calculation is diagonalized and, following Eq. (17), the orbital basis, both internal and external, is transformed. The first iteration of this procedure does a CI using the same CSFs, but constructed from the trans-formed orbital basis, producing different wave functions because the n-particle space is not invariant under orbital rotation. Iterations will be monitored for the consistency of $\{0,0;0\}$, the CI eigenvalues and eigenvectors of interest, the sum of internal orbital occupation numbers, and the degree to which the density matrix becomes diagonal. It is clear that if the density matrix in a transformed orbital basis is diagonal further iteration can produce no change in the desired wave functions. In general, however, the density matrix does not become diagonal through itera-tion, and the process does not converge. It would appear desirable to incorporate into the process a procedure which extrapolates orbital transformations to seek a maximum in the sum of internal orbital occupation numbers, similar to extrapolation procedures in traditional self consistent field calculations.

Computational experience[21] has shown cases, when single excita-tions into the externals are restricted to the subset of the class $\{0,1;1\}$ with zero matrix elements with a MCHF wave function expanded in $\{0,0;0\}$ by an extended Brillouin theorem, where the iterative process does converge. The density matrix becomes diagonal, the sum of the internal orbital occupation numbers becomes equal to the number of electrons, and the external occupation numbers become zero. Further, the wave functions resulting from this process are the MCHF wave functions, and the internal orbitals span the same space as the MCHF orbitals. The mathematical develop-ment showing that this should be, and the range of cases for which it is true, has not yet been worked out and represents a rather startling theoretical weak point which needs addressing.

The key to this iterative natural orbital procedure is that it provides a mechanism for mixing components of the initial external orbitals into the internals, in a way that increases the internal occupation number sum in the restricted CI calculation being iterated. It is anticipated, and shown in practice, that the

internal occupation numbers of these orbitals in more extended CI
calculations remain suitably high. The precise subspace of the
iterated CI chosen from {a,0;i} and {b,1;1} may not be critically
important, although all external orbitals must be used. Neither
may the degree to which a maximum occupation number sum is
approached be critical. These practical considerations require
testing in applications.

DETERMINATION OF EXTERNAL ORBITAL ASCNOs

For the same reason that it is desirable to obtain internal
orbitals which retain close to the maximum occupation number in
final CI calculations, it is desirable to find external orbitals
which, when ordered by occupation number, have near maximum partial
sums of the occupation numbers of the lead members in later CI
calculations. Orbital transformations producing external orbitals
with this property are said to have "packed" the orbitals, meaning
that the maximum amount of correlation information has been packed
into the leading members of the orbital set.

In looking to the practical requirement that CSFs be truncated
in the final functions, deletions from the complete CI n-particle
space should be made on the basis of their contribution to complete
CI wave function, which is unknown. However, these contributions
can be approximated through perturbation theory or through truncated
CI calculations. Inspection of which CSFs make the smallest contributions, within a given order space, may point the way to
criteria of orbital truncation, which means that in the n-particle
space all CSFs involving certain orbitals are dropped. It should
be clearly stated at the outset that the purpose in packing the
external orbital occupation numbers is the hope that this will
allow severe truncation of the external orbitals used for constructing CSFs in second and higher order spaces. Even with a
packed external orbital basis, the indications are that the level
of orbital truncation permissible in constructing first order CSFs
will be quite limited.[22] Cases where orbital truncation in the
construction of the first order space should occur will be when
only valence electrons are correlated. Those external orbitals
with strong radial overlap with the core orbitals will be deleted
from all order n-particle spaces.[23]

The desired packing of external orbitals can be achieved by
carrying out CI calculations in an n-particle space which includes
CSFs involving external orbitals belonging to the sets {0,1;1},
{1,1;1}, {0,2;1}. We will assume that CSFs are constructed from
an internal orbital set which is already near optimum, and a
starting external orbital set which spans the orthogonal complement
of the internal orbitals in the 1-particle space. These starting

external orbitals are usually available from the calculation which produced the internal orbitals, or are otherwise easily obtained. In actual calculations it may be possible to optimize internal and external orbitals in a single computation, but we separate these two aspects for the present discussion. We will further suppose that if one CSF from any of the above mentioned sets is included in the n-particle space, then all members with the same specific internal orbital excitation and all possible external orbital occupations, belonging to the same set, are included. Eigenvectors of a CI calculation in an n-particle space with this property are invariant under a rotation of the external orbitals. Suppose we diagonalize the external orbital subblock of the density matrix of a CI wave function of interest, and transform the external orbitals by the unitary transformation that performs the diagonal-ization. Then, rewriting the CI wave function in terms of these transformed external orbitals yields a function whose density matrix in the transformed orbital representation has a diagonal external orbital subblock.

To the extent that this is an approximately diagonal density matrix we argue that the external orbitals approximately have the property of maximum partial sums for the lead occupation numbers. Expansion in some subsets of the first order space can lead to density matrices with a zero connecting block between internals and externals. In these cases diagonalization of the full density matrix and subsequent orbital transformation does not mix internals and externals and the statement on maximum partial sums of the leading external orbital occupation numbers is exactly true.[19]

In practice, the CSFs included in the n-particle space for external orbital determination should involve all the external orbitals and should have the largest interactions with CSFs involving only internals. Also in practice, assuming starting near optimal internal orbitals, it makes little difference whether orbital transformations are based on diagonalization of the full density matrix, which generally mixes internals and externals, or are based on diagonalization of just the external subblock. We proceed, in the next section, to discuss various approximations which may be of practical use.

DESIGN OF CALCULATIONS

The design of actual calculations is naturally dependent on the dimension of the 1-particle space, which imposes the ultimate accuracy limitation, and on the dimension of the n-particle spaces.

In the trivial case where it is practicable to do a CI in the complete n-particle space which can be constructed, there is no calculation to design. One simply carries out the calculation without bothering to optimize internal orbitals, external orbitals, or to go through a process of CSF selection, since none of these factors can improve the computed result. A final transformation to a SCNO representation would be made to aid interpretation of results.

The next simplest case would be the case where it is practicable to work with the complete zeroth and first order n-particle spaces. With even the most approximate starting internal orbitals an iterated set of CI calculations in this space, in which at the end of each CI calculation the problem is set up again in a transformed orbital basis through diagonalizing an appropriate density matrix, will yield excellent internal and external orbitals. Iteration can be combined with extrapolation to seek a minimum energy expectation value or a maximum sum of internal orbital occupation numbers. The results of this calculation can be used to drop CSFs that are unimportant as judged through expansion coefficients or contribution to some property or other appropriate criterion. Finally, CSFs from higher order spaces, generated from external orbital sets, starting from a level of severe truncation, can be added systematically.

Now let us consider various possibilities for calculation where it is beyond our computational capacity to work at any one time with the full zeroth and first order spaces.

(i) Iterated natural orbital type CI calculations in an n-particle space containing the complete zeroth order space and the first order space complete except for the absence of the large {0,2;1} class. Starting from very approximate orbitals, this calculation will yield excellent internal orbitals for further use.[21,24] Moreover, in some calculations, where single excitations into external orbitals interact with {0,0;0} even for optimal internal orbitals, packing of the external orbitals may occur.

(ii) Whether or not packing of the externals results from calculation (i), the next step is to investigate {0,2;1}. We are assuming that working with the full space, or even the space involving excitation from only those internal orbitals crucial to the study, is beyond capacity. Two options are open; the first would be to look at the Rayleigh-Schrodinger first order perturbed wave function in the full space {0,2;1} or a more or less equivalent approximate CI in this full space, the second is to carry out accurate CI calculations in a space involving excitations from only one (or as many as possible) internal orbital pair into the full external set. This second approach is a calculation of the so called pseudo natural orbital (PSNO) type.[25,26]

(iii)   To investigate $\{0,2;1\}$ by Rayleigh–Schrodinger
perturbation theory one simply looks at the magnitude of either
$H_{KO}^2/(H_{KK}-H_{00})$, the contribution to the second order perturbation
energy, or $H_{KO}/(H_{KK}-H_{00})$, the coefficient of Φ_K in the first
order perturbed wave function, and drops CSFs whose contribution
is below a certain threshold.   In terms of the Hamiltonian matrix
for a CI calculation in this space, the required matrix elements
for evaluating these perturbation contributions do not include the
large set of interactions between members of $\{0,2;1\}$.

(iv)   An approximate CI calculation, which uses essentially
the same matrix elements required for the perturbation calculation, would be to find roots of the approximate CI matrix shown
schematically below.

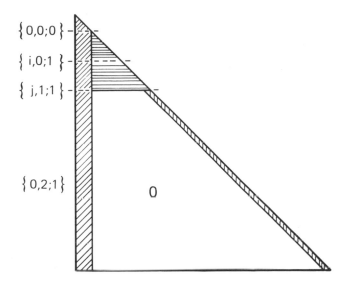

CSFs from $\{0,2;1\}$ would be dropped according to their contribution to the appropriate eigenvector or eigenvalue.[27]   We make
particular note that <u>single</u> excitations into the externals cannot
be dropped according to this criterion from this approximate CI,
or from the perturbation estimates of (iii).   This is most
dramatically illustrated by the case where $\{0,0;0\}$ contains only
a closed shell Hartree-Fock function. $\{0,1;1\}$ has no matrix
elements with $\{0,0;0\}$ and diagonalization of our approximate CI
matrix does not bring $\{0,1;1\}$ into the improved closed shell
wave function.   Diagonalization of the accurate CI matrix would,
however, bring $\{0,1;1\}$ in through interaction with $\{0,2;1\}$,
and in fact the contribution is an important one in the evaluation
of one electron properties.[28]

In the approximate CI matrix illustrated further approximation may be tried, but the key point is the setting to zero of all interactions within $\{0,2;1\}$ except the diagonal elements.

(v)   In cases with especially large basis sets it may be necessary to carry out calculations of type (iii) and (iv) in a stepwise manner, investigating only that part of $\{0,2;1\}$ arising from excitation out of a restricted set of internal orbitals.

(vi)   In the case that the calculation of (i) did not produce packing of the external orbitals it may be necessary to iterate the calculation of (iii) or (iv) using natural orbital transformations.

(vii)   In pseudo natural orbital calculations which offer an alternative to (iii) or (iv) the space $\{0,2;1\}$ is systematically investigated by carrying out a sequence of accurate CI calculations in a space involving excitations from as little as one internal orbital pair at a time.   These are laborious and time consuming and probably achieve results no more satisfactory than the methods of (iii) and (iv).

(viii)   Always the final step is to put together the results of all the previous calculations into accurate CI calculations in a truncated   n-particle space.   This final step has been arrived at by a combination of perturbation theory guidelines and variational calculations.   We emphasize that, even though perturbation theory has provided an important framework for setting up the calculations, our wave functions are not perturbation theory wave functions.   They are variational and provide a set of energy upper bounds.   It is in these final calculations that we start systematically investigating contributions from the second order space.   We insert the reminder that the second order space is not simply related to the usual notion of triple and quadruple replacements from a single CSF reference state.   Since $\{0,0;0\}$ may include more than one CSF, the space $\{0,2;1\}$ may well include CSFs that are triple and quadruple replacements from one of the members of $\{0,0;0\}$.   By the same token the second order space may contain CSFs that are single and double replacements from all the members of $\{0,0;0\}$ through our partitioning of degenerate spaces of a configuration.[16]

(ix)   So far, we have attempted to outline the variety of options open in systematic studies in configuration interaction. Actual choice is dependent on the problem at hand.   Whatever the choice, it would be desirable to compute the various levels of wave function (given levels of approximation within the   n-particle space) from systematically augmented   1-particle basis sets since it seems obvious that valuable convergence data could be so

obtained. This argues for (a) employing 1-particle basis sets
which contain all smaller 1-particle basis sets used in the study
as subsets, and (b) with smaller 1-particle basis sets carrying
out both the most accurate calculations possible and calculations
parallel to those done with the largest 1-particle basis sets.
Comparisons within (b) will again yield valuable convergence data.

RANGE OF APPLICATION AND COMPUTATIONAL ASPECTS

In the past, advanced research in computational chemistry
has demanded the most powerful computing machinery available;
this demand still exists and will exist as long as the properties
of molecules and their interactions are a topic of scientific
research. The applications making these computational demands will
continue to run from small systems, typified by those with less
than a dozen (valence) electrons, to the largest whose computational study can yield information not otherwise available. It
must be realized that, in spite of the advances made to the
present, there is still a great deal to be learned from atomic
systems, even those in the first row of the periodic table, while
the area of molecules with similar numbers of electrons is largely
untapped. It is applications at this end of the scale which, in
addition to providing needed data on specific systems, allow the
systematic testing of approximations and computational methods.
Tools tested and sharpened in these areas can be used with skill
and confidence in larger more complicated systems; this is why
advances in application to complex systems is strongly coupled to
advances in work of the highest possible accuracy in small systems.
In turn, that is why this particular topic belongs in a research
conference on large molecules, even though most of this paper has
been focused on requirements and techniques not yet adequately
tested in small systems.

Specifically addressing large molecule applications the
obvious first question is "what is a large molecule?" In the
context of this paper it asks what are the largest molecules that
can be treated by the ab initio methods that we have described,
aiming at an exact treatment within the 1-particle and n-particle
basis set constraints. Within these constraints, any approximations
made are negligibly small in regard to the determination of chemically interesting properties. If we accept the guidelines of
Table I for realistic sized 1-particle basis sets for molecular
prediction of quantitative worth, then atoms in the first row of
the periodic table require some fifteen basis functions of Slater
or contracted Gaussian type. In today's terms, thinking of
studies that can involve of order 100 hours of central processing
time on the largest available computers, while manipulating lists
of data containing millions of entries, this puts a large molecule

in the butadiene to benzene range. Calculations of quantitative
accuracy on this size system have not yet been carried out, although
with sufficient effort they could be. A few such calculations
would be of very great value in providing tests of more approximate
theories.

There are a number of directions which can be profitably
explored in reaching beyond the accurate ab initio large molecule
limit that we have just discussed.

(i) Use of lesser 1-particle basis sets will lower the
reliability of quantitative chemical prediction of properties such
as electric moments, but will be quite satisfactory for other
properties of great importance such as equilibrium nuclear
geometries. The paper by Pople in this volume addresses precisely
this point.

(ii) It may be possible to use basis sets tailored to a
specific chemical problem, with severe truncation in the region of
the molecule not strongly coupled to the behavior under study. A
limiting case of this type of approach is the treatment of regions
of a system through an effective potential. In fact, finding a
minimal representation which can describe specific phenomena is the
real challenge of research on molecules beyond the range of accurate
ab initio treatment.

(iii) Approximations within the constraints of a given
1-particle and n-particle basis offer other directions of investigation which may extend the range of computation. The most obvious
procedure is the approximation of large blocks of matrix elements,
particularly the crudest of such approximations which is to set all
matrix elements in certain blocks to zero. Such approximations
are often justifiable, especially in cases where one is looking at
property differences, and the same approximation is made in
evaluating the compared properties. The approximation can be valid
both for chemical data on a single system and in looking for trends
in data along a series of related systems. This type of approximation is the same one underlying all semi-empirical theories of
molecular structure. The measure of success is the vindication of
the approximation.

A final, and important, statement on large molecules is that
ab initio work, as well as experiment, is necessary in calibrating
and understanding the semi-empirical theories which offer the
only real hope in the foreseeable future of providing information
on really large systems. We are unlikely to ever reach beyond
the barrier to molecular ab initio computation contained in the
fact that the computational demands are proportional to some

power of the number of electrons in the range 3-6 depending on the computational weights of different steps.

Turning to purely computational aspects, which are common to the whole spectrum of application just discussed, the most important point to make is that the speed of current central processing units allows work with basis sets that give rise to large lists of intermediate data which must be kept on peripheral storage (magnetic disk or tape, for example). Storing and processing this data in turn taxes the capacity of the data channels to transfer information between this peripheral storage and the high speed storage used by the central processing unit. Processing the data taxes the ingenuity of the scientist designing computer programs in finding efficient algorithms. For example, in quantum chemical applications potential bottlenecks in the processing of matrix elements have been avoided by the use of sophisticated sorting techniques[30] using direct access peripheral storage, in order to complete processing in a fixed number of passes over the data rather than a number proportional to the length of the data.

The essential point is that these applications, falling into the class of large scale scientific computation, are not simply exercises in "number crunching", but from a computational point of view are extraordinarily difficult problems in the balanced use of the total resources of a large computer installation. A key element of future advances in quantum chemistry is the devoting of resources, along with the necessary cooperation, to see that the required computational tools are developed, tested, documented, and made available to research workers.

The obvious requirement of a largely computer independent language, with program versions tailored to specific installations, should be noted. In the United States, these factors strongly argue for a National Computation Center for Quantum Chemistry, which has been under discussion for some years, one of whose important commitments would be to making the tools useable and available.

REFERENCES

1.   W. Heitler and F. London, Z. Physik $\underline{44}$, 455 (1927).

2.   H. James and A. Coolidge, J. Chem. Phys. $\underline{1}$, 825 (1933).

3.   W. Kolos and L. Wolniewicz, J. Chem. Phys. $\underline{43}$, 2429 (1965).

4.   H. F. Schaefer III, "The Electronic Structure of Atoms and Molecules" Academic Press, New York 1972.

5.   See, for example, "IBMOL-5 Program User's Guide," E. Clementi
      and J. Mehl, IBM Research Report RJ 889, 1971 (unpublished)
      and "BISON, Part I, User's Manual and General Program
      Description," A. C. Wahl, P. J. Bertoncini, K. Kaiser, and
      R. M. Land, Argonne National Laboratory Research Report
      ANL-7271, 1968 (unpublished).

6.   P. O. Lowdin, "Advances in Chemical Physics," Vol II,
      Interscience Publishers, Inc. New York, 1959.

7.   J. C. Slater, "Quantum Theory of Molecules and Solids,"
      Vol. I and II, McGraw-Hill, New York (1963).

8.   C. F. Bunge, Theoret. Chim. Acta 16, 126 (1970).

9.   C. C. J. Roothaan, Rev. Mod. Phys. 23, 69 (1951).

10.   Compendia of optimized sets of Slater type elementary func-
      tions for atoms and small molecules may be found in:

10a. P. S. Bagus, T. L. Gilbert, and C. C. J. Roothaan, J. Chem.
      Phys. 56, 5195 (1972);

10b. E. Clementi, "Tables of Atomic Wave Functions," a supplement
      to IBM J. Res. Develop. 9, 2 (1965);

10c. A. D. McLean and M. Yoshimine, "Table of Linear Molecule
      Wave Functions," a supplement to IBM J. Res. Develop. 12,
      206 (1968).

11.   A classic case study of optimization of elementary functions
      for a small molecule is given by P. E. Cade, K. D. Sales,
      and A. C. Wahl, J. Chem. Phys. 44, 1973 (1966).

12.   A. D. McLean and M. Yoshimine, unpublished work.

13.   C. C. J. Roothaan and P. S. Bagus, "Methods in Computational
      Physics," Vol II, Academic Press, New York (1963).

14.   S. Huzinaga, "Approximate Atomic Functions," Volumes I
      and II, Research Report of the Dept. of Chemistry of the
      University of Alberta, 1971 (unpublished).

15.   See for example, J. C. Slater, "Quantum Theory of Atomic
      Structure," Vols. I and II, McGraw-Hill, New York (1960);
      U. Fano, Phys. Rev. 140, A67 (1965); R. K. Nesbet, J. Math.
      Phys. 2, 701 (1961); F. Sasaki, Int. J. Quant. Chem.
      (in press).

16. A. D. McLean and B. Liu, submitted for publication.

17. E. R. Davidson, "Advances in Quantum Chemistry," Vol. 6,
     Academic Press, New York (1972).

18. We recall that both γ' and γ are obtained from variationally
     determined functions.

19. C. F. Bunge, Phys. Rev. (in press).

20. G. Das and A. C. Wahl, J. Chem. Phys. 56, 1769 (1972).

21. The iterative natural orbital method was introduced and
     used in a somewhat different situation by E. R. Davidson
     and C. F. Bender, J. Phys. Chem. 70, 2675 (1966); an example
     of the use of the method as we have discussed it is given by
     H. F. Schaefer, J. Chem. Phys. 54, 2207 (1971).

22. B. Liu, Phys. Rev. Letters 27, 1251 (1971).

23. Unpublished work of the authors on several systems including
     CH, LiO, KrF, and KrF_2.

24. F. Sasaki and M. Yoshimine, unpublished results.

25. C. Edmiston and M. Krauss, J. Chem. Phys. 45, 1833 (1966).

26. A. W. Weiss, Phys. Rev. 162, 71 (1967).

27. Z. Gershgorn and I. Shavitt, Intern. J. Quantum Chemistry 1S,
     403 (1967).

28. S. Green, J. Chem. Phys. 54, 827 (1971).

29. R. K. Nesbet, "Advances in Chemical Physics," Volume 14, page
     1, Interscience Publishers (1969).

30. M. Yoshimine, J. Computational Phys. (in press).

APPROXIMATE METHODS IN QUANTUM CHEMISTRY

Michael C. Zerner

Dept. of Chemistry, University of Guelph

Guelph, Ontario, Canada

INTRODUCTION

The goal of any semi-empirical or approximate method is the achievement of a compromise between ease and accuracy. In molecular quantum chemistry ease generally means speed of obtaining results. Any method to be of practical use must execute at least as rapidly as methods which are more accurate. As obvious as this would seem, many suggested methods suffer from exactly this disadvantage. The criterion of accuracy is somewhat more difficult to define. Accurate with respect to what? If an approximate method confines itself closely to an exact theory, then the results should reproduce those obtainable from a correct treatment of that theory. If a method introduces pure parameters, then, perhaps, it is best guided to this purpose also. But if a method introduces semi-empirical parameters chosen from experiment, there exists the tempting idea that the model might extend beyond the confinements of the theory and best be compared directly with experiment. As attractive as a direct relation to experiment is, the idea is easy to abuse and has often led to a different method for different observables. Nevertheless, it is difficult to deny the utility of the Pariser-Parr-Pople (1) pi election model, and especially some of its refinements (2), in organizing thought about $\pi \to \pi^*$ spectra.

SOME CURRENT METHODS

I will review <u>some</u> of the more popular and promising semi-empirical and approximate methods, and must apologize for not

including more. Rather I refer those interested to a review
paper by Jug (3), a very complete bibliography included in a
paper by Nicholson (4), and a recent paper by Billingsley and
Bloor (5) where their ideas are compared with several others.
Similarly, I will confine myself to methods within the L.C.A.O. –
M.O. framework, not mentioning the very promising Scattered Wave
X̊-α method of Slater and Johnson (6), nor configuration inter-
action techniques such as that of Diner, Malrieu and Claverie (7).

Looking through the literature one still finds the simple
Hückel Molecular Orbital (HMO) method quite popular, especially
among experimental organic chemists. The great "eloquence" of
this method is its simplicity. Indeed, many chemists have developed a "feel" for the various parameters used (8), and several
universities have courses such as "Hückel Molecular Orbital
Theory Applied to Organic Chemistry". Unfortunately, sometimes
the method degenerates into a search for parameters, or is called
upon to rationalize phenomena beyond its reach.

In 1963 Lohr (9) and Hoffmann (10) introduced and system-
atized the Extended Hückel Method (EHM). This is an all valence
electron method that calculates the overlap, $\underset{\sim}{\Delta}$, between Slater
Type Orbitals (STO's), and that sets the off diagonal elements of
the Hamiltonian matrix,$\underset{\sim}{H}$, proportional to the average of the
appropriate diagonal terms times the overlap. The following
equations summarize the model.

$$\underset{\sim}{H}\,\underset{\sim}{C} = \underset{\sim}{\Delta}\,\underset{\sim}{C}\,\varepsilon$$

$$H_{ij} = \Delta_{ij}K_{ij}[H_{ij}+H_{ij}]/2$$

$$K_{ij} = [K_{ii} + K_{jj}]/2 \sim 1.75 - 2.00$$

$$H_{ij} = \text{Valence State Ionization Potential}$$

Perhaps no other method has had such an impact on experimental
chemistry, for it was through an examination of results from these
calculations that led Woodward and Hoffmann to postulate their
rules for concerted chemical reactions (11). Although it can be
argued that other methods give the same order of orbital symmetries,
this method is good enough.

Several modifications of the original method have been introduced. I mention but one of these.

In 1966 Rein, Fukuda, Win, Clarke and Harris, (12), Carrol,
Armstrong and McGlynn (13), and Zerner and Gouterman (14) introduced procedures to correct for the excessive charge built up at
electronegative centers in the original model. These methods

generally extrapolated parameters between appropriate positive
and negative ions in an iterative procedure until the "charges"
calculated for an atom were consistent with the parameters
assumed. Although these iterative methods gave results in which
the charge density appeared too smoothly distributed, such a
procedure was found essential in going directly from atomic
parameters to meaningful molecular results for transition metal
complexes (14).

In 1966 Pople and co-workers introduced their Complete
Neglect of Differential Overlap (CNDO) model (15), summarized in
the following equations:

1)   Rotational Invariance: $\gamma_{\mu\nu}^{AB} = \gamma_{AB} = (S_A S_A | S_B S_B)$

2)   Core Integrals: $U_{\mu\mu}^{AA} \equiv (\mu|-\nabla^2/2-Z/R|\mu) = -I_A - (Z_A-1)\gamma_{AA}$ or

$-(I_A+A_A)/2 -(Z_A-1/2)\gamma_{AA}$

3)   ZDO: $(\mu\nu| = \delta_{\mu\nu}(\mu\mu|$

4) $V_{AB} \equiv (S_A|Z_B/R_B|S_A) = Z_B\gamma_{AB}$

5) $H_{\mu\nu}^{AB} \equiv \Delta_{\mu\nu}(\beta_A + \beta_B)/2$

In many ways the introduction of this method was important, for
it attempted to tie a systematic, approximate method to an exact
formalism. The work stressed the necessity of choosing parameters that maintained essential invariants – a criticism of
several previously postulated techniques. To compensate for the
neglected core orbitals, one center core integrals ($U_{\mu\mu}$) were
approximated from ionization potentials, I, (CNDO/1) or from ionization potentials and electron affinities, (I+A)/2 (CNDO/2).
After this, the basis set was assumed to be orthogonal, and ideas
of Zero-Differential Overlap (ZDO) were invoked. The Intermediate Neglect of Differential Overlap (INDO) technique differs
from CNDO by including one center exchange integrals (16).

Several variants of these ideas have also been introduced, of
which I briefly mention three.

There has been an attempt by several investigators to gather
more information from CNDO by assuming the calculation relates to
a symmetrically orthogonalized set ($\chi'=\chi\Delta^{1/2}$); they thus de-
orthogonalize the resulting solution. This approach I think is
unlikely to be successful, for a symmetrically orthogonalized ab-
initio Fock matrix, properly core orthogonalized, bears little
resemblance to the CNDO/2 Hamiltonian matrix. For this comparison
core integrals obtained from ionization potentials appear to be
better. This idea would also seem to give eigenvalues in better

Fock Matrix for N_2 (A.U.)

	2s	2pσ	2pπ	2s	2pσ	2pπ	2s	2pσ	2pπ
2s	-0.65			-0.91			-0.81		
A 2pσ	0.24	-0.16		0.17	-0.42		0.17	-0.26	
2pπ	0	0	-0.15	0	0	-0.27	0	0	-0.13
2s	-0.57	-0.63	0	-0.64	-0.58	0	-0.50	-0.51	0
B 2pσ	-0.63	-0.61	0	-0.58	-0.35	0	-0.51	-0.44	0
2pπ	0	0	-0.43	0	0	-0.63	0	0	-0.48
	$\Delta^{-1/2}(F + V) \Delta^{-1/2}$ ab-initio			F(CNDO/2) U from 1/2 (I + A)			F(CNDO/2) U from I		

agreement with ab-initio calculations, but the simple interpretation given the dipole moment by Pople and co-workers is destroyed.

A second modification has been a spectroscopic parameterization begun by Del Bene and Jaffe (17). Taking advantage of the information gained from pi orbital only calculations, this reparameterization has led not only to accurate predictions of $\pi \rightarrow \pi^*$ spectra, but appears reliable in indicating the locations of $n \rightarrow \pi^*$.

The final modification I mention is by Dewar and co-workers – MINDO/2 (Modified INDO) (18). The theory as originally postulated has been considerably changed, and parameters occuring have been adjusted to fit experimental bond lengths and heats of formation. If one is willing to subtract constants for bonds involving hydrogen atoms, the method is remarkably accurate for bond lengths. The method also is accurate in reproducing heats of formation. This would seem surprising, as the connection between enthalpies of formation and the results of molecular calculations is not direct. Rather appeal has been made here directly to the concept of bond energy additivity.

Several other investigators have pursued the simplification inherent in utilizing an orthogonalized basis set. The more exact of these are typified by methods suggested by Cook, Hollis and McWeeny (19) and by Brown, Burden and Williams (20). The approach is summarized below.

1) $H_{\mu\nu} = (\mu|-\nabla^2/2 - \Sigma_A Z_A/R_A|\nu)$
   $\underset{\sim}{H}' = \underset{\sim}{\Delta}^{-1/2} \underset{\sim}{B}^+ \underset{\sim}{A}^+ \underset{\sim}{H}' \underset{\sim}{A} \underset{\sim}{B} \underset{\sim}{\Delta}^{-1/2}$

2) $(\mu'_a \nu'_b| = \delta_{\mu\nu} (\mu'_a \mu'_a|$ ZDO

2´) $(\mu'_a \nu'_b| = \delta^{ab} (\mu'_a \nu'_a|$ NDDO

The one electron Hamiltonian is calculated exactly for STO's. $\underset{\sim}{H}$ is then core orthogonalized, $\underset{\sim}{A}$, perhaps hybridized $\underset{\sim}{B}$, as suggested by Cook, Hollis and McWeeny, and Lowdin orthogonalized, $\underset{\sim}{\Delta}^{-1/2}$. Advantage is then taken of the observation that in the orthogonal-

ized set many integrals involving differential overlap can be
neglected.   The following table presents some of these integrals
for N_2 (ξ = 1.95, ρ = 3.25).

		SYM.				SYM.
	STO's	ORTH.			STO's	ORTH.
$(2s_a2s_a\|2s_a2s_a)$	0.709	0.738	$(2s_a2p_{\sigma a}\|2s_b2s_b)$		0.121	0.084
$(2s_a2s_a\|2s_b2s_b)$	0.452	0.437	$(2s_a2p_{\sigma a}\|2p_{\sigma b}2p_{\sigma b})$		0.135	0.094
$(2s_a2s_a\|2s_a2s_b)$	0.277	-0.016	$(2s_a2p_a\|2s_a2p_a)$		0.157	0.138

   The first group of integrals indicates that one center
Coulomb integrals generally increase slightly, two center Coulomb
integrals decrease, while those involving differential diatomic
overlap become small, and usually negative. The second group
indicates that integrals which involve one center differential
overlap must <u>not</u> be neglected. The hybridization in the Cook,
Hollis and McWeeny scheme might be considered as somewhat awkward
in a general method – especially if one searches for bond angles.
The scheme proposed by Brown et. al. includes an estimate of the
non-vanishing integrals in the orthogonalized set. This is rather
cumbersome: the simple empirical scaling procedure adopted by
Cook, Hollis and McWeeny incorporated into Brown's procedure
appears to give results very similar to those reported by Brown.

   In 1966 Manne introduced his "Valence" method (21), in which
he carefully considered the explicitly ignored core electrons,
calculated the kinetic energy, and applied the simple Mulliken
approximation (22) for potential energy integrals. To insure

$$(\mu\nu| = \Delta_{\mu\nu}\{\mu\mu| + (\nu\nu| \ \}/2$$

$$(\mu\nu|\sigma\sigma) = \Delta_{\mu\nu}\{(\overline{\mu\mu}|\overline{\sigma\sigma}) + (\overline{\nu\nu}|\overline{\sigma\sigma})\}/2$$

rotational invariance, $\overline{\mu}$ indicates A.O. μ is of S symmetry. The
problems of this approach are shortcomings in the Mulliken approximation itself, some of which are indicated below.

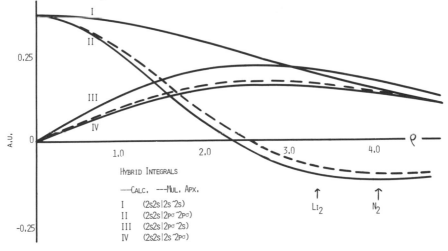

HYBRID INTEGRALS

—CALC. ---MUL. APX.

I   (2s2s|2s¯2s)
II   (2s2s|2pσ¯2pσ)
III   (2s2s|2pσ¯2s)
IV   (2s2s|2s¯2pσ)

Note particularly that integrals III and IV are predicted to be
the same and that integrals involving pσ diatomic differential
overlap can be underestimated by large factors. Integrals
involving overlap squared are, of course, progressively worse.

Most of the more accurate approximate schemes involve
Ruedenberg's generalization (23) of the simple Mulliken idea.
Differential overlap is expanded in terms of A.O.'s on the two
centers involved. The coefficients, C, in the expansion are

$$\pi^{AB}_{(ij)} = \chi^A_i \chi^B_j = \sum_p C_{kp} \Omega^A_p + \sum_q C_{kq} \Omega^B_q$$

$$\{\Omega^A\} = \{1s1s, 1s2s, 1s2p_x \ldots 2p_z 2p_z \ldots\}$$

choosen by a least square fit to match overlap, as in the PDDO
(Projection of Diatomic Differential Overlap) method of Newton
and co-workers (24), or to reproduce hybrid integrals as
originally suggested by Harris and Rein (25), and utilized by
Billingsley and Bloor in their LEDO (Limited Expansion of Diatomic Overlap) method (5).

Both of these methods are very accurate in reproducing the
results of the exact calculations to which they are modeled. The
criticism, if there is one, is met in expanding single integrals
in terms of hundreds or thousands of simpler ones. Because of
this, one might expect execution time to approach that of the
STO - nG method (26), which expands STO's in terms of n Gaussian
functions by a fitting procedure (27). Since this latter method
is variational, it might be preferred.

                              THE CORE

In developing an approximate M.O. method, it seems reasonable
to start with simplifying assumptions which are soundest, leaving
the more questionable approximations for last where they can be
most easily modified. Dropping explicit consideration of the core
has been one of the common features in all but the most demanding
methods. There are several reasons for believing this omission
can be successful. Chemists seldom invoke consideration of core
orbitals in explaining phenomena. In addition, early studies on
diatomic molecules in which outer shell orbitals were orthogonalized to the inner shell produced small matrix elements connecting
the two sets.

The first attempt to systematically eliminate consideration
of the core is that of Phillips and Kleinman in 1956 (28), using
a method generalized and discussed in a review paper on pseudopotentials by Weeks, Hazi and Rice (29). Manne in his "Valence"
method used a partitioning technique and derived a Brillouin type

perturbation expansion representing the core potential (21). Both these have the advantage of not relying on the concept of a "frozen" core. Their utilization, however, is somewhat complicated: Manne's is very accurate, but nearly as complicated as the original secular equation itself.

A simpler, if less accurate, approach begins by Schmitt orthogonalizing the valence orbitals to the core (30).

$$\chi_\alpha' = \chi_\alpha \qquad \alpha, \beta \ldots = \text{core A.O.'s}$$

$$\chi_i' = \chi_i - \sum_\alpha \Delta_{i\alpha} \chi_\alpha \qquad i, j \ldots = \text{valence A.O.'s}$$

The overlap and Fock matrices are then developed in terms of the old matrices plus corrections.

$$\Delta_{ij}' = \Delta_{ij} - 2\sum_\alpha \Delta_{i\alpha} \Delta_{\alpha j} + \sum_{\alpha,\beta} \Delta_{i\alpha} \Delta_{\alpha\beta} \Delta_{\beta j}$$

$$F_{ij}' = F_{ij} - \sum_\alpha (\Delta_{i\alpha} F_{\alpha j} + F_{i\alpha} \Delta_{\alpha j}) + \sum_{\alpha,\beta} \Delta_{i\alpha} F_{\alpha\beta} \Delta_{\beta j} \equiv F_{ij} + V_{ij}$$

These matrices are than simplified by the following three approximations:

a) $\Delta_{ij}' = \Delta_{ij}$

b) $F_{i\alpha}' = 0$

c) $F_{\alpha\beta}' = \delta_{\alpha\beta} F_{\alpha\alpha}$

The first approximation concerning overlap between valence orbitals is generally within 1-2%. Although the orthogonalization greatly reduces elements of the Fock matrix between core and valence orbitals, $F_{i\alpha}'$, they are not, in general, zero. In the worst diatomic case examined, however, that of N_2, the error is 0.009 A.U. in a total energy of 109A.U. These approximations yield for the "pseudo-potential" V

$$V_{ij} = -\sum_\alpha \Delta_{i\alpha} F_{\alpha\alpha} \Delta_{\alpha j}$$

an expression involving only inner-shell outer-shell overlap, and the diagonal elements of the Fock matrix for core orbitals, $F_{\alpha\alpha}$. $F_{\alpha\alpha}$ can be calculated, or can be taken from an average of model molecular calculations. Realizing that $F_{\alpha\alpha}$ values are nearly the same for atoms as they are for molecules, they can be taken from atomic calculations. Since $F_{\alpha\alpha}$ (atom) $\approx \varepsilon_\alpha$ (atom) the easiest, and apparently most successful, values are obtained from ionization potentials of core electrons as determined by ESCA (31).

Applying these ideas, the Fock matrix can then be written in terms of the one electron Hamiltonian defined for "core charges", $\underset{\sim}{H}(Z^V)$, the pseudo-potential, and the two electron matrix $\underset{\sim}{G}$,

$$F_{ij}' = H_{ij}(Z^V) + G_{ij}(VV) + V_{ij}$$

where $\underset{\sim}{G}(VV)$ indicates consideration of only valence orbitals. The total energy is now expressible in terms of the valence energy, E^V, plus one center atomic terms.

$$E_T = E^V + \{\sum_\alpha \eta_\alpha H_{\alpha\alpha}(\text{atom}) + 1/2 \sum_{[\alpha,\beta]\epsilon A} \eta_\alpha \eta_\beta [\alpha\alpha|\beta\beta) - 1/2(\alpha\beta|\alpha\beta)]$$
$$- 1/4 \sum_{[\alpha,i]\epsilon A} \eta_\alpha \eta_i (\alpha i|\alpha i)\}$$

(η_α is number of electons in A.O. α for the isolated atom; $[\alpha,\beta]\epsilon A$
indicates both orbitals in summation simultaneously belong to the
same atom.) This expression suggests that most system properties
of interest should be reflected in the valence energy. This
treatment also dictates that the nuclear repulsion energy to be
used is the classical one between core charges, $\Sigma(A<B)Z_A^V Z_B^V/2$, and
that no scaling is required.

The table below shows the result of this core separation on
first row atoms. The differences in calculated valence shell
energies and those obtained from experiments (32) are principally
those of starting with a calculation far from the Hartree Fock

Valence Calculations on Atoms, from (30)

	E^V(a.u.)	E_{exp}^a (a.u.)	E_T	E_{calc}^b
Li	-0.199	-0.198	-7.426	-7.418
Be	- .973	-1.012	-14.595	-14.556
B	-2.520	-2.623	-24.534	-24.495
C	-5.240	-5.438	-37.651	-37.619
N	-9.472	-9.808	-54.286	-54.265
O	-15.064	-15.915	-74.525	-74.533
F	-23.272	-24.209	-98.905	-98.934

a)   From Moore's Tables, (32)
b)   Minimum basis set calculations using exponents from Slater's
     rules, (33).

limit, and electron correlation. The total energies calculated
from the "valence energy only" approach are in good agreement with
those from ab-initio calculation. The next table shows that the
valence calculations mimic to some degree the trends of ab-initio
calculations on which they are modeled.

Valence Calculations on Homonuclear Diatomics[a], from (30)

	E^V(a.u.)	D_e^V(ev)	D_e(ev)[b]
Li_2	-0.3935	+0.11	-0.15
N_2	-18.9864	-1.16	-1.19
F_2	-46.5284	+0.40	+0.30

a)   Minimal basis using Slater Orbitals (33)
b)   From Ransil (34)

THE ONE ELECTRON MATRIX

The one electron Hamiltonian is not difficult to evaluate
exactly, the only time consuming part being the evaluation of

three center nuclear attraction integrals. A particular method,
however, may require adjustments to the one electron matrix to
compensate for approximations introduced in estimating the two
electron matrix.

A common feature of many semi-empirical methods has been the
selection of parameters for the diagonal elements of the one
electron matrix from ionization potentials, or from ionization
potentials and electron affinities (CNDO/2). The more systematic
procedures use these atomic parameters plus the appropriate
coulomb and exchange terms to estimate the one-center one-electron
core integral, U.

$$U_{\mu\mu} = (\mu|-\nabla^2/2 - Z/R|\mu)$$

The major rationale for using such a procedure is to include the
inner-shell outer-shell "repulsion" in a simple fashion, as well
as some of the correlation energy, etc.

An alternant procedure for obtaining core integrals semi-
empirically is directly from atomic spectra. This technique
escapes the criticism of orbital expansions and contractions met
in going from atoms to negative and positive ions. The table
below compares atomic energies as calculated from ionization
potentials (INDO/1), ionization potentials and electron affinities
(INDO/2) and from core integrals obtained by Karlsson and Zerner
from a least squares fit of atomic configurations (35).

Atomic Valence Energies (A.U.)

	INDO/1	INDO/2	K-Z(35)	EXP.
Li	-.198	-.232	-.198	-.198
Be	-1.039	-1.143	-1.009	-1.012
B	-2.689	-2.907	-2.623	-2.623
C	-5.641	-6.052	-5.425	-5.426
N	-10.220	-10.824	-9.808	-9.810
O	-16.706	-17.398	-15.911	-15.895
F	-24.668	-26.431	-24.209	-24.209

a) From Moores tables (32), plus term promotion energies.

The INDO/2 idea seems quite incapable of reproducing atomic
energies. Although it is tempting to begin a molecular calculation
with parameters that accurately reproduce atoms, numerical investigations indicate that "underestimates" of core integrals obtained
from ionization potentials are often of the same size as are the
ignored, or grossly underestimated, two center exchange terms of
the diagonal Fock Matrix. Of course, the <u>distance</u> dependence is
quite different.

THE TWO ELECTRON MATRIX

All the real difficulty in molecular quantum mechanics is met in calculating the two electron matrix $\underset{\sim}{G}$, which, for a closed shell L.C.A.O.-M.O. calculation is given as

$$G_{ij} = J_{ij} - K_{ij} = \sum_{\alpha,\lambda} P_{\sigma\lambda}(ij|\sigma\lambda) - 1/2\sum_{\sigma,\lambda} P_{\sigma\lambda}(i\sigma|j\lambda)$$

where $\underset{\sim}{J}$ is the Coulomb contribution, $\underset{\sim}{K}$ the exchange part, and $\underset{\sim}{P}$ the Fock-Dirac density.

Two well known difficulties arise which collectively become worse for large molecules. Firstly, many of the integrals are difficult to evaluate; secondly, there is an n^4 summation implied in forming $\underset{\sim}{G}$, where n is the number of basis functions.

Examining first the Coulomb matrix, a great simplification is effected by utilizing the Mulliken idea for differential overlap. We obtain

$$J_{\mu\nu}^{AB} = \sum_{\alpha,\lambda} P_{\sigma\lambda}(\mu\nu|\sigma\lambda) = S_{\mu\nu}^i [J_{\mu\mu}^{AA} + J_{\nu\nu}^{BB}]/2$$

and observe that all two center terms can be obtained from "appropriate" one center elements.

The question now asked is "can a simple overlap be found from which the off diagonal elements of $\underset{\sim}{J}$ can be formed given the exact diagonal terms?" Several simple "Projection overlaps" are suggested, three of which are listed below.

1) Orbital Overlap
$$S_{\mu\nu} = (\mu|\nu)$$

2) Local Potential
$$S_{\mu\nu}^V = \frac{2[(\mu|Z_A/R_A|\nu) + (\mu|Z_B/R_B|\nu)]}{[(\bar{\mu}|Z_A/R_A + Z_B/R_B|\bar{\mu}) + (\bar{\nu}|Z_A/R_A + Z_B/R_B|\bar{\nu})]}$$

3) Hybrid Integrals
$$S_{\mu\nu}^L = [\frac{2}{Z_A + A_B}] \{\frac{Z_A(\mu\nu|\mu\mu)}{[(\bar{\mu}\bar{\mu}|\bar{\mu}\bar{\mu}) + (\bar{\mu}\bar{\mu}|\bar{\nu}\bar{\nu})]} + \frac{Z_B(\mu\nu|\nu\nu)}{[(\bar{\mu}\bar{\mu}|\bar{\nu}\bar{\nu}) + (\bar{\nu}\bar{\nu}|\bar{\nu}\bar{\nu})]}\}$$

where $\mu \epsilon A$, $\nu \epsilon B$ and $\bar{\mu}$, $\bar{\nu}$ indicate A.O.'s μ and ν are given S symmetry. The last projection overlap weights the hybrid integrals by approximately the weight they receive in calculating $\underset{\sim}{J}$. The following table is an example of how well these overlaps perform for diatomic nitrogen. The "most appropriate" diagonal Coulomb elements are averages within a subshell. For estimating $\underset{\sim}{J}$, $\underset{\sim}{S}^L$ gives consistently the best results. In most cases $\underset{\sim}{S}^V$ gives similar results, the exception being N_2 as shown in the table (note $J_{\pi\pi'}$).

Coulomb Terms for N_2 (A.U.)(a)

	$S_{\mu\nu}$	$S_{\mu\nu}^V$	$S_{\mu\nu}^L$	Exact
J_s	5.74	5.76	5.81	5.82
$J_{\sigma\sigma}$	5.96	5.98	6.03	6.05
$J_{\pi\pi}$	5.64	5.66	5.71	5.74
J_{ss}'	2.59(b)	2.69(b)	2.74(b)	2.76
$J_{\sigma\sigma}'$	1.84(b)	2.31(b)	2.35(b)	2.29
$J_{\pi\pi}'$	1.62(b)	1.43(b)	1.59(b)	1.59
J_{so}'	2.53(b)	2.81(b)	2.84(b)	2.79
$\Delta\%$(c)	5.1%	1.9%	0.8%	

a)   Over Valence Shell Orbitals Only, Using Ransil Density (34)

b)   from $J_{\mu\nu}^{AB} = S_{\mu\nu}[J_{\overline{\mu\mu}} + J_{\overline{\nu\nu}}]/2$

c) $\Delta\% \equiv [\Sigma(\Delta J)^2/\Sigma J^2]^{1/2}$

In the above table I have gone one step further and also calculated the diagonal terms using these overlaps. The simplest method capable of reliably calculating the diagonal elements of J from S^L that I have found is summarized below.

$$J_{\mu\mu}^{AA} = \Sigma P_{\sigma\lambda}(\mu\mu|\sigma\lambda)$$

$$= [\sigma,\lambda]\varepsilon A^{\Sigma} P_{\sigma\lambda}(\mu\mu|\sigma\lambda) + \{\Sigma_{\sigma} M_{\sigma\sigma}(\mu\mu|\overline{\sigma}\overline{\sigma}) - \sigma\varepsilon A^{\Sigma} P_{\sigma\sigma}(\mu\mu|\sigma\sigma)\} +$$

$$+ 2\Sigma P_{\sigma\lambda}(\overline{\mu\mu}|\sigma\lambda)$$
$$[\sigma<\lambda]\varepsilon B$$
$$B \neq A$$

$$M_{\sigma\sigma} \equiv \Sigma P_{\sigma\lambda} S_{\lambda\sigma}$$

The first term in the expression for $J_{\mu\mu}$ involves only one center integrals and is easily evaluated exactly. The second term indicates that two center coulomb integrals may be evaluated with only modest loss of accuracy treating all orbitals of other atoms as if they were of S type. The last term, with two different orbitals on another atom, cannot be ignored without loss of con- siderable accuracy. Worse, this term is often negative, and there is little hope that it will cancel with an underestimate of ex- change terms. There appears to be little loss of accuracy, however, if this last term is evaluated treating μ as if it were of S symmetry.

I know of no method simpler than the PDDO and LEDO ideas that can be used to reliably estimate the exchange matrix. Application of the Mulliken approximation, even including one center differen- tial overlap, seriously underestimates these terms. Evaluating K in a symmetrically orthogonalized set, not correcting the int- egrals, and neglecting only diatomic differential overlap underestimates these terms just as badly as does the simple Mulliken approximation. If the integrals are estimated in the orthogonal set by a simple scaling procedure as, for example, that recommended by Cook, Hollis and McWeeny (19), the diagonal elements of $\underset{\sim}{K}$ can be corrected, but the two center elements are still badly underestimated.

To obtain the two center exchange terms from the appropriate diagonal elements requires an approximation such as suggested by Ruedenberg (23).

$$\chi_\mu^A(1)\chi_\nu^B(2) = S_{\mu\nu}^i[\chi_\mu^A(1)\chi_\mu^A(2) + \chi_\nu^B(1)\chi_\nu^B(2)]/2$$

Although the details of such an approximation are rather poor, a numerical investigation is surprising, as shown in the following table, where $G_{\mu\nu}^{AB} = S_{\mu\nu}^V[G_{\mu\mu}^{AA} + G_{\nu\nu}^{BB}]/2$, and $G_{\overline{\mu\mu}}$ is that obtained from a core orthogonalized ab-initio calculation.

Two Electron Matrix for N_2 (A.U.)[a]

	2s	2pσ	2pπ
2s´	2.20(2.18)	2.30(2.27)	0(0)
2pσ´	2.30(2.27)	1.88(1.87)	0(0)
2pπ´	0(0)	0(0)	1.18(1.20)

Two Electron Matrix for LiF [a]

F /Li	2s	2pσ	2pπ
2s	1.28(1.30)	2.05(2.08)	0(0)
2pσ	0.25(0.24)	0.32(0.28)	0(0)
2pπ	0(0)	0(0)	0.47(0.43)

a) Numbers in parenthesis are from ab-initio calculations, $\underset{\sim}{F} + \underset{\sim}{V} - \underset{\sim}{H}(Z^V)$

The problem that remains, if one is willing to suffer the inaccuracies indicated in the above table, is the calculation of the diagonal elements of $\underset{\sim}{K}$. Roughly two-thirds of the value of $K_{\mu\mu}$ in a valence calculation stem from one-center contributions that are not difficult to treat. If rather unsatisfying, purely empirical equations can readily be formulated that are fairly successful in correcting the two center part of $G_{\mu\nu}$ (36). Some results of such a procedure are given below.

N_2

	$2\sigma_g$	$2\sigma_u$	$3\sigma_g$	$1\pi_u$	$1\pi_g$	R_e (Å)	ω_e (cm^{-1})
L-STO(a,b)	−1.45	−0.73	−0.54	−0.58	0.27		
APX. H.F.(c)	−1.47	−0.78	−0.64	−0.62		1.122	2730
THIS APX.(d)	−1.38	−0.75	−0.60	−0.62	0.22	1.12	~2400
EXP.	−1.38	−0.69	−0.57	−0.61		1.094	2358

CO

	3σ	4σ	5σ	6σ	1π	2π	μ(Debye)
L-STO(a,b)	-1.49	-0.73	-0.48	0.93	-0.58	0.26	0.730
APX. H.F.(f)	-1.52	-0.80	-0.55		-0.64		0.274
THIS APX.	-1.46	-0.75	-0.55	1.11	-0.63	0.21	0.20
EXP.(e,g)	-1.42	-0.74	-0.54		-0.64		0.112

(a) Eigenvalues in A.U.   (c) Ref. (37), Exp. R_e
(b) From Ransil, Ref. (34)   (d) Ref. (36)
(e) μ from Ref. (40)   (f) Ref. (39), Exp. R_e

(g) Eigenvalues from Koopmann's Apx., Ref. 31: R_e, ω_e from
     Hertzberg (38).

FORWARD

   Much attention has been devoted to the development of approximate molecular methods by many investigators. The attraction
of calculating the electronic structure of large systems in
seconds or minutes rather than hours or days (if at all) is
irresistable. The results of such calculations, when used with
care, have proven very useful indeed in suggesting rationale for
experimental behavior. In addition, such model building has also
led to a physical "feel" for the mathematical expressions encountered in molecular quantum mechanics.

   In the past there has been some friction between ab-initio
and semi-empirical investigators. This, I think, is dying. The
former group is realizing that truly enormous calculations are
required to obtain acceptable results for large molecules: the
latter group is confining itself more and more to theoretically
sound principles. It is also becoming apparent that exact
calculations within the Hartree Fock approximation do not agree
quantitatively with experiment. Often they will not agree as well
as the results of approximate methods. In either case, useful
chemical information is gathered only after an examination of the
results within the context of the model used, and its shortcomings.
Argument by analogy and comparison is essential in explaining
phenomena within our most well established molecular theories, as
it is within our least sophisticated.

ACKNOWLEDGEMENT

   I am pleased to acknowledge that this work was supported in
part by a National Research Council of Canada Grant.

REFERENCES

1. R.G. Parr, Quantum Theory of Molecular Electronic Structure (Benjamin, New York, 1963).
2. I. Fischer-Hjalmars and M. Sundbom, Acta. Chem. Scand. 22, 607 (1968).
3. K. Jug, Theoret. chim. Acta 14, 91 (1969).
4. B.J. Nicholson, Adv. in Chem. Phys. 18, 249 (1971).
5. F.P. Billingsley II and J.E. Bloor, J. Chem. Phys. 55, 5178 (1971).
6. J.C. Slater and K.H. Johnson, Phys. Rev., B5, 844 (1972).
7. S. Diner, J.P. Malrieu and P. Claverie, Theoret. chim. Acta 13, 1, 13 (1968).
8. A. Streitwieser, Molecular Orbital Theory for Organic Chemists (Wiley, New York, 1961).
9. L.L. Lohr and W.N. Lipscomb, J. Chem. Phys. 38, 1607 (1963).
10. R. Hoffmann, J. Chem. Phys. 39, 1397; 40, 2047, 2474, 2480, 2745, (1964).
11. R.B. Woodward and R. Hoffmann, The Conservation of Orbital Symmetry (Academic Press, 1969).
12. R. Rein, N. Fukuda, H. Win, G.E. Clarke and F. Harris, J. Chem. Phys. 45, 4743 (1966).
13. Carrol, Armstrong and McGlynn, J. Chem. Phys. 44, 1865 (1966).
14. M. Zerner and M. Gouterman, Theoret. chim. Acta 4, 44 (1966).
15. J.A. Pople, D.P. Santry and G.A. Segal, J. Chem. Phys. 43, S129 (1965); J.A. Pople and G.A. Segal, ibid., S136 (1965); ibid., 44, 3289 (1966).
16. J.A. Pople, D.L. Beveridge and P.A. Dobosh, J. Chem. Phys. 47, 2026 (1967).
17. Janet Del Bene and H.H. Jaffé, J. Chem. Phys. 48, 1807, 4050 (1968); 49, 1221 (1968).
18. M.J.S. Dewar and E. Haselbach, J. Am. Chem. Soc. 92, 590 (1970); (1970); N. Bodor, M.J.S. Dewar, A. Harget and E. Haselbach, J. Am. Chem. Soc. 92, 3854 (1970).
19. D.B. Cook, P.C. Hollis and R. McWeeny, Mol. Phys. 13, 553 (1967).
20. R.D. Brown, F.R. Burden and G.R. Williams, Theoret. chim. Acta 18, 98 (1970).
21. R. Manne, Theoret. chim. Acta 6, 299 (1966); J. Chem. Phys. 46, 4645 (1967).
22. R.S. Mulliken, J. chim. Phys. 46, 497 (1949).
23. K. Ruedenberg, J. Chem. Phys. 19, 1433 (1951).
24. M.D. Newton, N.S. Ostlund and J.A. Pople, J. Chem. Phys. 49, 5192 (1968); M.D. Newton, J. Chem. Phys. 51, 3917 (1969); M.D. Newton, W.A. Lathan, W.J. Hehre and J.A. Pople, J. Chem. Phys. 51, 3927 (1969).
25. F.E. Harris and R. Rein, Theoret. chim. Acta 6, 73 (1966).
26. W.J. Hehre, R.F. Stewart and J.A. Pople, J. Chem. Phys. 51, 2657 (1969); W.J. Hehre, R. Ditchfield, R.F. Stewart and J.A. Pople, J. Chem. Phys. 52, 2769 (1970).

27. R.F. Stewart, J. Chem. Phys. 52, 431 (1970).
28. J.C. Phillips and L. Kleinman, Phys. Rev. 116, 287 (1959).
29. J.D. Weeks, A. Hazi and S.A. Rice, Adv. in Chem. Phys. 16, 283 (1969).
30. M. Zerner, Mol. Phys. 00, 0000.
31. K. Seigbahn, C. Nordling, G. Johansson, J. Hedman, P.F. Heden, K. Hamrin, U. Gelium, T. Bergmark, L.O. Werme, R. Manne and Y. Baer, ESCA Applied to Free Molecules (North-Holland, (1969).
32. C.E. Moore, Atomic Energy Levels, U.S. Dept. of Commerce, National Bureau of Standards Circ. No 467 (1949).
33. J.C. Slater, Phys. Rev. 36, 57 (1930).
34. B.J. Ransil, Rev. Mod. Phys. 32, 245 (1960).
35. G. Karlsson and M. Zerner, to be published.
36. M. Zerner, to be published.
37. P.E. Cade, K.D. Sales and A.C. Wahl, J. Chem. Phys. 44, 1973 (1966).
38. G. Hertzberg, Spectra of Diatomic Molecules (D. Van. Nostrand Co., Inc., (1950).
39. W.M. Huo, J. Chem. Phys., 43, 624 (1965).
40. C.A. Burrus, J. Chem. Phys., 28, 427 (1958).

LCAO-MO CLUSTER MODEL FOR LOCALIZED STATES IN COVALENT SOLIDS

G. D. Watkins and R. P. Messmer

General Electric Corporate Research and Development

Schenectady, New York 12301

I. THE PROBLEM

In this contribution we will be concerned with a theoretical description of the effect of isolated defects and impurities in an otherwise perfect crystalline covalent solid. One of the main effects of the defect or impurity is to create localized electronic states about itself as distinguished from the delocalized Bloch states which extend throughout the perfect crystal. These localized states have associated with them, energy levels in the gap between the valence and conduction bands (in "band language") or in the gap between the bonding and anti-bonding orbitals of the solid (in "bond language").

A good theoretical description can be given impurity levels which have associated with them energy levels which are quite shallow, i.e., within ~0.1 eV of the band edges. This description is based on the Kohn-Luttinger effective mass treatment or modifications thereof (1). However, many, perhaps even the majority, of defects and impurities have energy levels "deep" into the gap, i.e., $\gtrsim .1$ eV from band edges and which have more highly localized wave functions associated with them. For these "deep" levels there is no generally satisfactory theoretical treatment. One of the complicating features of the deep level problem, as deduced from experiment, is the importance of lattice relaxations and distortions around the defect or impurity (2).

From the above, we can see that a proper theoretical treatment of the deep level problem must meet the following conditions:

a)   it must locate the electronic levels, introduced by the
      impurity or defect, with respect to the band edges;
b)   it must provide localized wave functions for the
      electrons in the deep level which can be compared to
      experimental information, e.g. EPR studies;
c)   it must provide for the possibility of investigating
      lattice distortions and relaxation around the defect;
d)   it must be the basis of a practical computational
      scheme.

We will now introduce the model which we have proposed to
treat this problem and briefly summarize the computational results
for three different defects.

II. THE METHOD

The most commonly used approach to the deep level problem is
based on the Koster-Slater (3) theory or modifications thereto;
however this yields such a gigantic computational problem that
very few calculations have actually been carried out on other than
simple models, and the calculations on real systems have never
included lattice relaxations and distortions (4,5). The basic
idea of the method is to start with the solutions for the perfect
crystal and by a perturbation theory technique to describe the
localized defect states in terms of these starting basis functions.
Thus one must know the solution of the perfect crystal problem,
before one can start to solve the defect problem.

The viewpoint we have taken is to start with the defect and
continue to add neighbors to build up the crystal. Hopefully,
with not too many atoms a reasonable approximation to the solid
can be made. In fact, as we shall see shortly, this is indeed the
case. Then using this cluster of atoms, or "large molecule" we
will apply a simple molecular orbital technique to describe the
electronic structure.

As a demonstration of the approach, we will use the 35-atom
cluster shown in Fig. 1 as a representation of the diamond lattice,
in which we wish to investigate defects. The molecular orbital
technique we employ is the simple Extended Hückel Theory (EHT) (6).
This was discussed in some detail in an earlier contribution to
this meeting (7) and hence we will only describe our specific
procedure. We use the original Hoffmann version of EHT (6) (i.e.,
no charge iterative modifications which cause problems in our
case); the molecular orbital ϕ_i is written as

$$\phi_i = \sum_\mu c_{i\mu} \chi_\mu,$$

i.e., a linear combination of atomic orbitals, which leads to a set of equations

$$\sum_{\nu}(F_{\mu\nu}-\varepsilon_i S_{\mu\nu})c_{i\nu} = 0 \qquad\qquad i = 1,....N \quad , \qquad\qquad (1)$$

whose solution yields the orbital energies ε_i and the wave function coefficients $\{c_{i\nu}\}$. The matrix elements are:

$$F_{\mu\mu} = -I_\mu \text{ (atomic ionization potential of orbital } \mu)$$

$$F_{\mu\nu} = -\frac{K}{2}(I_\mu + I_\nu)S_{\mu\nu} \quad (K = 1.75 \text{ is chosen})$$

$$S_{\mu\nu} = \langle\chi_\mu|\chi_\nu\rangle$$

where χ_μ are Slater orbitals, and the orbital exponents are chosen by using Slater's rules (8).

The solution of Equations (1) for the cluster of 35 carbon atoms gives the result shown in Fig. 2a, where we see a primitive conduction and valence band separated by a gap. We have obtained the elastic constants, C_{11}, C_{12} and C_{44} for this cluster (using the EHT approximation to the total energy which is the sum over the occupied one electron molecular orbital energies), and the results (9) are in remarkable agreement with the experimental values. Hence we can be somewhat optimistic that when we apply

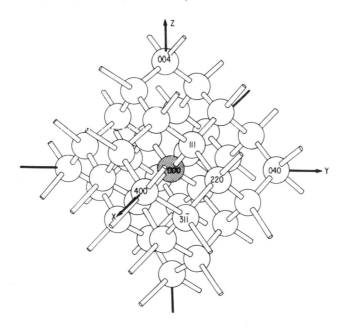

Fig. 1   The 35 atom "diamond" cluster used for LCAO-MO calculations.

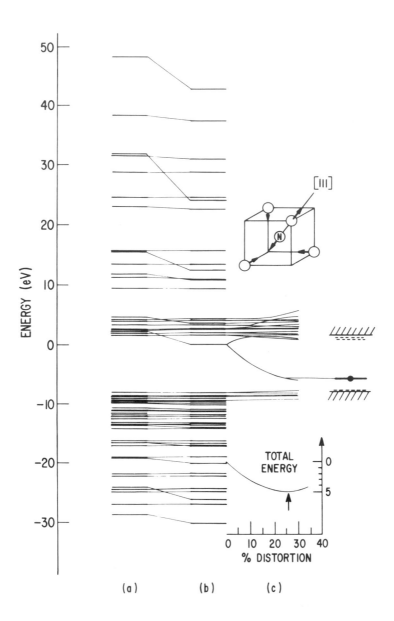

Fig. 2 (a) The energies of the one electron MO's for a 35 carbon
atom "diamond". (b) The energies with nitrogen as the central
atom, no relaxation. (c) Effect of trigonal Jahn-Teller
distortion involving the nitrogen and its four nearest neighbors
(see insert).

the method to investigate lattice relaxations and distortions around a defect, reasonable results will be obtained.

III.   DEFECTS IN DIAMOND-EXAMPLES

We now wish to review the results obtained using this cluster for the following three defects in diamond: (A) substitutional nitrogen atom impurity (9), (B) a lattice vacancy (10), and (C) a carbon atom interstitial (11).

A.   Substitutional Nitrogen

When the center carbon atom of the 35-atom cluster of Fig. 1 is replaced by a nitrogen, the results shown in Fig. 2b are obtained. A triply degenerate (t_2 symmetry) level appears in the gap and is occupied by one electron. The wave function is quite localized on the nitrogen and its four carbon neighbors. We recognize immediately that this degenerate ground state configuration may undergo a Jahn-Teller distortion by coupling to E and T_2 lattice vibrations, which can remove this degeneracy. Calculations employing both tetragonal (E) and trigonal (T_2) local distortions show that the total energy is lower for the T_2 distortion and thus predict a trigonal Jahn-Teller distortion. This result is shown in Fig. 2c.

The details of these calculations have been given elsewhere (9,12) and we only note here that the calculations predict: the position of the energy level in the gap, a trigonal Jahn-Teller distortion, a highly localized wave function for the electron which is primarily centered on the nitrogen and the single carbon nearest neighbor atom along the trigonal distortion axis, hyperfine parameters, and nitrogen vibrational frequencies, all of which are in good agreement with experiment. We believe that these calculations represent the first time that a theoretical computational scheme has met the criteria given in section I for a proper treatment of the deep level problem.

B.   Lattice Vacancy

Our calculations on the vacancy have shown (10) that a level in the gap close to the valence band edge results and that the wave function associated with this energy level is ~54% localized on the four carbon atoms neighboring the vacancy. Jahn-Teller distortions were found to be significant, and an estimate of ~0.5 eV was given as the stabilizing effect of these distortions.

Although no unambiguous experimental information is available about the vacancy in diamond, there is a great deal known about the vacancy in silicon through EPR studies (13). Our calculations are generally consistent with what is known about the vacancy in silicon and in addition have provided valuable physical insight as to the origin of certain subtle features observed in the magnetic hyperfine interactions which were not previously understood.

C.   Carbon Interstitial

Previous calculations on the host-atom interstitials in diamond type lattices have assumed the most stable configuration to be that in which the atom is centered either in the tetrahedral or hexagonal interstitial position of the lattice (15-19). Then the activation energy for migration was calculated by taking the difference in energies for the tetrahedral and hexagonal positions (16,17), which were along the assumed migration path.

In our calculations however we have found that the tetrahedral and hexagonal sites are much higher in energy than other "interstitialcy" configuration in which the extra carbon atom nestles into a bonding configuration with its neighbors. Using these calculations as a guide, it was possible to propose a specific mechanism for "athermal" migration of the interstitial (11), which can explain for the first time the experimental evidence that interstitials are able to migrate long distances in a low temperature radiation damage experiment.

Thus with these three rather diverse examples of defects in diamond, we can see that these calculations have provided a good deal of physical insight and the ability to explain experiments. There are, however, several aspects of the method which can be questioned: 1) what are the effects of the surfaces of the cluster?, 2) what effect does the size of the cluster have on the results? and 3) how well does EHT represent the electronic structure (band structure) of the bulk crystal? These questions are taken up in the next section.

IV.   CRITIQUE OF THE METHOD

A.   Effect of Surfaces

The main effect of the surfaces of the cluster are to introduce $N_b/2$ extra levels into the "valence band" and consequently reduce the conduction band by $N_b/2$ levels, where N_b is the number of dangling bonds on the surfaces of the cluster (see Fig. 1).

These extra states are not "surface states" in the usual sense,
in that for the most part they are not well localized on the
surface, but rather tend to penetrate well into the cluster.   It
is for this reason that we have chosen (9-12) to treat these
levels for the 35-atom cluster as part of the valence band.   The
detailed rationale behind this has been discussed previously (12,20).
There is no question, however, that this situation is undesirable
and should be eliminated if possible.   In fact, by choosing
periodic boundary conditions, as discussed in section V, this
problem of surface effects can be eliminated.

B.   Effect of Cluster Size

The effect of the cluster size on the distribution of energy
levels is shown in Fig. 3. We see that the valence band converges
much more rapidly toward its limiting value than does the conduction band.   The infinite limit is determined by an EHT band
structure calculation using the same parameters (21).   One reason
for the better convergence of the valence band is the presence of
the additional $N_b/2$ levels which would have been in the conduction
band in the absence of the surfaces.

The details of how the electronic structure of a defect level
in the gap changes as a function of cluster size is discussed
elsewhere (12); suffice it to say here that although some of the
quantitative aspects are somewhat affected, none of the qualitative
features or conclusions of section III are changed.

The most obvious effect of the finite cluster size is that
the band gap is much too large (see Fig. 3) as compared to the
infinite case.   It would certainly be most desirable to have the
band gap match that of the infinite crystal.   This too can be
achieved by periodic boundary conditions.

C.   EHT Band Structure

To assess the efficacy of EHT for describing the electronic
structure of the perfect infinite lattice, we have carried out EHT
band structure calculations on diamond (21), using the same
parameters as used in the cluster calculations.   The result of
these calculations is shown in Fig. 4a and compared to an ab
initio OPW calculation of Herman et al. (Fig. 4b).   We note that
all four of the valence bands and the two conduction bands (defined
by L_3, Γ_{15} and X_3) agree rather well.   The success of EHT in the
defect calculations described in section III must stem in part
from this fact.   We note however, that the other two conduction
bands are very poorly described.

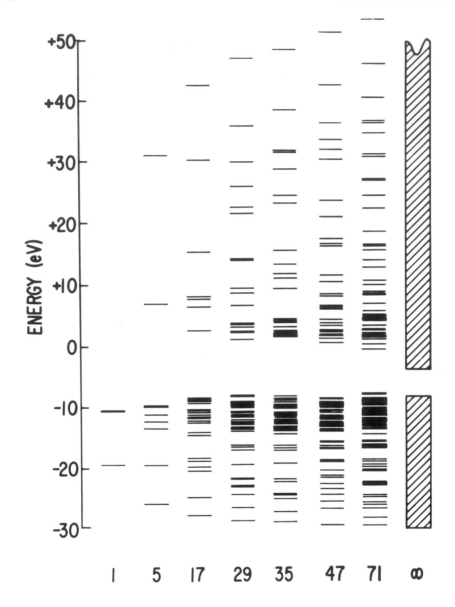

Fig. 3 The energies of the one electron MO's for a diamond cluster as a function of cluster size, starting with the central atom (1) and adding successive shells of neighbors up to and including the sixth nearest neighbor shell (71). The result for the infinite cluster comes from a band structure calculation using the identical EHT parameters.

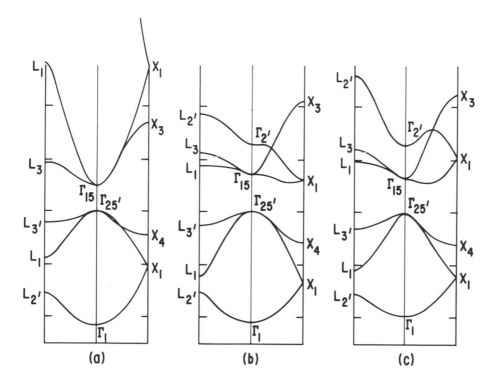

Fig. 4 Comparison of the diamond band structure (a) calculated using normal EHT parameters with (b) first principle results of Herman et al. (ref. 22). (c) Band structure with modified EHT parameters.

         In the spirit of the empirical pseudo-potential method we can make small adjustments to some of the EHT parameters, as previously shown by one of the authors (21), to obtain a band structure for diamond in better agreement with the ab initio result and experiment. Such a result is shown in Fig. 4c. For this, the carbon s-orbital exponent was increased by 24% (Slater orbital coefficient 2.0 instead of 1.625) and the value of K for p-p matrix elements was increased to 2.0. All other parameters were unchanged.

         A comparison of Figs. 4b and 4c clearly shows that EHT, empirically adjusted, can provide quite a good description of the electronic structure of the infinite crystal. In the next section we take advantage of this fact, and at the same time eliminate the surface effect and finite size effect by using the Molecular Unit Cell Approach (MUCA) to the problem.

V.   THE MOLECULAR UNIT CELL APPROACH

The LCAO-Molecular Unit Cell Approach has previously been applied by Messmer et al. to the problem of chemisorption (23). The idea is quite simple: if by chemical or physical intuition we feel that a particular problem can be treated by considering the electronic structure of a cluster of atoms and yet wish to have no problems with surfaces, we simply take a suitably shaped cluster of atoms and apply periodic boundary conditions. That is, we use this particular cluster of atoms as the unit cell for a band structure calculation. We call this approach the Molecular Unit Cell Approach (MUCA).

In the defect problem this means of course, that we will be forming a periodic array of defects. If, however, the molecular unit cell can be chosen large enough the defects will be sufficiently far apart that they can be considered as isolated.

We realize at the outset of course that to compute a band structure for a very large unit cell over all of $\vec{k}$-space is a tremendous task which must be avoided, if we are to have a reasonable computation scheme. However, as the unit cell becomes larger, fewer points in $\vec{k}$-space should be necessary to characterize the properties of the system. In effect, the $\vec{k}$ dependence simply determines the relative phase of the periodically reproduced wavefunctions in the adjoining cells. As these cells become farther removed, their relative phase should have less and less effect on the properties of a defect in the central cell. A fundamental part of MUCA is therefore that it will be sufficient to sample only a few $\vec{k}$-points with the large molecular unit cell to characterize the properties of the system.

We want as large a unit cell as possible. This must be counterbalanced on the other hand by the practical limitations of computing requirements. For the diamond case, a good compromise appears to be achieved by employing a 64 atom molecular unit cell. This cluster is cubic and by unit translation along the cubic axes, fills all of space, a necessary requirement to serve as a unit cell for a band structure calculation. For such a unit cell, the periodically reproduced atomic sites are 16th nearest neighbors (7.12Å separation) and are therefore substantially removed from each other.

For the perfect diamond lattice, a calculation at a single

point in the Brillouin zone of the 64 atom unit cell is equivalent to sampling 32 points of the Brillouin zone of the primitive (two atom) unit cell. This connection is illustrated in Fig. 5 for the $\Gamma(\vec{k} = 0)$ and $X(\vec{k} = 001)$ points in the large unit cell. The solution at any specific $\vec{k}$-point still gives 64 x 4 = 256 discrete molecular orbital states, just as it would with a finite cluster of 64 atoms. However, now with the periodic boundary conditions these states can be directly related to specific states of the infinite crystal, with no surfaces. The solution at the Γ point represents a particularly attractive starting point for the defect calculations because, as can be seen in Fig. 5, it samples states over the full range of the conduction and valence bands, and the forbidden gap obtained is <u>exactly</u> the solution for the infinite crystal.

In addition, because many points are being sampled in the true Brillouin zone for each $\vec{k}$ point in the larger molecular unit cell, we anticipate that average properties, such as the total energy of the system, may be determined with reasonable accuracy by the solution at a single $\vec{k}$ point. This is illustrated in Table I where the elastic properties are calculated from the total EHT energy at $\vec{k} = 0$ vs. the size of the molecular unit cell.

The small change in the elastic constants in going from the 64 to the 512 atom unit cell (equivalent to sampling eight $\vec{k}$-points in the 64 atom cell) clearly indicates that the 64 atom unit cell is already large enough to give a satisfactory description of the elastic properties from the solution at a single $\vec{k}$-point. We note further the very good agreement between the calculated and the experimentally observed values. The calculated values were

Table I. Elastic Constants of Diamond (10^{12} dynes/cm^2) Calculated with the Modified EHT parameters at k = 0 vs. the size of the Molecular Unit Cell.

Atoms/unit cell	C_{11}	C_{12}	C_{44}
2	5.06	14.47	6.09
8	9.58	3.13	6.83
<u>64</u>	<u>10.14</u>	<u>0.81</u>	<u>5.72</u>
512	10.27	.62	5.47
EXPERIMENT[a]	10.76	1.25	5.76

[a]H.J. McSkimin and W.L. Bond, Phys. Rev. <u>105</u>, 116 (1957).

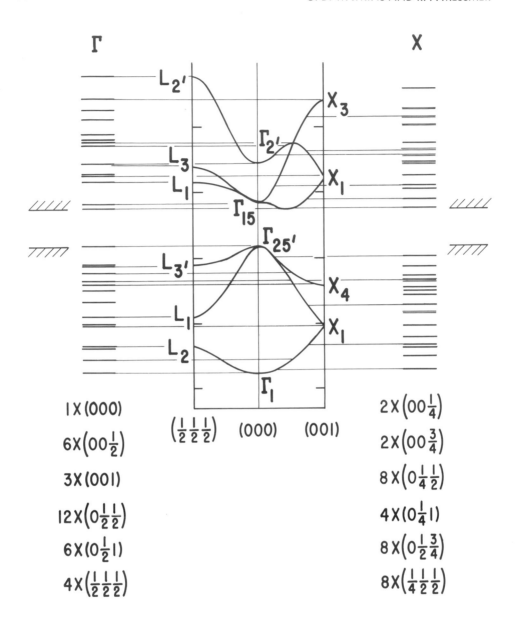

Fig. 5 The connection between the one electron MO energies for a 64 carbon atom diamond molecular unit cell at $\Gamma(\vec{k} = 0)$ and $X(\vec{k} = 001)$ and the band structure for the infinite lattice. The points sampled and their degeneracy are indicated.

obtained using the modified EHT parameters that led to the improved band structure of Fig. 4. Table I therefore also confirms that these modified EHT parameters still provide realistic elastic properties for the system.

Using this new approach, we now present some preliminary results for the lattice vacancy in diamond. When the central atom is removed, the solution at the Γ point yields a localized state of t_2 symmetry at ~0.6 eV above the valence band edge. Approximately half of the wavefunction is found to be localized on the four nearest carbon atoms. The result is therefore qualitatively similar to that described earlier for the 35 atom cluster calculation with the unmodified EHT parameters (10). The result at the X point, however, points up an interesting problem. Here the defect state is still highly localized on the nearest carbon atoms but its electrical level position is 1 eV lower than it is at the Γ point. This means that there remains a sizeable interaction between the periodically reproduced defects. There is, in effect, an impurity band and it has a 1 volt dispersion.

We are forced therefore to the disappointing conclusion that the 64 atom unit cell is still not large enough to effectively contain the localized state of the vacancy. Inspection of its wavefunction reveals the reason. Approximately 50% of the wavefunction is highly localized on the four nearest neighbors, but the remainder is spread throughout the cell with little attenuation vs. distance from the vacancy. It is these tails that are apparently producing the strong coupling between cells. This is a peculiar property of the vacancy. The nitrogen donor, previously discussed, is much more highly localized and a similar calculation for it reveals a much smaller dispersion. For it, the 64 atom unit cell is, in effect, sufficiently large.

The vacancy therefore remains a somewhat illusive defect for theory. It serves as an instructive example however in that it points to the extreme difficulty inherent in the deep level problem. In this present MUCA approach, we have come closer by far than any previous approach to treating both the localized and the extended character of such a defect. It should serve as a dramatic warning that results on a small cluster should be interpreted cautiously at least for a strongly covalent material such as the elemental semiconductors.

VI. CONCLUSIONS

We conclude that a cluster approach to the deep level problem is an extremely promising one. The Molecular Unit Cell Approach represents a significant extension of the method in eliminating surfaces and truncation effects associated with the finite cluster and allowing a better contact to the extended states of the crystal. Extended Huckel Theory does a very good job for the elemental semiconductor diamond (and is also useful for silicon (24)), particularly when the parameters are empirically adjusted to match the band properties.

EHT should not be used however for the more ionic materials such as the III-V and II-VI semiconductors. Here a more sophisticated MO technique that includes self-consistency is required. The new scattered wave Xα method described by Johnson at this meeting (25) represents a promising MO method for this purpose. Any SCF method requires more computing time than EHT. However, one compensating feature is that in the more ionic crystals, the bands are narrower and it may be possible to get by with smaller clusters.

REFERENCES

1. W. Kohn, in Solid State Physics, edited by F. Seitz and D. Turnbull (Academic Press Inc., N.Y. 1957) Vol. 5, p. 258.

2. G. D. Watkins, in Radiation Effects in Semiconductors, edited by F. L. Vook, (Plenum Press, N.Y., 1968), p. 67.

3. G. F. Koster and J. C. Slater, Phys. Rev. 95, 1167; ibid. 96, 1208 (1954).

4. J. Callaway and A. J. Hughes, Phys. Rev. 156, 860 (1967).

5. N. J. Parada, Phys. Rev. B3, 2042 (1971).

6. R. Hoffmann, J. Chem. Phys. 39, 1397 (1963).

7. M. Zerner, contribution to this symposium.

8. J. C. Slater, Phys. Rev. 36, 57 (1930).

9. R. P. Messmer and G. D. Watkins, Phys. Rev. Letters 25, 656 (1970).

10. R. P. Messmer and G. D. Watkins, Radiation Effects 9, 9 (1971).

11. G. D. Watkins, R. P. Messmer, C. Weigel, D. Peak, and J. W. Corbett, Phys. Rev. Letters 27, 1573 (1971).

12. R. P. Messmer and G. D. Watkins, Phys. Rev. B, submitted for publication.

13. G. D. Watkins, J. Phys. Soc. Japan 18, Suppl. II, 22 (1963).

14. G. D. Watkins, in Radiation Damage in Semiconductors (Dunod, Paris, 1965), p. 97.

15. T. Yamaguchi, J. Phys. Soc. Japan 18, 386, 923 (1963).

16. K. H. Bennemann, Phys. Rev. 137, A1497 (1965); 139, A482 (1965).

17. R. R. Hasiguti, J. Phys. Soc. Japan 21, 1927 (1966).

18. S. P. Singhal, Phys. Rev. B4, 2497 (1971).

19. K. Schroeder, Bull. Amer. Phys. Soc. 16, 397 (1971).

20. G. D. Watkins and R. P. Messmer, Phys. Rev. B4, 2066 (1971).

21. R. P. Messmer, Chem. Phys. Letters 11, 589 (1971).

22. F. Herman, R. L. Kortum, and C. D. Kuglin, Intern. J. Quantum Chem. 1S, 533 (1967).

23. R. P. Messmer, B. McCarroll and C. M. Singal, J. Vacuum Sci. Technol. 9, 891 (1972).

24. R. P. Messmer and G. D. Watkins, to be published.

25. K. H. Johnson and J. W. D. Connolly, contribution to this symposium.

DESCRIPTION OF MOLECULES IN TERMS OF LOCALIZED ORBITALS

Klaus Ruedenberg

Ames Laboratory-USAEC and Department of Chemistry

Iowa State University, Ames, Iowa 50010

INTRODUCTION

Historically the Hartree approximation originated from the wishful conjecture that it ought to be possible to reduce the many-electron problem to a pseudo-one-electron problem with a self-consistency requirement. It was a most welcome turn of events when V. Fock succeeded in deriving the Hartree-Fock equations from the many-electron Schrödinger variation principle by constraining the solution to be an antisymmetrized orbital product, characteristic of an "independent particle model." However, Fock also pointed out that this rigorous point of view not only leads to the novel exchange terms in the energy and the self-consistent equations but, moreover, revealed a fundamental flexibility and ambiguity in the self-consistent-field orbitals: the SCF orbitals of the pseudo-one-electron equations are not the only possible solutions of the posed problem; there exist an infinite number of different equivalent orbital sets which represent equally good and valid solution possibilities. Nonetheless, the instinctive drive towards the quasi-one-electron formalism continued and, for a considerable time, Fock's observation seems to have been repressed as a disturbing complication. Yet, eventually this reality proved to be a fact of remarkable usefulness and became the basis for a fruitful reconciliation between the seemingly incompatible requirements of spectroscopic and chemical intuition.

Even though the understanding of individual molecules is an important aspect of chemistry, actual chemical research proceeds through comparison of molecular systems. Important information

is gained from measuring energy differences of various systems,
such as reaction energies, ionization energies or excitation
energies. Even in the case of properties that pertain to isolated
systems, such as dipole moments, the chemist is particularly interested in comparing them for series of related compounds in
order to establish trends and regularities. This method has an
interesting consequence for the theory of quantum mechanical observables: The comparison of related molecular systems can yield
experimental information pertaining to theoretical quantities of
subsystems, such as orbital contributions, which, because of the
indistinguishability of electrons, would not qualify as observables in the sense of being expectation values of an operator for
any one isolated system.

     In this "research by comparison" two fundamentally different
approaches can be distinguished: On the one hand, the investi-
gator may compare the same molecule in various quantum states.
The entire field of spectroscopy is based on this type of mea-
surement. On the other hand, he may compare corresponding states,
e.g., ground states, of structurally related molecules in order
to gain insights of importance for understanding chemical reactions. Until recently it was a common belief that different
theoretical models were needed for bringing order into the variety
of phenomena observed in these two areas. Since Robert S.
Mulliken's fundamental papers on the band spectra of diatomics,
there has been no question that the molecular orbital theory was
the suitable approach for making sense of molecular spectroscopy;
and since Linus Pauling wrote "The Nature of the Chemical Bond,"
it was taken for granted that the most fruitful comparison of
ground states of different molecules could be obtained through
valence bond theory.

     The development of localized-orbital aspects of molecular
orbital theory can be regarded as a successful attempt to deal
with these two kinds of comparisons from a unified theoretical
standpoint. It is based on the flexibility in the choice of the
molecular orbitals first observed by Fock: The same many-electron
Slater determinant can be expressed in terms of various sets of
molecular orbitals. In the classical spectroscopic approach one
particular set, the canonical set, is used. On the other hand,
for the same wavefunction an alternative set, the set of localized
molecular orbitals, can be found which is especially suited for
comparing corresponding states of structurally related molecules.
Thus, one many-electron molecular-orbital wavefunction can be
cast into several forms, which are adapted for use in different
comparisons: for a comparison of the ground state of a molecule
with its excited states the canonical representation is most
effective; for a comparison of a particular state of a molecule
with corresponding states in related molecules, the localized

representation is most effective.   In this way the molecular orbital theory provides a unifed approach to both types of problems.

What has been said applies to approximate as well as to ab-initio molecular orbital wavefunctions, i.e., those obtained by solving the self-consistent-field equations exactly.   Hence, the localized orbital approach also offers an attractive tool for bridging the gap between rigorous quantitative calculations and qualitative chemical intuition.   The localized orbital picture corresponds closely to intuitive chemical thinking.

INVARIANCE OF SCF WAVEFUNCTION

The closed shell Hartree-Fock wavefunction

$$\psi = \mathcal{A}\{(u_1\alpha)^{(1)} (u_1\beta)^{(2)} \cdots (u_N\alpha)^{(2N-1)} (u_N\beta)^{(2N)}\}$$

is invariant under the orthogonal transformation

$$v_k = \sum_{i=1}^{N} u_i T_{ik}$$

among the occupied orbitals.   Invariant are furthermore the density matrix

$$\rho = 2 \sum_n u_n(x) u_n(x') \, ,$$

the one-electron energy

$$H_1 = 2 \sum_n (u_n|h|u_n) \, ,$$

the Coulomb energy

$$C = 2 \sum_n \sum_m \langle u_n u_m |g| u_n u_m \rangle \, ,$$

the exchange energy

$$X = \sum_n \sum_m \langle u_n u_m | g | u_m u_n \rangle,$$

the total energy

$$E = H_1 + C - X$$

and the SCF equations

$$\mathfrak{F}(\rho) u_n = \sum_m u_m \lambda_{nm}.$$

SPECTROSCOPIC ORBITALS

The traditional SCF orbitals ϕ_n, the so called "canonical" or "spectroscopic" orbitals, result when the arbitrary choice $\lambda_{ij} = 0$ is made for $i \neq j$. The canonical SCF equations

$$\mathfrak{F} \phi_n = \epsilon_n \phi_n$$

have pseudo-one-electron character, result from a pseudo-one-electron variation principle and usually belong to irreducible representations of the symmetry group that commutes with the Hamiltonian. The canonical orbital energies are directly re-lated to the energy differences observed in spectroscopic and ionization experiments.

CHEMICAL ORBITALS

In forming localized molecular orbitals, (LMO's), the ob-jective is to confine each molecular orbital to as small a space in the molecule as possible and, at the same time, to have these confined molecular orbitals as far removed from each other as possible. The more the orbitals can be confined and the more they can be separated from each other, the less likely they are to change when distant parts of the molecule are modified. Thus, the more the orbitals are localized in this sense, the more they can be expected to be transferable among molecules having related structures. It is this transferability of localized orbitals which makes them appropriate tools for comparing corresponding

states of related molecules and for pinpointing differences
between them.

Quantitative similarities of molecules can easily be recognized if it is possible to define quantities for molecular parts
which are additive as well as transferable. Such quantities can
be derived from transferable molecular orbitals because any one-
electron property, such as dipole moment, quadrupole moment,
kinetic energy, is a sum of the corresponding contributions from
all molecular orbitals in a system, if such orbitals are chosen
mutually orthogonal. Thus, for each transferable orthogonal
molecular orbital there exists, e.g., a transferable orbital
dipole moment. Since chemists value additive decompositions of
molecular properties into transferable orbital contributions,
there exists an interest in mutually orthogonal localized molecular orbitals. If it proves possible to establish such decompositions of molecular observables in terms of transferable contributions from localized orbitals, then, by measuring the observable in question for sufficiently many molecules, it would be
possible to deduce quantitative values for the contributions from
individual localized orbitals.

In order to construct localized orbitals for molecules, it
is necessary to define a "measure for the degree of localization"
of an arbitrary set of molecular orbitals. The "localized orbitals" are then defined as that set of orthogonal molecular orbitals for which the measure of localization has the maximum
value. In 1950 Lennard-Jones and Pople suggested to use the self-
energy-sum

$$L = \sum_n \langle u_n u_n | g | u_n u_n \rangle$$

as localization measure, which has several attractive features.
However, various other criteria have been advanced, notably by
Gilbert and Adams. Those and other criteria are discussed in a
recent review by Weinstein, Pauncz, and Cohen.

Localized orbitals are usually, but not always, unique.
Usually they belong to reducible representations of the symmetry
group. Quite often they are different from the spectroscopic
orbitals, although occasionally some localized orbitals may coincide with some canonical orbitals.

LOCALIZED ORBITALS IN DIATOMICS

In diatomic molecules, all localized orbitals belong to one of three categories: atomic inner shells, atomic lone-pair valence orbitals, and diatomic bond orbitals. Among the latter, pure sigma bonds and sigma-pi hybrid bonds can be distinguished. All localized diatomic orbitals are bonding or nonbonding, never strongly antibonding. Lone-pair orbitals do, however, have slightly antibonding character corresponding to "nonbonded repulsions." The description of the various diatomic bonding situations through localized orbital structures is considerably more distinctive and suggestive of interpretation than by means of canonical orbitals. The canonical orbitals emphasize the geometric similarity of all diatomics, whereas the localized orbitals exhibit the physical differences between the various diatomics.

LOCALIZED ORBITALS IN POLYATOMICS

In saturated polyatomics the localized orbitals reproduce the intuitive chemical bonds in terms of bond orbitals. In water, for example, one finds two bond orbitals and two lone pairs in two mutually orthogonal planes and in methane one finds four CH bond orbitals. In all saturated hydrocarbons, there will be one localized orbital for each CH and CC bond.

Since localized orbitals are orbitals of maximal localization, they are the only molecular orbitals that can serve to define and compare the degree of localization of electronic systems. For example, one finds that a small but noticeable degree of delocalization exists, for saturated bonds in hydrocarbons, into a bond that is two bond distances away. For unsaturated π-electrons, the localized orbitals are much less localized than for the σ-electrons (they extend about twice as far), and this "local delocalization" of the π-bonds of maximal localization is the origin of the stabilization described by the "delocalization energy."

Localized orbitals have been particularly helpful in elucidating certain electronic structures with multicenter bonding. Lipscomb and coworkers have successfully used localized orbitals to clarify the electronic structures of many boron hydrides. Thompson has shown that Gillespie's rules for the conformation of polyatomics must be thought of as applying to localized molecular orbitals and that they are equivalent to Walsh rules for canonical orbitals. This suggests that localized orbitals may play a considerable role in elucidating the reasons for molecular geometries.

LOCALIZED ORBITALS IN SOLIDS

It is possible to apply the aforementioned ideas to the solid state. Localized groundstate SCF orbitals in a solid are obtained by a suitable orthogonal transformation among the orbitals below the Fermi level. Thus, they are different from the Wannier functions, which are obtained by an orthogonal transformation among all Bloch orbitals in one band. As an example, consider a one-dimensional system with a half-filled band, whose Bloch orbitals are simply given by plane waves. In the usual treatment the periodic Born-von Karman boundary condition is assumed to hold at the crystal surfaces. In this case, the localized orbitals for a very large linear lattice are found to be (starting at one surface)

$$\lambda_s^{\,o}(x) = \frac{1}{\sqrt{2}} \frac{\sin\frac{1}{2}\pi y_s}{\frac{1}{2}\pi y_s} \,,$$

where the coordinate x has its origin at the surface and is measured in units of the internuclear distance. Furthermore

$$y_s = x - 2s \,, \quad \text{with } s = 1, 2, 3, \cdots$$

so that the $\lambda_s^{\,o}(x)$ are spaced over approximately two bond lengths (assuming each atom contributes one electron). It is however also possible to form the localized orbitals while assuming the more realistic boundary condition that all orbitals vanish at the surface. In this case one finds

$$\lambda_s(x) = \lambda_s^{\,o}(x) + \frac{1}{\sqrt{2}} \frac{\sin\frac{1}{2}\pi y_s}{\frac{1}{2}\pi (y_s + 4s)} \,.$$

This result shows that, close to the surface (i.e., $y_s + 4s \approx 2s$), all localized orbitals differ from $\lambda_s^{\,o}$ approximately by the periodic function $(\sin\frac{1}{2}\pi y_s / \sqrt{2}\,\pi s)$. For the localized orbitals that are moreover located near the surface, i.e., for which s is a small integer, one finds therefore a marked difference between $\lambda_s(x)$ and $\lambda_s^{\,o}(x)$. Thus, the localized orbitals provide a means to assess the deficiencies introduced by the Born-van Karman condition near the surface.

REFERENCES

The ideas outlined in the present note are elaborated in considerable detail in a recent review by England, Salmon, and Ruedenberg.[1] Another review by Weinstein, Pauncz, and Cohen[2] discusses a variety of theoretical questions that are raised by the problem of localization; in particular they review the various proposed localization criteria. These two reviews complement each other in a gratifying manner, and between them, contain references to most other important articles on the subject.

1.   W. England, L. S. Salmon, and K. Ruedenberg: Fortschritte der chemischen Forschung, Vol. 23, p. 31. New York: Springer-Verlag 1971.

2.   H. Weinstein, R. Pauncz, and M. Cohen: Advances in Atomic and Molecular Physics, Vol. 7. New York: Academic Press 1971.

4. Computational Methods II

INTRODUCTORY REMARKS

P. S. Bagus

IBM Research Laboratory

San Jose, California 95114

The computational methods familiar to quantum chemists can be conveniently divided into two groups: ab initio methods and semi-empirical methods.

In the ab initio approaches, one first selects a one-particle basis for the system under study (molecular orbitals) and an n-particle basis (configurations). One then proceeds to solve Schroedinger's equation _exactly_ in the space defined by these bases. Now, _exactly_ is a very comfortable word, but we must not lose sight of the fact that ab initio methods involve approximations, often very considerable approximations. The most important of these are in the selection of one- and n-particle bases. The levels of approximation which have been discussed by previous speakers, or-dered roughly in increasing complexity and accuracy, include:

i) Minimal basis set Hartree-Fock

ii) Double-zeta (or double-zeta plus polarization) basis set Hartree-Fock

iii) Limited configuration interaction in a limited or extended one-particle basis

iv) Large scale configuration interaction in an extended one-particle basis.

Thus, ab initio calculations depending on the property being con-sidered and the level of approximation used may or may not give "accurate" results. The key point is that one may remove these approximations and perform accurate calculations within the same

theoretical framework. This has always been true, in principle,
and is now, very importantly with improvements in our knowledge
of theory and numerical analysis and in our computing power, becoming true in fact. Furthermore, the theoretical framework allows
us by computation to determine the optimum values of parameters
used in the calculation as, for example, the optimum values of
exponents in elementary basis sets of Slater or Gaussian functions.

In semi-empirical methods, one traditionally begins with the
same basic chemical models that are used in ab initio methods, but
one introduces approximations into the resulting equations rather
than solving them exactly. These approximations are parametrized
with reference to experimental data, the parametrization being
chosen to fit either the results of the equivalent ab initio model
or to fit experimental results.

The speakers in this session, Professors Johnson and Harrison,
will discuss computational methods which involve approximations
which are -- at least to quantum chemists -- quite new.

THE SCF-Xα SCATTERED-WAVE METHOD[*]

K. H. Johnson and J. G. Norman, Jr.[†]
Department of Metallurgy and Materials Science
Massachusetts Institute of Technology
Cambridge, Massachusetts

and

J. W. D. Connolly
Department of Physics
University of Florida
Gainesville, Florida

ABSTRACT

The recently developed SCF-Xα scattered-wave theory
of molecular electronic structure and localized states
in solids is reviewed. Calculations of the electronic
structures and cohesive properties of small and medium
sized molecules are described. The ionization energies
and optical properties are interpreted, using the
transition-state procedure, which accounts for the
effects of spin-orbital relaxation. Emphasis is on
applications to more complex systems, and as illustrative examples, results for metal complexes and biological
macromolecules are described. In the latter applications,
the electronic structure of a macromolecule is built up
from the SCF-Xα charge distributions of suitably chosen
component polyatomic clusters. The final section is
devoted primarily to a summary of applications to localized states in solids, including those associated with
crystal defects, excitons, magnetism, and chemisorption.

I.   INTRODUCTION

This paper is a comprehensive review of recent progress at M.I.T. and the University of Florida in applying the self-consistent-field Xα scattered-wave (SCF-Xα-SW) method (1,2) to polyatomic molecules and solids. This technique permits accurate calculations of the electronic structures of complex molecules and localized electronic states in solids, from first principles, but requires only moderate amounts of computer time. Two review articles devoted to the Xα and multiple-scattered-wave formalisms, respectively, are in press (3,4). A number of other papers dealing with the SCF-Xα-SW method and its applications have also been published, or are close to publication (5-15).

II.   THE SCF-Xα SCATTERED-WAVE THEORY

The method is based, first of all, on the arbitrary division of matter into component clusters of atoms. A cluster may be an isolated polyatomic molecule in the gaseous phase, in a crystalline environment, or in aqueous solution. It may also be a finite group of atoms in a bulk crystal or at a crystalline surface, or it may be part of a macromolecule. The cluster, in turn, is partitioned into three fundamental types of regions: I. <u>atomic</u> – the regions within nonoverlapping spheres surrounding the constituent atoms; II. <u>interatomic</u> – the region between the atomic spheres and an outer sphere surrounding the entire cluster; III. <u>extramolecular</u> – the region outside the cluster.

The main objective is to solve the one-electron Schrödinger equations (written here in Rydberg units)

$$[-\nabla_1^2 + V_C(1) + V_{X\alpha}(1)]\, u_i(1) = \varepsilon_i u_i(1) \qquad (1)$$

for the spin orbitals $u_i(1)$ and energy eigenvalues ε_i of the cluster, subject to the boundary conditions on the cluster. Each spin orbital can be associated with either the spin-up or spin-down Pauli function. Thus the electronic charge density

$$\rho(1) = \Sigma(i) n_i u_i^*(1) u_i(1) \qquad (2)$$

where n_i is the occupation number of the ith spin orbital, can be separated into spin-up $\rho\!\uparrow(1)$ and spin-down $\rho\!\downarrow(1)$ parts. In expression (1), $V_C(1)$ is the electro-

static potential energy at position 1 due to the total
electronic and nuclear charge, determined classically.
The quantity $V_{X\alpha}(1)$ is Slater's (16-18) Xα statistical
approximation to exchange correlation, namely

$$V_{X\alpha}\uparrow(1) = -6\alpha[\frac{3}{4\pi}\rho\uparrow(1)]^{1/3} \tag{3}$$

with a similar expression for spin down. This potential
is dependent only on the local electronic charge density
$\rho\uparrow(1)$ and on the scaling parameter α. The value $\alpha = 1$
yields the exchange potential originally derived by
Slater (19) in 1951. The value $\alpha = 2/3$ leads to the
exchange approximation derived independently by Gaspar
(20) and by Kohn and Sham (21). Actually a value of α
systematically chosen between these two limits is
generally better for most applications to atoms, molecules, and solids (see below).

   With expressions (2) and (3), the one-electron
Schrödinger equations (1) define a self-consistent-
field problem which can be solved iteratively, in
analogy to the more traditional Hartree-Fock method.
It can be shown, however, that the SCF-Xα method, in
conjunction with the multiple-scattered-wave procedure
for solving the one-electron equations, has several
advantages over the Hartree-Fock method and is therefore
not just an approximation to the latter approach.

   The numerical solutions of equations (1) are
implemented for a polyatomic cluster by first spherically averaging the potential $V_C(1) + V_{X\alpha}(1)$ inside each
of the atomic regions I. The potential is generally
assumed to be a constant throughout the interatomic
region II, equal to the volume average of $V_C(1) +$
$V_{X\alpha}(1)$ over this region. However, spherical averages
can also be carried out in region II inside supplementary nonoverlapping "interstitial" spheres, e.g. in
the regions of "lone-pair" electrons (see below). For
localized molecular orbitals, the potential is usually
spherically averaged in the extramolecular region III
with respect to the defined center of the cluster.
Alternative boundary conditions can also be imposed on
region III to describe the effects of different local
environments on the cluster. The initial potential is
based on a superposition of free-atom or free-ion
potentials in the SCF-Xα approximation, using atomic
self-consistent-field computer programs of the type
originally developed by Herman and Skillman (22). The
scaling parameter α in expression (3) is optimized for

each component atom using the procedure suggested by Slater (3,18). This procedure consists of matching the Xα statistical total energy of the atom to the Hartree-Fock total energy in the case of a closed-shell atom, or to the average of the multiplet energies for the ground-state configuration in the case of an open-shell atom. Using this scheme, Schwarz (23) has recently determined the α parameters for the atoms hydrogen through niobium. These values are then adopted for the corresponding atomic regions I of the molecular cluster. Appropriately weighted averages of the component atomic α values are used in regions II and III. This procedure ensures as accurate a description of the cluster in the separated-atom limit as possible, within the framework of the SCF-Xα-SW theory.

The partitioning of the space of the cluster into local regions of spherically averaged and volume averaged potential permits one to introduce a rapidly convergent composite partial-wave representation of the solutions of equations (1). Inside each atomic sphere j (and each interstitial sphere, if present), the spin orbitals are expanded in the form

$$u_i^{I}(1) = \Sigma(L) C_L^{j} R_\ell^{j}(\varepsilon;r) Y_L(\vec{r}) \tag{4}$$

where $L = (\ell,m)$ is the partial-wave (angular-momentum) index. The functions $R_\ell^{j}(\varepsilon;r)$ are solutions of the radial Schrödinger equation, the functions $Y_L(\vec{r})$ are spherical harmonics, and the coefficients C_L^{j} are to be determined. The radial functions are generated by outward numerical integration of the radial Schrödinger equation for each partial wave ℓ and each trial energy ε. In the extramolecular region III, the orbitals are expanded with respect to the center of the cluster in the representation

$$u_i^{III}(1) = \Sigma(L) C_L^{0} R_\ell^{0}(\varepsilon;r) Y_L(\vec{r}), \tag{5}$$

where the functions $R_\ell^{0}(\varepsilon;r)$ are obtained by inward numerical integration of the radial Schrödinger equation. For the intersphere region II, Schrödinger's equation (1) reduces to the ordinary scalar wave equation

$$(\nabla_1^2 + \varepsilon - \bar{V}_{II}) u_i^{II}(1) = 0 \tag{6}$$

where $\bar{V}_{II}$ is the volume average of $V_C(1) + V_{X\alpha}(1)$ over that region. The exact solutions of Eq. (6) for the energy range $\varepsilon < \bar{V}_{II}$ can be expanded in the multicenter representation

$$u_i^{II}(1) = \Sigma(L)i^{-\ell}A_L^{\ 0}j_\ell(i\kappa r_0)Y_L(\vec{r}_0)$$

$$- \Sigma(j)\Sigma(L)i^{-\ell}A_L^{\ j}h_\ell^{(1)}(i\kappa r_j)Y_L(\vec{r}_j) \qquad (7)$$

where $\kappa = (\bar{V}_{II} - \epsilon)^{\frac{1}{2}}$ (8)

is the "wave propagation constant," $j_\ell(i\kappa r_0)$ is a spherical Bessel function, and $h_\ell^{(1)}(i\kappa r_j)$ is a spherical
Hankel function of the first kind. For the energy range
$\epsilon > \bar{V}_{II}$, the solutions of Eq. (6) are

$$u_i^{II}(1) = \Sigma(L)A_L^{\ 0}j_\ell(\kappa r_0)Y_L(\vec{r}_0)$$

$$+ \Sigma(j)\Sigma(L)A_L^{\ j}n_\ell(\kappa r_j)Y_L(\vec{r}_j), \qquad (9)$$

where $\kappa = (\epsilon - \bar{V}_{II})^{\frac{1}{2}}$ (10)

and $n_\ell(\kappa r_j)$ is a spherical Neumann function.

The first term in each of the expressions (7) and
(9) may be interpreted as a superposition of "incoming"
spherical waves which have been "scattered" by the
extramolecular region of spherically averaged potential
and which are directed toward the center of the cluster.
The second term may be interpreted as a superposition of
"outgoing" spherical waves which have been "scattered"
by the atomic regions of potential. The arguments of
the spherical Hankel functions in Eq. (7) are imaginary.
Thus for localized spin orbitals of the cluster in the
energy range $\epsilon < \bar{V}_{II}$, one is not dealing with progressive
waves in the usual sense of scattering theory, but
rather spherical waves which in region II decay exponentially away from the atoms. The occurrence of spherical Neumann functions with real arguments in Eq. (9)
suggests "standing waves" in the intersphere region for
the energy range $\epsilon > \bar{V}_{II}$.

The intersphere wave functions (7) and (9) and
their first derivatives are required to be continuous
with the solutions (4) and (5) of Schrödinger's equation
in the atomic and extramolecular regions, respectively.
This leads to equations relating the coefficients $A_L^{\ j}$
and $C_L^{\ j}$ and the coefficients $A_L^{\ 0}$ and $C_L^{\ 0}$. Thus the
cluster spin orbitals undergo a smooth transition from
atomic-like behavior inside each atomic sphere to exponentially decreasing behavior away from each atom.
This is very much like the behavior of molecular orbit-

als represented as traditional linear combinations of
analytic atomic orbitals (the LCAO method). However,
the scattered-wave representation suffers from none of
the convergence problems and other computational difficulties associated with the LCAO method.

By relating the waves incident on each region of
the cluster to the waves scattered by all the other
regions, it is possible for one to derive a set of
compatibility relations among the coefficients $A_L{}^j$ and
$A_L{}^0$. These relations are in the form of a set of secular
equations which have nonvanishing solutions only for
certain values of the energy ε, the eigenvalues. The
secular equations are similar in form to those found in
the Korringa-Kohn-Rostoker (KKR) (24,25) method of
crystal band theory, in that there are two principal
kinds of matrix elements. There are those elements
which depend only on the geometrical arrangement of
atoms in the cluster (the "structure factors") and those
elements which depend only on the nature of the atoms,
through the logarithmic derivatives of the radial wave
functions at the various sphere radii separating regions
I, II, and III (the "scattering factors"). In the event
that the cluster is a repeating or periodic unit cell of
a crystal and the boundary condition of wave function
localization in the extramolecular region III is replaced
by the Bloch condition of band theory, the secular equations can be transformed exactly to the KKR form (see
Ref. 2).

The eigenfunctions of the one-electron equations
(1) for an initial superposed-atom potential $V_C(1)$ +
$V_{X\alpha}(1)$ are the starting point for a complete SCF-Xα-SW
calculation. The initial set of spin orbitals $u_i(1)$ is
substituted into expression (2) to obtain the electronic
charge density throughout the cluster. This charge
density is then substituted into Poisson's equation of
classical electrostatics to determine a new Coulomb
potential $V_C(1)$ and into Eq. (3) to determine a new
exchange-correlation potential. As before, these potentials are spherically averaged in regions I and III and
volume averaged in region II. A weighted average of the
new potential and the initial potential serves as input
for the first iteration. A new set of spin orbitals is
calculated by the scattered-wave method, and the entire
computational procedure is repeated until self consistency is attained.

From the self-consistent cluster spin orbitals and

electronic charge density, one can compute the expectation value of the Xα statistical total energy (1,3,18),
namely

$$\langle E_{X\alpha} \rangle = \Sigma(i) n_i \int u_i^*(1) f_1 u_i(1) dv_1$$

$$+ \tfrac{1}{2}\int \rho\uparrow(1) [\int \rho(2) g_{12} dv_2 + \tfrac{3}{2} V_{X\alpha}\uparrow(1)] dv_1$$

$$+ \tfrac{1}{2}\int \rho\downarrow(1) [\int \rho(2) g_{12} dv_2 + \tfrac{3}{2} V_{X\alpha}\downarrow(1)] dv_1 \qquad (11)$$

The term containing the one-electron operator f_1 represents the average of the kinetic and potential energies
of the electrons in the electrostatic field of the
nuclei.   The term involving the two-electron operator
g_{12} represents the average Coulomb interaction of the
electrons with the electron cloud and includes the interaction of an electron with itself.   The term in g_{12} also
includes the average internuclear Coulomb repulsion
which, for fixed nuclear positions, is a constant to be
added to the total electronic energy at the end of the
calculation.   The terms involving $V_{X\alpha}\uparrow(1)$ and $V_{X\alpha}\downarrow(1)$
(see Eq. 3) remove the electron self interaction and
account for exchange effects.   These quantities are
different for spin-up and spin-down electrons, so that
the two terms are written separately.

     The eigenvalues of the one-electron equations (1)
can be shown (1,3,18) to be related to the statistical
total energy expression (11) as the first derivatives
of the latter quantity with respect to occupation numbers,
i.e.

$$\varepsilon_{iX\alpha} = \frac{\partial \langle E_{X\alpha} \rangle}{\partial n_i} \qquad (12)$$

This is in distinct contrast to the Hartree-Fock method,
where the one-electron energies are given by the difference between the total energies calculated respectively when the ith spin orbital is occupied and when it is
empty, i.e.

$$\varepsilon_{iHF} = \langle E_{HF}(n_i=1) \rangle - \langle E_{HF}(n_i=0) \rangle \qquad (13)$$

In general, $\langle E_{X\alpha} \rangle$ is not a linear function of n_i, so
that the values of $\varepsilon_{iX\alpha}$ will usually be different from
the values of ε_{iHF}.   This is true even if the spin-
orbital eigenfunctions and the total energies of the
two methods are identical.

There are several important consequences of expressions (11) and (12) which make the SCF-Xα-SW technique generally superior to the Hartree-Fock method. First of all, $<E_{X\alpha}>$ for the molecule or cluster converges approximately to the sum of the free-atom total energies when the internuclear distances are increased to the separated-atom limit, even in the spin-restricted case. This is not true for $<E_{HF}>$. Secondly, the SCF-Xα-SW theory satisfies Fermi statistics, thereby ensuring the proper ordering and occupation of electronic energy levels, while the Hartree-Fock theory does not (1,3). Finally, excited spin orbitals relevant to ionization and optical transitions, including the effects of spin-orbital relaxation, are readily within the scope of the SCF-Xα-SW method, using Slater's transition-state concept (1,3). No analogous concept, other than Koopmans' theorem (which neglects spin-orbital relaxation), exists in the Hartree-Fock theory. On the other hand, like the Hartree-Fock approach, the SCF-Xα-SW method can be made to satisfy both the virial and Hellmann-Feynman theorems to reasonable accuracy. These consequences are elaborated upon in greater detail in connection with the examples described in the following sections.

III.   APPLICATIONS TO SMALL AND MEDIUM SIZED MOLECULES

As emphasized in Refs. 1-4, the SCF-Xα-SW method has been designed primarily for the description of polyatomic molecules and clusters composed of many-electron atoms, where traditional ab initio methods of quantum chemistry and theoretical solid-state physics are difficult and costly to implement. Such examples are discussed in later sections of this paper. Nevertheless, the method has also been applied to a wide range of small and medium sized molecules, in order to determine how accurate the SCF-Xα-SW theory is for systems where both experimental data and the results of ab initio Hartree-Fock SCF-LCAO calculations are available for comparison. Some of this work has already been published or will be published soon. A partial listing of examples with references includes: SO_4^{2-} (5,6); ClO_4^- (5); SF_6 (8); C_2H_6 (10); N_2, CO, NO, and CF_4 (12); H_2O (13); CH_4 (14); H_2 and Li_2 (26); CO_2, C_3O_2, N_2O, C_4H_4S, and C_6H_6 (27); NH_3 (27,28); H_2O_2 (28); P_4 and P_8 (29).

A.   Calculation of Ionization Energies

In the Hartree-Fock method, the ionization energies of an atom, molecule, or solid are usually calculated on the basis of Koopmans' theorem. This theorem states that the ionization energy associated with the excitation of an electron from a particular spin orbital is equal to the negative of the orbital energy. Unfortunately, Koopmans' theorem ignores the orbital relaxation which accompanies ionization, and thus is generally a poor approximation for most systems. In order to include this relaxation within the Hartree-Fock framework, it is necessary for one to perform separate total energy calculations for the neutral system and the ion, subtracting the two total energies to obtain the ionization energy. This is too costly a computational procedure to implement in general, even for relatively small or medium sized molecules of the type listed above.

On the other hand, in the SCF-Xα-SW method the difference between the statistical total energies of the neutral system and the ion is equal, to very good approximation, to the energy of a spin orbital from which one-half a unit of electronic charge has been removed (1,3). This modified spin orbital, calculated self consistently, is called a transition state, since it represents a one-electron state "halfway" between the initial and final states. The transition states automatically include the effects of spin-orbital relaxation, and the negative values of their energies may be identified with the ionization energies.

For example, in Fig. 1 there is a comparison of the SCF-Xα-SW transition-state energies for the valence orbitals of the sulfur hexafluoride (SF_6) molecule (8) with the experimental ionization energies measured by the method of electron spectroscopy for chemical analysis (ESCA) (30). Also shown, for comparison, are the theoretical ionization energies based on a Koopmans' theorem interpretation of semiempirical "complete-neglect-of-differential-overlap" (CNDO) (30) and ab initio SCF-LCAO (31) molecular-orbital calculations. It is evident in Fig. 1 that the ordering of the SCF-Xα-SW transition-state energies is almost identical with the ordering of orbital energies obtained in the ab initio SCF-LCAO calculation, but it is in substantial disagreement with the CNDO ordering. Furthermore, the SCF-Xα-SW results are in significantly better quantitative agreement with the ESCA data than are the ab initio results. It should also

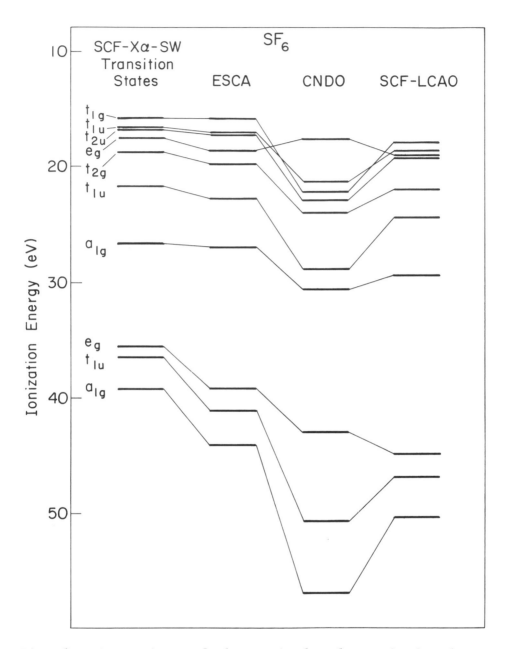

Fig. 1. Comparison of theoretical valence ionization energies for the SF_6 molecule with the experimental values measured by the method of electron spectroscopy for chemical analysis (ESCA).

be noted that the present calculation required a total
of only six minutes of time on an IBM 360/65 computer,
including the SCF determination of all the transition-
state energies (32). This may be contrasted with 20 to
30 hours on the same type of machine required in the ab
initio calculation (31).

The SCF-Xα-SW transition-state procedure has been
applied to all of the molecules listed above, and the
results are consistently in better agreement with ESCA
data than the results of both ab initio and semiempirical
LCAO calculations, yet they have required relatively
small amounts of computer time. The reader is referred
to the references listed above for details of these
applications.

B.   Calculation of Optical Excitations

Another advantage of the SCF-Xα-SW method over the
Hartree-Fock LCAO method is that the former approach
leads to a reasonably accurate description of the unoccupied or virtual spin orbitals to which electrons in
the ground-state orbitals may be excited by optical or
other means. The latter procedure typically leads to
virtual orbitals of positive energy which have no
quantitative physical connection to the final states of
an optical excitation. This is due to the fact that in
the Hartree-Fock theory the occupied and unoccupied
orbitals are treated differently. An electron in an
occupied orbital moves in the average potential field
of N-1 electrons, as it should, but an electron in a
virtual orbital moves in the potential of all N electrons.

In the SCF-Xα-SW method the two types of electrons
move in the same local potential field of N-1 electrons,
so that the unoccupied spin orbitals are treated as
accurately as the occupied ones. In calculating the
excitation energy involved in promoting an electron from
an occupied orbital to an empty one it is, of course,
important for one to use the transition-state procedure,
which accounts for the effects of spin-orbital relaxation (1,3). Thus it is necessary to calculate the
difference between the energy of an initial spin orbital
whose electron occupation number has been reduced by
one half and the energy of a virtual orbital with occupation number equal to one half. In general, for optical
excitations from occupied orbitals just below the Fermi
level to empty orbitals immediately above the Fermi

level, the relaxation effects are considerably less than those which accompany ionization.

For example, in applying this procedure to the benzene (C_6H_6) molecule (27) one finds that the two lowest unoccupied orbitals (e_{2u} and b_{2g}) are π^* states, as they should be. In contrast, <u>ab initio</u> SCF-LCAO calculations (33) predict that the lowest virtual orbital is a σ^* state (e_{2g}), in complete disagreement with experimental optical data (34). The transition-state procedure does not yield the detailed multiplet structure of excited states, but it does provide an approximation to the average over multiplet states arising from a given electronic configuration. When this average is calculated for the experimental multiplet states of C_6H_6, it is found to be in excellent agreement with the π-π^* one-electron energy difference calculated by the transition-state procedure. These results are summarized in Table I.

Table I.   Comparison of theoretical and experimental $\pi \rightarrow \pi^*$ optical excitation energies (in eV) for benzene (C_6H_6).

Multiplet	Multiplet Energy[a]	Weight	Average Multiplet Energy	SCF-Xα-SW $e_{1g}(\pi) \rightarrow e_{2u}(\pi^*)$ Transition Energy
$^3B_{1u}$	3.66	3		
$^3E_{1u}$	4.69	6		
$^1B_{2u}$	4.89	1		
$^3B_{2u}$	5.76	3	5.06	4.85
$^1B_{1u}$	6.14	1		
$^1E_{1u}$	6.75	2		

[a]See Ref. 34.

A part of the multiplet structure can be determined by the SCF-Xα-SW method if the transition-state calculations are carried out in spin-unrestricted (spin-

polarized) form. In this case it is possible to perform
separate averages over multiplet states of the same
multiplicity. For example, in the calculation of the
ionization energies of the open-shell molecule nitric
oxide (NO), the method was able to predict reasonably
well the splittings between the singlet and triplet
excited states of the NO^+ ion (12).

More detailed information about the multiplet
structure of the excited states of a molecule or cluster,
within the framework of the SCF-Xα-SW theory, can be
obtained only by using the spin-orbital wave functions
as a basis for a full multiplet calculation. That is,
just as in atomic theory, one will have to use the
SCF-Xα-SW spin orbitals to construct determinantal
functions and linear combinations of them, to describe
the multiplets. This has not yet been implemented,
although plans are under way to do so. The scattered-
wave representation of the spin orbitals is very rapidly
convergent and constitutes a complete solution of the
Schrödinger equation throughout the space of the molecule. Therefore, it is reasonable to expect that multiplet calculations by this approach will be practical,
perhaps more easily implemented than multiplet calculations based on Hartree-Fock SCF-LCAO molecular orbitals.

C. Calculation of Molecular Total Energies and Conformations

As described in Section II, the SCF-Xα-SW statistical total energy (see Eq. 11) of a molecule or cluster
goes approximately to the sum of the free-atom total
energies in the limit of infinite internuclear distances,
provided that the atomic sphere radii are adjusted in
proportion to these distances. This ensures that as the
atoms go to infinite distances from each other, the
atomic sphere radii become infinite and the problem
reduces to that for the isolated atoms. This is true
for homonuclear molecules even in the spin-restricted
formulation of the theory. However, the restricted
Hartree-Fock total energy is well known not to reduce
to the separated free-atom limit (35).

For example, in the case of the hydrogen molecule
(H_2), the Hartree-Fock determinantal wave function implies
that the probability of finding two electrons on the same
atom is equal to the probability of finding them on different atoms. This is, of course, incorrect in the limit
of infinite internuclear distance and leads to a signifi-

cant error in the total energy. This error is equal to the self-energy term for a single hydrogen atom (i.e., $\sim$0.62 Rydberg). The spin-restricted statistical total energy $<E_{X\alpha}>$, on the other hand, gives the correct dissociation limit in principle, since the potential energy and electronic charge density go smoothly over to those for the two separate hydrogen atoms. Effectively, the exchange-correlation scaling parameter α (see Eq. 3) is defined so that the self-energy term is cancelled out, thus giving the correct total energy at infinite internuclear distance. An actual SCF-Xα-SW calculation for the H_2 molecule (26), using an α value appropriate for the hydrogen atom (namely α_H = 0.978), yielded an equilibrium internuclear distance of 1.32 a.u. and a total energy of -2.467 Ry, compared with the experimental values of 1.42 a.u. and -2.348 Ry. A smaller value of α would give results closer to the experimental results. Using a reduced value of α for H_2 can possibly be justified from the fact that α must decrease in the limit of zero internuclear distance to the value for the helium atom (namely α_{He} = 0.773). However, no procedure for determining the variation of α with internuclear distance has yet been developed. Actually, the H_2 molecule is an unfavorable case for calculating the statistical total energy. For any other molecule, where the electronic charge density is higher, the variation of α from infinite to zero internuclear distance is much smaller and thus the error in assuming α to be a constant is not as serious.

The spin-restricted SCF-Xα-SW method has also been applied to the lithium (Li_2) molecule, for which the variation of α with respect to internuclear distance should be small (α_{Li} = 0.781 and α_C = 0.759). As in the case of the H_2 molecule, the statistical total energy dissociates to the correct Hartree-Fock limit, whereas the restricted Hartree-Fock total energy does not. Because the equilibrium internuclear distance for Li_2 is quite large ($\sim$5 a.u.), the large dissociation error causes the restricted Hartree-Fock total energy to give a poor result, only approximately 15% of the experimental dissociation energy (36). The SCF-Xα-SW calculation (using α_{Li} = 0.781), having the correct dissociation properties, yields a much more reasonable result, approximately 75% of the experimental dissociation energy and an equilibrium internuclear distance only slightly larger than the experimental value. This comparison is shown in Fig. 2. For both H_2 and Li_2 the virial theorem is nearly exactly satisfied at the calculated equilibrium internuclear distances.

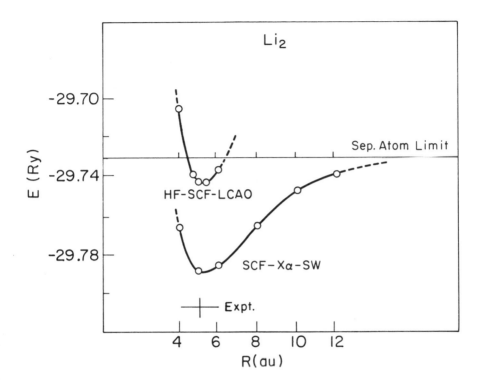

Fig. 2.   Comparison of the spin-restricted Xα statisti-
           cal total energy of the Li₂ molecule as a
           function of internuclear distance, calculated
           by the scattered-wave method, with the spin-
           restricted Hartree-Fock total energy calculated
           by the ab initio LCAO method. The experimental
           dissociation energy and equilibrium internuclear
           distance are also shown. The SCF-Xα-SW energy,
           in contrast to the HF-SCF-LCAO energy, goes
           properly to the sum of the energies of the
           separated atoms as the internuclear distance is
           increased to infinity.

Preliminary results for the statistical total
energies of certain other diatomic molecules, such as
carbon (C_2) and nitrogen (N_2), are not as satisfactory,
although the one-electron transition-state energies are
in excellent agreement with ESCA ionization energies
(26).   This is probably due to the fact that the total
energy expression (11) is presently being calculated on
the basis of the muffin-tin approximation to the electronic charge distribution.   In other words, the computed
SCF-Xα-SW charge distribution, which is not muffin-tin
like at the final stage of the SCF procedure (e.g., see
Fig. 4), is spherically averaged within the atomic and
extramolecular regions and volume averaged, i.e.,
constant, within the interatomic region.   This is, of
course, a much more stringent assumption for calculating
the total energy than it is for the one-electron energies,
especially for molecules like C_2 and N_2 where, in
contrast to H_2 and Li_2, the presence of p orbitals on
both atoms leads to a nonspherical charge distribution
along the bond.   Thus it is important to calculate the
statistical total energy from the complete non-muffin-
tin charge distribution which is automatically obtained
in the SCF-Xα-SW computational procedure, and modifications to the existing computer programs to accomplish
this are in progress.

Nevertheless, for many larger molecules the statistical total energies calculated in the simple muffin-tin
approximation are generally as good as or better than the
results of ab initio Hartree-Fock LCAO investigations
requiring a much greater computational effort.   For
example, the total energies of water (H_2O) (13), methane
(CH_4) (14), and ammonia (NH_3) (26) as functions of the
bond distances have yielded equilibrium bond lengths,
dissociation energies, and stretching force constants
which are in excellent agreement with experiment.   Again
the virial theorem is nearly exactly satisfied at the
equilibrium bond lengths.   The statistical total energy
of CH_4 as a function of C-H bond distance is shown in
Fig. 3.   The computer-generated valence charge distribution for CH_4 is shown in Fig. 4, illustrating that the
SCF-Xα -SW method is indeed capable of yielding non-
muffin-tin charge distributions and "directed" chemical
bonds.

The dissociation limit of CH_4 deserves special
comment, since it behaves differently from the case of
a homonuclear diatomic molecule.   The limit of the
statistical total energy in this case is not exactly
equal to the free-atom Hartree-Fock limit, even though

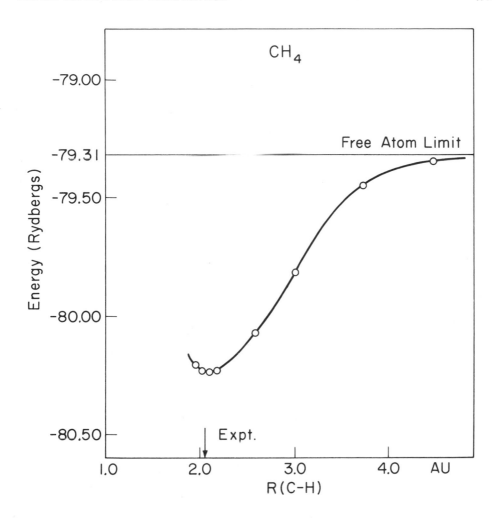

Fig. 3.   The spin-restricted Xα statistical total energy
of the CH$_4$ molecule as a function of symmetrically increasing C-H internuclear distance, calculated by the scattered-wave method.   Also indicated is the experimental equilibrium C-H internuclear distance.   The dissociation limit in
this case is determined by Fermi statistics.   In
the limit of large C-H distance, there is a
charge transfer of approximately 0.2 electron
from the carbon atom to the hydrogen atoms, with
a lowering of the statistical total energy by
approximately 0.02 Ry from the sum of free-atom
Hartree-Fock energies.

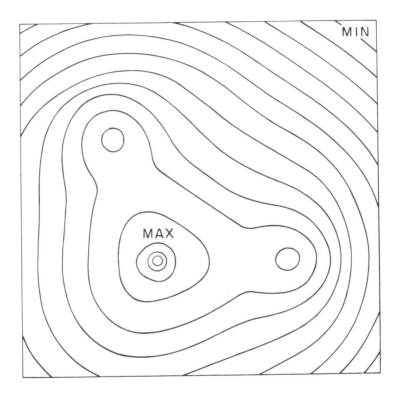

Fig. 4.   Contour map of the valence electronic charge density (i.e. that associated with six electrons in the $1t_2$ orbital) in an H-C-H plane of a CH_4 molecule. The C-H internuclear distance is the calculated equilibrium value of 2.109 a.u. which corresponds to the minimum statistical total energy shown in Fig. 3. The charge density corresponding to the maximum contour shown is 0.330 electron per cubic Bohr radius. The values of succeeding contours decrease by a factor of two to a minimum charge density of 0.805×10^{-4} electron per cubic Bohr radius.

the exchange-correlation α-parameters are defined so
that $<E_{X\alpha}>$ is equal to the Hartree-Fock values for the
individual carbon and hydrogen atoms (i.e., α_C = 0.759
and α_H = 0.978). The dissociation limit in this case
is determined essentially by Fermi statistics. It can
be shown (1,3) that the one-electron energies, defined
in Eq. (12), rigorously satisfy Fermi statistics, i.e.,
for the statistical total energy to be a minimum, there
cannot be a partially occupied or fully occupied
orbital above an empty or partially occupied orbital.
In the separated-atom limit of the CH_4 molecule, one
finds that the one-electron energy corresponding to the
hydrogen 1s level is below that of the carbon 2p level.
Since these are both partially occupied in the isolated
atoms, the atomic configuration cannot represent the
minimum statistical total energy in the dissociation
limit. In order for this to be achieved, there must be
electronic charge transfer from the carbon 2p level to
the hydrogen 1s levels. This must lower the 2p energy
and raise the 1s energy until the two energies are exactly
equal, at which point the total energy will be a minimum.
The actual calculations for the CH_4 molecule (14) show
that in the limit of very large C-H bond distance this
is actually the case. The molecular orbitals which
are predominantly carbon 2p ($1t_2$) and hydrogen 1s ($1t_2$
and $2a_1$) all approach a common energy value intermediate
between their isolated free-atom energy levels. It turns
out that there is a total charge transfer of approximately
0.2 electron from the carbon atom to the four hydrogen
atoms, with a lowering of the total energy by approximately 0.02 Ry.

There are several classes of problems in molecular
conformation, requiring an understanding of total-energy
differences, for which the SCF-Xα-SW method seems to be
reasonably well suited. For example, the method has been
used to calculate the internal rotation barrier in ethane
(C_2H_6) (10). It is well known that most ab initio and
approximate SCF-LCAO methods (even the semiempirical extended Hückel technique) lead to a rotation barrier of
C_2H_6 in good agreement with experiment. Therefore, it is
worthwhile to test the SCF-Xα-SW procedure on the C_2H_6
problem to see whether comparable results are obtained.
The calculations were carried out for the more stable
staggered conformation, using the experimental bond
distances and angles. The same bond distances were used
in the calculations for the eclipsed geometry. The atomic
and extramolecular sphere radii were chosen to be touching,
in the usual fashion, and the sphere radii were not

optimized to achieve the lowest statistical total energy. The difference between the statistical total energies of the staggered and eclipsed conformations was calculated to be 2.90 kcal/mole. This may be compared with the experimental rotation barrier of 2.93 kcal/mole (37). Optimization of sphere radii and geometries would certainly not improve the agreement between theory and experiment. Indeed an indication that stable and meta-stable equilibria were approximately obtained for the staggered and eclipsed geometries, respectively, is that the virial theorem was closely satisfied in both cases.

Calculations of the inversion barrier in ammonia (NH_3) and main rotation barrier in hydrogen peroxide (H_2O_2) have also been made (28). In these cases it is essential to include supplementary regions of spherically averaged $X\alpha$ potential in the interatomic regions of the lone-pair electrons. The magnitudes of the barriers are sensitive to the lone-pair sphere radii (particularly for H_2O_2), although setting them approximately equal to the hydrogen sphere radii leads to barriers in reasonable agreement with experiment. In order to avoid the intro-duction of lone-pair spheres, it is necessary for one to compute the statistical total energy from the complete non-muffin-tin charge distribution.

Another very interesting problem is the stereo-chemical stability of the tetrahedral phosphorus mole-cule (P_4) relative to that of the hypothetical cubic form (P_8). This is an excellent case where there is interplay between molecular and solid-state concepts. It is well known that the most chemically reactive and least stable crystalline form of phosphorus is white phosphorus, in which the atoms have local tetrahedral (P_4) coordination (38). The most stable crystalline form is black phos-phorus, which consists of corrugated sheets, each phos-phorus atom being bonded to three neighbors approximately at 101° angles (38). Because the phosphorus atom has three 3p electrons, it should be energetically favorable for the atom to form right-angle (p_x, p_y, p_z) bonds in the molecule as well as in the solid, suggesting that a P_8 molecule should be more stable than a P_4 molecule (39). Indeed, the so-called "strain energy" and possible occurrence of "bent bonds" in the tetrahedral P_4 molecule have been long-standing issues of controversy (40). Thus it is worthwhile to apply the SCF-$X\alpha$-SW method to both the P_4 and P_8 molecules, and to determine the dif-ference between their statistical total energies as a measure of their relative stabilities.

This has been carried out (29), and it has been
found that the equilibrium statistical total energy of
P$_4$, namely $\langle E_{X\alpha} \rangle$ = -1362.272 Hartrees, is somewhat lower
than the value of $\langle E_{HF} \rangle$ = -1362.004 Hartrees determined
by Brundle et al. (41) using the <u>ab initio</u> SCF-LCAO
method. The P$_4$ ionization energies, calculated by the
transition-state procedure, are also in excellent agreement with photoelectron data. Of most significance is
the fact that the statistical total energy of the P$_8$
molecule, namely $\langle E_{X\alpha} \rangle$ = -2724.787 Hartrees, is lower
than the sum of the statistical total energies of two
P$_4$ molecules, i.e. $2\langle E_{X\alpha} \rangle$ = -2724.544 Hartrees, suggesting that P$_8$ is indeed more stable than P$_4$. If this
difference in energy (0.243 Hartree) is divided among
the twelve possible P-P near-neighbor bonds, P$_8$ turns
out to be more stable than P$_4$ by approximately 13 kcal
per bond. This figure is significantly greater than
the $\sim$6 kcal estimate for the P-P bond "strain energy"
in P$_4$ (40). There is very little discussion in the
literature about the possible occurrence of the P$_8$ molecule in nature. The fact that the black crystalline
form of phosphorus is very difficult to prepare (in
comparison with the white form), requiring several days
of high pressure and elevated temperature plus a catalyst,
suggests that the chemical kinetics may also be unfavorable for the preparation of P$_8$ in molecular (vapor)
form. An attempt is now being made to determine the
experimental conditions which are essential for the
preparation of P$_8$ (39).

IV.   APPLICATIONS TO METAL COMPLEXES

Much of inorganic chemistry, organic chemistry,
and solid-state physics is concerned with the behavior
of metal atoms in complex molecular and crystalline
environments. Among the many examples are:

(1)   isolated metal complexes in the vapor phase
      (e.g. TiCl$_4$).
(2)   metal-metal bonded molecules [e.g. Mn$_2$(CO)$_{10}$].
(3)   heavy-metal complexes (e.g. transuranium
      complexes).
(4)   complex metal anions in ionic crystals and
      aqueous solutions (e.g. MnO$_4^-$ in KMnO$_4$).
(5)   antiferromagnetic insulating crystals
      (e.g. NiO).
(6)   "one-dimensional" crystals with metal-metal
      bonded columnar sequences [e.g. Magnus'

green salt $Pt(NH_3)_4PtCl_4]$.

(7)   metal-atom clusters (e.g. $Mo_6Cl_8^{4+}$).
(8)   metal impurities in semiconductors (e.g. "luminescent" impurities in II–VI compounds).
(9)   molecules and surfaces important in catalysis (e.g. transition-metal carbonyls).
(10) organometallic complexes, crystals, and polymers (e.g. metal phthalocyanines).
(11) biologically active prosthetic groups of certain proteins and enzymes (e.g. hemoglobin and cytochrome iron porphyrins).
(12) transition-metal reaction mechanisms (e.g. ligand exchange).

It has been shown that with the SCF-Xα-SW method, it is possible for one to calculate with reasonable accuracy the electronic energy levels, charge distributions, electronic excitations, and spin polarization of polyatomic molecules and clusters, using only moderate amounts of computer time. Because the computational effort does not increase inordinately with the number of electrons per atom, this technique is therefore ideally suited for the description of electronic structure in metal complexes such as those listed above.

The permanganate ion (MnO_4^-) is an excellent illustrative example. The nature of the electronic structure of this transition-metal complex has been as important an issue in inorganic chemistry as the electronic structure of benzene has in organic chemistry. In 1938 and 1939, Tetlow (42) carried out the first investigation of the visible and near-ultraviolet optical absorption of MnO_4^- ions in host crystal lattices such as $NaClO_4$ and $LiClO_4$. The principal absorption peak observed at 2.3 eV is responsible for the characteristic purple color of $KMnO_4$ crystals. Thirteen years later, Wolfsberg and Helmholz (43) carried out the first theoretical calculations of the electronic structure of MnO_4^-, using a semiempirical LCAO molecular-orbital method. Since then, there has been further experimental work on the optical properties of MnO_4^- (44), and there have been many attempts to determine the electronic structure of MnO_4^- by both semiempirical (45–51) and ab initio (52–54) LCAO methods. Unfortunately, these calculations are widely inconsistent with respect to ordering and magnitudes of orbital energies, and none of them is in quantitative agreement with the measured optical properties.

In an attempt to resolve this long-standing controversy over the nature of the electronic structure and
assignment of optical transitions for MnO_4^-, we have
carried out calculations for the ion with the SCF-Xα-SW
method (2). The similarities of the optical absorption
of MnO_4^- in various crystalline environments (42,44)
suggest that the neighboring cations in a crystal like
$KMnO_4$ have little effect on the chemical bonding of an
MnO_4^- cluster, other than providing a stabilizing electrostatic field. The unit cell of the orthorhombic
$KMnO_4$ lattice, for example, can be divided into four
distinct tetrahedrally coordinated MnO_4^- molecules and
four K^+ ions (55). In the present applications of the
SCF-Xα-SW method, the stabilizing field was approximated
by surrounding the MnO_4^- cluster with a spherical shell
of charge +1 which does not overlap the nearest K^+ ions
and neighboring MnO_4^- molecules in the $KMnO_4$ unit cell.

Xα exchange-correlation scaling parameters of α =
0.712 and α = 0.744, determined by Schwarz (23) for
free Mn and O atoms respectively, were used for the
potentials in the corresponding atomic regions I of
the MnO_4^- cluster. A weighted average of α = 0.738
(four parts O to one part Mn) was adopted for the interatomic region II. The value of α = 0.744 appropriate
for O was used in the extramolecular region III.

All of the 58 electrons of MnO_4^- were included in
the SCF-Xα-SW calculation. The SCF procedure converged
in 15 iterations, requiring a total time of only 8
minutes on an IBM 360/65 computer (32). The higher
levels which are involved in chemical bonding of MnO_4^-
and which are relevant to the measured visible and
near-ultraviolet optical properties are shown in Figure
5. The levels are labeled according to the various
irreducible representations of the tetrahedral (T_d)
symmetry group. The corresponding SCF-Xα free-atom
energy levels are included for comparison. MnO_4^- is a
closed-shell system. The highest fully occupied valence
orbital is the one labeled $1t_1$ in Figure 5. This is a
nonbonding orbital localized almost entirely on the
oxygen ligands and composed principally of oxygen 2p-like
partial waves. The $6t_2$ and $6a_1$ orbitals are also essentially nonbonding and O 2p-like, except for small admix-
tures of Mn 3d- and Mn 4s-like partial waves, respectively. The $1e$ and $5t_2$ orbitals correspond to π and σ
symmetrized combinations of O 2p- and Mn 3d-like partial
waves, and they are largely responsible for the bonding
of MnO_4^-. Contour maps for the $5t_2$ and $1e$ orbital wave

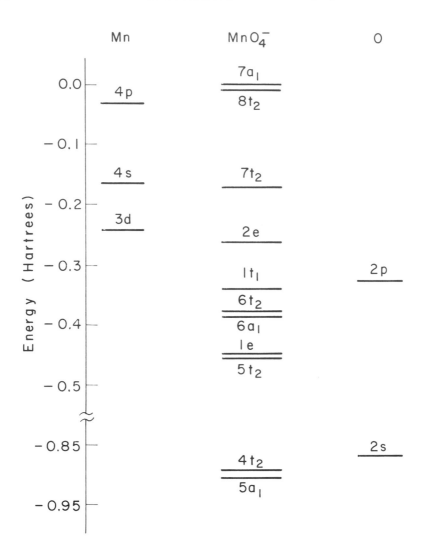

Fig. 5.   The SCF-Xα-SW electronic energy levels of an
MnO₄⁻ cluster in the approximate stabilizing
electrostatic field of a typical crystalline
environment (e.g. KMnO₄). The levels are
labeled according to the irreducible representations of the tetrahedral (T_d) symmetry group.
The highest occupied level in the ground state
is $1t_1$. Also shown, for comparison, are the
corresponding SCF-Xα energy levels of the free
atoms.

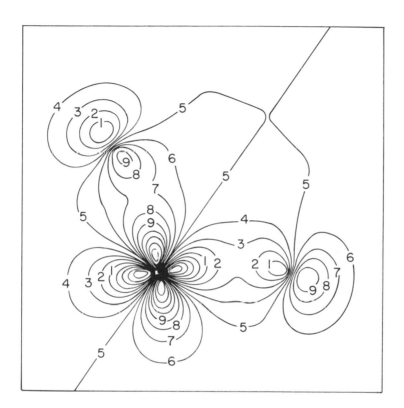

Fig. 6. Contour map of a normalized $5t_2$ orbital wave
function in an O-Mn-O plane of an MnO_4^-
cluster, showing the presence of Mn3d-O2p
"σ bonding." The Mn-O internuclear distance
is 3.00 a.u. The value of contour No. 1 is
-0.2, the value of contour No. 9 is +0.2,
and the contour interval is 0.05. The wave
function has a node along contour No. 5.

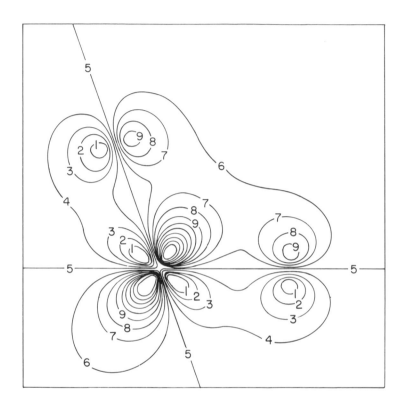

Fig. 7.   Contour map of a normalized 1e orbital wave
          function in an O-Mn-O plane of an MnO_4^-
          cluster, showing the presence of Mn3d-O2p
          "π bonding." The Mn-O internuclear distance
          is 3.00 a.u. The value of contour No. 1 is
          -0.2, the value of contour No. 9 is +0.2,
          and the contour interval is 0.05. The wave
          function has a node along contour No. 5.

functions of MnO_4^- in the O-Mn-O plane are shown in
Figures 6 and 7, respectively. $O2p\sigma$-Mn3d bonding is
clearly visible in Figure 6, as is $O2p\pi$-Mn3d bonding in
Figure 7. It should be emphasized that these maps have
not been generated from wave functions based on linear
combinations of atomic orbitals of the type traditionally
used in Hartree-Fock molecular-orbital theory. They have
been generated directly from the numerical SCF-Xα partial-
wave solutions of Schrödinger's equation, the wave functions and their first derivatives being joined throughout
the various regions of the MnO_4^- cluster by multiple-
scattered-wave theory.

Table II.   Comparison of theoretical and experimental
             optical excitation energies (in eV) for an
             MnO_4^- cluster.

Transition	SCF-Xα-SW Transition Energies	Experiment[a] $(^1A_1 \rightarrow {}^1T_2)$
$1t_1 \rightarrow 2e$	2.3	2.3
$6t_2 \rightarrow 2e$	3.3	3.5
$1t_1 \rightarrow 7t_2$	4.7	4.0
$5t_2 \rightarrow 2e$	5.3	5.5

[a]See Ref. 44.

     The first two unoccupied levels 2e and $7t_2$ illus-
trated in Figure 5 are principally Mn 3d-like orbitals.
The ordering of these levels is consistent with that
expected for tetrahedral geometry on the basis of
ligand-field theory. The positions of these levels with
respect to the occupied valence levels are critical for
the interpretation of the measured optical properties.
The optical absorption spectrum of MnO_4^-, measured for
a solid solution of $KMnO_4$ in $KClO_4$, consists of three
intense bands with maxima at 2.3 eV, 4.0 eV, and 5.5 eV,
and a "shoulder" at 3.5 eV (44). In Table II we list
the calculated energy differences between the initial and
final SCF-Xα transition-state orbitals for each orbitally
allowed optical transition of MnO_4^-. The present theoretical optical transition energies are in better

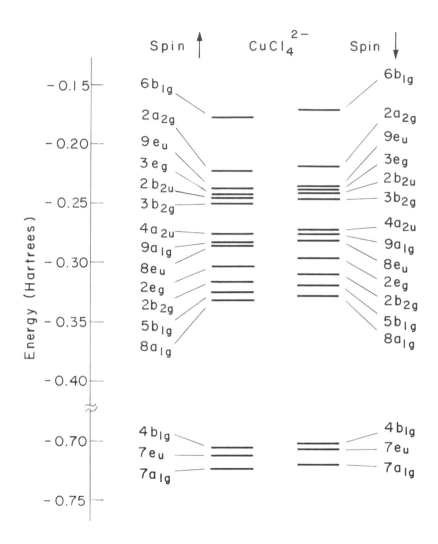

Fig. 8.　The spin-polarized SCF-Xα-SW electronic energy
levels of a CuCl₄²⁻ cluster in the approximate
stabilizing electrostatic field of a (NH₄)₂CuCl₄
crystal.　The levels are labeled according to
the irreducible representations of the square-
planar (D₄h) symmetry group.　The highest
occupied level in the ground state is 6b₁g↑.

quantitative agreement with experiment than are the
results of semiempirical (43,45-51) and ab initio (52-54)
LCAO calculations. The transition-state calculation
brings the $1t_1 \rightarrow 2e$ theoretical transition energy into
"exact" agreement with the measured 2.3 eV absorption
energy, although this accuracy may be fortuitous. Thus
we interpret this absorption peak as due predominantly
to electronic "charge transfer" between the $1t_1$ nonbonding 2p-like orbital localized on the O ligands and the
empty 2e 3d-like orbital localized on the Mn atom.

    Applications of the SCF-Xα-SW method to most of the
areas of transition-metal bonding listed above are in
progress. Among the isolated metal complexes (in the
gas phase or in crystalline environments) for which
calculations have already been completed are: MnO_4^-
(see discussion above and Ref. 2); $TiCl_4$ (56); NiF_6^{4-}
(26); $CuCl_4^{2-}$ (11,57); $PtCl_4^{2-}$ (57); $Ni(CO)_4$ (11);
$Cr(CO)_6$ and $Fe(CO)_5$ (58).

    In the treatment of open-shell complexes, such as
NiF_6^{4-} and $CuCl_4^{2-}$, it is essential for one to carry out
the calculations in spin-unrestricted form, so that the
effects of spin polarization are accounted for. For
example, the valence spin-orbital energy levels of a
"square-planar" $CuCl_4^{2-}$ cluster in a $(NH_4)_2CuCl_4$ crystalline environment are shown in Figure 8. The spin-split-
tings of the levels, although relatively small, are key
to understanding both the measured paramagnetism and
observed optical properties (e.g. yellow color) of
$(NH_4)_2CuCl_4$ crystals (11,57).

    In applications to heavy-metal complexes, it is
important for one to include, as completely as possible,
the relativistic effects on the electronic structure.
Many questions remain as to exactly how a truly rigorous relativistic theory of the electronic structure of
polyatomic molecules should be formulated. Nevertheless,
within the framework of the Dirac or Pauli formulation,
it is far easier to include relativistic corrections in
the SCF-Xα-SW computational procedure, than it is in
the Hartree-Fock SCF-LCAO method. Such corrections are
currently being programmed, and preliminary relativistic
calculations on PbTe complexes have already been successfully carried out (15,59).

    V. APPLICATIONS TO BIOLOGICAL MACROMOLECULES

    One of the most promising and exciting areas for

applying the SCF-Xα-SW method is the electronic structure
of biological macromolecules. The basic assumption is
that one can systematically build up the electronic structure of a macromolecule by first calculating the self-
consistent electronic structures of component polyatomic
clusters. The SCF-Xα charge distributions of these
clusters are then used as the starting point for determining the interactions between clusters and calculating
the electronic structure of the entire macromolecule.

One class of applications to biomolecules which is
in progress (and which is also one of the areas of metal
bonding cited in the preceding section) is the electronic
structure of the biologically active iron porphyrins in
heme proteins and enzymes (e.g. hemoglobin and cytochrome).
This work has already been outlined in an earlier publication (2), and the reader is directed to this reference
for details of how the SCF-Xα-SW method is being implemented on this problem.

Another very interesting and important biomolecular
problem which is under investigation is the nature of
the chemical bonding of phosphate-containing molecules
to enzymes. Calculations are in progress (29) on a model
for the 5'-phosphate binding site in the staphylococcus
aureus nuclease – Ca^{2+}-thymidine 3',5'-diphosphate macromolecule (60). The actual nuclease site consists of
guanidinium moieties of arginine residues 35 and 87, and
the 5'-phosphate (60). The normal biological function
of the nuclease is to catalyze the hydrolysis of nucleic
acids. The Ca^{2+}-thymidine 3',5'-diphosphate group complexes with the nuclease in such a way so as to inhibit
its normal biocatalytic mechanism.

For the purpose of the calculations, the binding
site of the nuclease has been approximated by two guanidinium ions $[C(NH_2)_3^+; Gu^+]$ and a hydrogen phosphate ion
(HPO_4^{2-}). These ions are stereochemically oriented to
permit hydrogen bonding between the guanidinium hydrogen
atoms and phosphate oxygen atoms, in a fashion similar
to that observed in the nuclease (60) and, more quantitatively, in the high-resolution X-ray diffraction structure of methylguanidinium phosphate (61).

Of principal interest in the electronic-structure
calculations is the nature of the charge distribution
in hydrogen-bonded guanidinium phosphate (Gu_2HPO_4)
relative to that in the isolated $C(NH_2)_3^+$ and HPO_4^{2-}
ions. In particular, it is hoped that the calculations

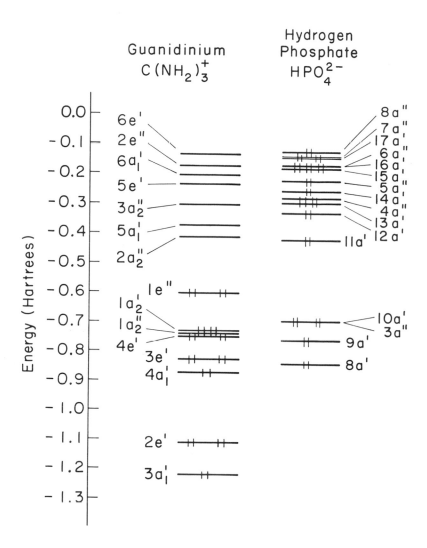

Fig. 9.   The SCF-Xα-SW electronic energy levels for
           the guanidinium and hydrogen phosphate ion
           clusters, as the starting point for building
           up the electronic structure of hydrogen-bonded
           guanidinium phosphate. The latter is a model
           for the 5'-phosphate binding site in a
           staphylococcus aureus enzyme-inhibitor complex.
           The number of electrons occupying each energy
           level is indicated.

will confirm the intuitive suggestion (39,60) that the
hydrogen bonding from the guanidinium ion serves to
polarize the P-O bonds, thereby rendering them more
susceptible to the nucleophilic attack for phosphate
hydrolysis, and thus providing a partial explanation
for the mechanism of action of the nuclease.

Although SCF-Xα-SW calculations for the electronic
structure of the complete nuclease-phosphate model have
not yet been finished, results have been obtained for
the separate $C(NH_2)_3^+$ and HPO_4^{2-} ions. These results
are shown in Fig. 9. The electronic structure of the
guanidinium ion is itself very interesting. Its charge
distribution is consistent with a significant amount of
π-electron delocalization, which explains the very
short C-N bond distance. Comparison of the electronic
energy levels of $C(NH_2)_3^+$ with those of HPO_4^{2-} (see
Fig. 9) indicates that there are low-lying empty
molecular orbitals on the positive ion which span
almost exactly the same energy range as the highest
filled orbitals of the anion. This suggests the like-
lihood of significant P-O bond polarization when the
separate ions are linked via hydrogen bonds. Comparison
of the charge distributions in hydrogen phosphate (HPO_4^{2-})
and the ordinary phosphate ion (PO_4^{3-}) shows a slightly
greater electron density on phosphorus in the protonated
ion. It will be interesting to see how the effects of
hydrogen bonding from $C(NH_2)_3^+$ to HPO_4^{2-} will compare
with those due to protonation of PO_4^{3-}. These calculations are not only relevant to the nature of the nuclease-
phosphate binding, but also to the much broader problem
of the binding of phosphate-containing macromolecules
(e.g., DNA, RNA, ATP, and various enzyme cofactors) to
enzymes and proteins, a major issue in molecular biology.

VI.   SUMMARY AND CONCLUSIONS

A few of the many types of calculations on molecules which are practicable with the SCF-Xα-SW method
have been described in the preceding sections. The
chief advantage of this technique over more conventional
methods of quantum chemistry is that it yields a reasonably accurate description of the electronic structures
of many-electron polyatomic molecules with relatively
little computational effort. It is not an ab initio
method, as traditionally defined in the Hartree-Fock
SCF-LCAO approach, since one works with a statistical

expression (Eq. 11) for the total energy, not the "exact"
many-electron Hamiltonian. It is also not a semiempirical
method, since no parameters based directly on experimental
data are incorporated at any point of the computational
procedure. On the other hand, because a consistent set
of initial nonempirical approximations (the Xα exchange-
correlation potential and the division of molecule space
into different regions of potential) are introduced in
the SCF-Xα-SW theory, it may be called a first-principles
technique. Unlike the ab initio SCF-LCAO method, the
present approach does not suffer from the problem of
choosing the proper molecular-orbital basis set (e.g.,
Gaussian vs. Slater type atomic orbitals), because the
Xα-type one-electron Schrödinger equation (Eq. 1) is
solved "exactly" throughout the space of the molecule in
a numerical partial-wave representation. Thus one does
not encounter the so-called "N^4 problem," i.e., the fact
that in the LCAO method the number of multicenter integrals increases with the fourth power of the number of
electrons in the molecule, thereby increasing the required
computational effort markedly with the number of electrons. In the SCF-Xα-SW approach, there are no multicenter integrals and the computer time does not increase
excessively with either the number of atoms in the molecule or with the number of electrons per atom.

Of course, the main criticism of the SCF-Xα-SW
approach is its reliance on the muffin-tin representation of the potential. However, it has been shown (26-
29) that even for highly non-spherically-symmetrical
molecules, such as the linear system carbon suboxide
(C_3O_2), the method generally leads to a more accurate
description of the one-electron properties (e.g.,
orbital orderings, optical excitations, ionization
energies, and charge distributions) than does the ab
initio SCF-LCAO technique, and with significantly less
computational effort than in the latter approach. It
appears, therefore, that the muffin-tin approximation
is an asset, not a liability, as far as the one-electron
aspects of the electronic structure are concerned. As
we have described in Section IIIC, the muffin-tin approximation often leads to equilibrium statistical total
energies, bond distances, force constants, and internal
barriers which are at least as satisfactory as the
results of more elaborate Hartree-Fock LCAO calculations,
and which satisfy the virial theorem to reasonable
accuracy. Moreover, the statistical total energy exhibits
the proper behavior as the molecule dissociates, as long
as the atomic sphere radii are increased in proportion

to the internuclear distances. In general, however, to
obtain a reasonable minimum equilibrium total energy as
a function of molecular conformation (e.g., bond angle),
it is necessary for one to include at least the first-
order interatomic non-muffin-tin corrections to the
charge density. This can frequently be accomplished,
without any modifications to the computer programs,
through the introduction of supplementary interatomic
regions of spherically averaged potential (e.g., in the
regions of lone-pair electrons). More generally, it is
necessary to use the complete charge density (such as
that shown in Fig. 4 for CH_4) as a basis for calculating
non-muffin-tin corrections to the total energy. Modifications of the computational procedure to include these
corrections are in progress.

There are, of course, situations where it will be
important to go beyond the one-electron description
provided by the SCF-Xα-SW theory. For example, in the
analysis of optical excitations, ionization energies,
and crystal-field splittings for open-shell molecules
and clusters, the transition-state procedure yields at
most the differences between the average energies of
multiplets of differing multiplicity. For more detailed
information about multiplet structure and oscillator
strengths, it is necessary to determine linear combinations of determinantal functions of the calculated spin-
orbital wave functions. Since the method leads to a
rapidly convergent complete set of spin orbitals, they
are ideally suited for the construction of determinantal
functions.

Because of space limitations, only some aspects of
treating the general problem of localized electronic
states of solids have been described in this paper,
particularly in Section IV in connection with the electronic structures of metal complexes in crystalline
environments. For example, in explaining the principal
features of the optical properties of a $KMnO_4$ crystal,
which has four K^+ ions and four MnO_4^- ions per unit cell,
it was not necessary to carry out a complete band-structure calculation for the crystal. It was necessary only
to calculate the detailed electronic structure and
optical excitations (via the transition-state procedure)
for a single MnO_4^- cluster in the average stabilizing
electrostatic field of the crystal. There are many
other examples of problems involving localized states in
solids which are accessible with the SCF-Xα-SW method.

Among these are:

(1)   localized crystal defects (e.g., "deep-level"
      impurities and vacancies).
(2)   localized optical excitations in crystals (e.g.,
      excitons).
(3)   localized magnetic properties (e.g., Heisenberg
      exchange integral; local moments).
(4)   chemisorption (e.g., chemisorption of atoms and
      small molecules at catalytic interfaces).

Applications of the SCF-Xα-SW method to all these
areas are in progress. For example, the exciton problem
and a procedure for calculating the Heisenberg exchange
integral are discussed in Ref. 1. Calculations of the
deep levels associated with "luminescent" impurities in
II-VI compounds (e.g., Mn in ZnS) are described in Ref.
2. A determination of vacancy and interstitial levels
in lead telluride (PbTe), including relativistic effects,
is outlined in Ref. 15. Also under way are theoretical
studies of the electronic structure of interstitial
carbon in iron lattices (62) and of the chemisorption of
CO, NO, H_2, and O on transition-metal and transition-
metal-oxide surfaces (56,63).

It is sufficient to summarize here that a certain
amount of useful information about the effects of a
localized perturbation (such as a defect or local excitation) on the electronic structure of an otherwise perfect
crystal or interface can be extracted from only a knowledge of the electronic structure of a sufficiently large
but finite cluster of atoms centered on the perturbation.
Focussing one's attention on a finite cluster of atoms,
rather than the entire solid, is not a new idea, and a
similar approach to the problem of impurities and
vacancies in diamond lattices has been described by
Watkins and Messmer (64). The main difference is that
Watkins and Messmer use a semiempirical LCAO method to
calculate the electronic energy levels, while in our
applications the SCF-Xα-SW method is used.

The chief difficulty in placing confidence on the
results of such calculations is the proper treatment of
the termination effects or boundary conditions on the
cluster. Within the framework of the SCF-Xα-SW method,
there are three principal ways to account for these
effects, to first approximation. For reasonably ionic
crystals, one can use the same procedure as described in
Section IV for a MnO_4^- cluster in a crystalline environ-

ment, namely introducing a spherical shell of charge
around the cluster to approximate the average electrostatic field (Madelung constant) of the rest of the
lattice.   Reasonably small clusters generally suffice
in this situation.   For predominantly covalent solids,
larger clusters should be used [of the sizes described
by Watkins and Messmer (64)], and the "saturating
effects" of the remainder of the solid should be
accounted for, unless "surface states" due to "dangling
bonds" are to be tolerated.   This can be accomplished
by saturating the atoms at the periphery of the cluster
with hydrogen-like atoms or groups of atoms.   Finally,
it may be meaningful under certain circumstances to view
the cluster as a periodic entity (i.e., a complex unit
cell) in the bulk solid or at the interface.   A similar
viewpoint has been expressed by Watkins and Messmer (64).
Then it is possible to replace the cluster boundary
condition of wave function localization by the Bloch
condition.   It has been shown (2) that in this limit,
the SCF-Xα-SW theory reduces to a band-structure formalism
similar to that of the KKR method.

ACKNOWLEDGMENTS

    We are very grateful to Professor J. C. Slater for
originally suggesting the SCF-Xα-SW approach.   His many
helpful discussions and constant encouragement throughout the course of this research have greatly contributed
to its successful completion.   We are also grateful to
Dr. F. C. Smith, Jr. for his unfailing expertise in
developing and improving the computer programs used in
this work.   We thank Professor F. A. Cotton for introducing us to the very interesting and challenging
problems of molecular phosphorus and enzyme-phosphate
binding, and for making available computer support.   Our
thanks also go out to our colleagues at other institu-
tions, who have developed an interest in the SCF-Xα-SW
method and have contributed to the results reported in
this paper.   Among these are Dr. U. Wahlgren of the
University of Stockholm, Sweden, Drs. C. Nordling, U.
Gelius, and H. Siegbahn of the University of Uppsala,
Sweden, and Dr. R. P. Messmer of the General Electric
Research and Development Center.   Finally, we are grateful
to Dr. Frank Herman and his colleagues at the IBM San Jose
Research Laboratory for organizing a most stimulating
symposium and for inviting us to participate in it.

REFERENCES

*Research sponsored by the Air Force Office of Scientific Research, United States Air Force (AFSC), Contract No. F44620-69-C-0054, by the National Science Foundation, Grants No. GP-21312 and No. GP-16464, by the Advanced Research Projects Agency, Contract No. DAHC15-67-C-0222, and by the Institute of General Medical Sciences of the National Institutes of Health, Grant No. GM13300 (made available by arrangement with Professor F. A. Cotton).

†Visiting from the Department of Chemistry, M.I.T.; address after August 1, 1972, Department of Chemistry, University of Washington, Seattle, Washington.

1.   J. C. Slater and K. H. Johnson, Phys. Rev. B5, 844 (1972).

2.   K. H. Johnson and F. C. Smith, Jr., Phys. Rev. B5, 831 (1972).

3.   J. C. Slater, in Advances in Quantum Chemistry, edited by P.-O. Löwdin (Academic, New York, 1972), Vol. 6.

4.   K. H. Johnson, in Advances in Quantum Chemistry, edited by P.-O. Löwdin (Academic, New York, in press), Vol. 7.

5.   K. H. Johnson and F. C. Smith, Jr., Chem. Phys. Letters 7, 541 (1970).

6.   K. H. Johnson and F. C. Smith, Jr., Intern. J. Quantum Chem. 5S, 429 (1971).

7.   K. H. Johnson and F. C. Smith, Jr., Chem. Phys. Letters 10, 219 (1971).

8.   J. W. D. Connolly and K. H. Johnson, Chem. Phys. Letters 10, 616 (1971).

9.   K. Schwarz and J. W. D. Connolly, J. Chem. Phys. 55, 4710 (1971).

10.   U. Wahlgren and K. H. Johnson, J. Chem. Phys. 56, 3715 (1972).

11.   K. H. Johnson and U. Wahlgren, Intern. J. Quantum Chem. (in press).

12. J. W. D. Connolly, Intern. J. Quantum Chem. (in press).

13. J. W. D. Connolly and J. R. Sabin, J. Chem. Phys. (in press).

14. J. B. Danese, Intern. J. Quantum Chem. (in press).

15. G. W. Pratt, Jr., J. Nonmetals (in press).

16. J. C. Slater, T. M. Wilson, and J. H. Wood, Phys. Rev. $\underline{179}$, 28 (1969).

17. J. C. Slater, J. B. Mann, T. M. Wilson, and J. H. Wood, Phys. Rev. $\underline{184}$, 672 (1969).

18. J. C. Slater and J. H. Wood, Intern. J. Quantum Chem. $\underline{4S}$, 3 (1971).

19. J. C. Slater, Phys. Rev. $\underline{81}$, 385 (1951); $\underline{82}$, 538 (1951).

20. R. Gaspar, Acta Phys. Hung. $\underline{3}$, 263 (1954).

21. W. Kohn and L. J. Sham, Phys. Rev. $\underline{140}$, A1133 (1965).

22. F. Herman and S. Skillman, Atomic Structure Calculations (Prentice-Hall, Englewood Cliffs, N. J., 1963).

23. K. Schwarz, Phys. Rev. B$\underline{5}$, 2466 (1972).

24. J. Korringa, Physica $\underline{13}$, 392 (1947).

25. W. Kohn and N. Rostoker, Phys. Rev. $\underline{94}$, 1111 (1954).

26. J. W. D. Connolly (unpublished results).

27. J. W. D. Connolly, C. Nordling, U. Gelius, and H. Siegbahn (unpublished results).

28. U. Wahlgren and K. H. Johnson (unpublished results).

29. J. G. Norman, Jr., K. H. Johnson, and F. A. Cotton (unpublished results).

30.   K. Siegbahn, C. Nordling, G. Johansson, J. Hedman,
      P.-F. Hedén, K. Hamrin, U. Gelius, T. Bergmark,
      L.-O. Werme, R. Manne, and Y. Baer, ESCA Applied
      to Free Molecules (North-Holland, Amsterdam, 1969).

31.   U. Gelius, B. Roos, and P. Siegbahn, Chem. Phys.
      Letters 4, 471 (1970); U. Gelius, private communica-
      tion for the molecular-orbital energies.

32.   Recent modifications to the SCF-Xα-SW computational
      procedure now permit the calculations to be done in
      less than half this time.

33.   G. Berthier, A. Y. Meyer, and L. Draud, Jerusalem
      Symposium on Quantum Chemistry and Biochemistry,
      edited by E. D. Bergmann and B. Pullman (Jerusalem
      Academic Press, 1971), Vol. 3, p. 174.

34.   D. R. Kearns, J. Chem. Phys. 36, 1608 (1962).

35.   J. C. Slater, Quantum Theory of Molecules and
      Solids, Vol. 1 (McGraw-Hill, New York, 1963).

36.   G. Das and A. C. Wahl, J. Chem. Phys. 44, 87
      (1966).

37.   S. Weiss and G. E. Leroi, J. Chem. Phys. 48, 962
      (1968).

38.   F. A. Cotton and G. Wilkinson, Advanced Inorganic
      Chemistry, 2nd ed. (Interscience, New York, 1966),
      p. 489.

39.   F. A. Cotton (private communication).

40.   L. Pauling and M. Simonetta, J. Chem. Phys. 20, 29
      (1952).

41.   C. R. Brundle, N. A. Kuebler, M. B. Robin, and H.
      Basch, Inorg. Chem. 11, 20 (1972).

42.   J. Tetlow, Z. Phys. Chem. B40, 397 (1938); B43,
      198 (1939).

43.   M. Wolfsberg and L. Helmholz, J. Chem. Phys. 20,
      837 (1952).

44.   S. Holt and C. J. Ballhausen, Theoret. Chim. Acta
      7, 313 (1967).

45. C. J. Ballhausen and A. D. Liehr, J. Mol. Spectry. 2, 342 (1958).

46. R. F. Fenske and C. Sweeney, Inorg. Chem. 3, 1105 (1964).

47. A. Viste and H. B. Gray, Inorg. Chem. 3, 1113 (1964).

48. L. Oleari, G. de Michelis, and L. diSipio, Mol. Phys. 10, 1111 (1966).

49. R. D. Brown, B. H. James, M. F. O'Dwyer, and K. R. Roby, Chem. Phys. Letters 1, 459 (1967).

50. J. P. Dahl and C. J. Ballhausen, in Advances in Quantum Chemistry, edited by P.-O. Löwdin (Academic Press, New York, 1968), Vol. 4, p. 170.

51. J. P. Dahl and H. Johansen, Theoret. Chim. Acta 11, 8 (1968).

52. I. H. Hillier and V. R. Saunders, Proc. Roy. Soc. (London) A320, 161 (1970).

53. I. H. Hillier and V. R. Saunders, Chem. Phys. Letters 9, 219 (1971).

54. P. D. Dacre and M. Elder, Chem. Phys. Letters 11, 377 (1971).

55. R. C. L. Mooney, Phys. Rev. 37, 1306 (1931).

56. K. H. Johnson (unpublished results).

57. R. P. Messmer, U. Wahlgren, and K. H. Johnson (unpublished results).

58. L. Noodleman and K. H. Johnson (unpublished results).

59. D. Choo and G. W. Pratt, Jr. (unpublished results).

60. A. Arnone, C. J. Bier, F. A. Cotton, V. W. Day, E. E. Hazen, Jr., D. C. Richardson, J. S. Richardson, and A. Yonath, J. Biol. Chem. 246, 2302 (1971).

61. F. A. Cotton, V. W. Day, E. E. Hazen, Jr., and Mrs. S. Larsen, J. Amer. Chem. Soc. (to be published).

62. R. Kaplow, W. Choo, and K. H. Johnson (unpublished results).

63. R. P. Messmer and K. H. Johnson (unpublished results).

64. G. D. Watkins and R. P. Messmer (this volume).

THE ORBITAL CORRECTION METHOD

Walter A. Harrison

Applied Physics Department

Stanford University, Stanford, California

It may be helpful at the outset to state briefly the method and the findings which we are to discuss. The orbital correction method is an approximate scheme for computing the total energy of molecules as a function of the arrangement of the constituent atoms. It is based upon a self-consistent-field treatment of the electron-electron interaction and an expansion technique analogous to perturbation theory. The expansion parameter in this case is the difference between the true one-electron state and an assumed starting state, which is essentially a linear combination of atomic orbitals. The simplification allowed by this expansion permits the treatment of arbitrarily large systems and gives a simple framework for the treatment of all properties of molecular systems. Preliminary application to the central hydrides illustrates the fact that many properties can be meaningfully treated in zero order (no orbital correction), while the inclusion of the orbital correction is absolutely essential in others.

A solid-state theorist working on molecules comes with a series of opinions based upon his experience in solid-state theory. Since these are not shared by many of those working in molecular physics, it may be most helpful with respect to this conference to enumerate those views which led to the development of the orbital correction method. I will then review briefly the method and some results which clarify the nature of the method.

The first point carried over from solid-state physics is that it may be very desirable to focus specifically on the total energy of the system as a function of the position of the constituent atoms. This is relevant to studies of the stable configuration, the vibration frequency, and barriers against various conformal changes. This may be contrasted with studies of the cohesive energy itself which seek high accuracy in the total energy for a specific configuration. In solid-state theory entirely different methods are used for the two problems. It also may be contrasted with studies of the excitation spectrum, corresponding to band structure in solids. Such spectra have been calculated for all classes of solids but the only solid-state systems for which the total energy as a function of arrangement has been solved in any completeness are the simple metals; the method used has been the pseudopotential approximation. Interestingly enough, these systems correspond in molecular terminology to open-shell structures.

The second point carried over from solids is the relevance of the one-electron approximation in the calculation of total energy. In solids one in fact has no choice. Because of the intrinsic complexity of the problem, not even Hartree-Fock theory is possible and one must use some form of exchange potential such as that discussed earlier by K. Johnson. Just exactly which of the possible schemes are used is not of major importance because of the inherent inaccuracy of the expansion techniques we will use in calculating the energy. Our experience that meaningful treatments can be made both in solids and in free atoms using such exchange potentials would strongly suggest that meaningful treatments can be made also in the context of molecules.

The third point carried over from work in solids is the importance of a perturbation expansion in the calculation. In solids such an expansion is absolutely necessary since quantities such as the total energy cannot be calculated without such an approximation, except possibly in the case of the perfectly ordered crystal. The situation is different in the physics of relatively simple molecules. In those cases, at least within the one-electron approximation, it is possible to diagonalize matrices rather than use perturbation theory in obtaining the energy; thus there is not the compelling necessity for perturbation theory. There remain, however, at least two reasons for trying such an approximate approach: First, the great simplicity of a perturbation expansion may provide an understanding of molecular systems which would be difficult to obtain from a calculation in which the details are buried in a computer diagonalization. A second possibility is that the great simplicity from the perturbation calculation would permit us to avoid many of the rather crude approximations to the one-electron potential which are required in other calculation techniques, giving added accuracy and flexibility. Probably only experience will tell whether these two motivations are well founded.

We note immediately that the perturbation technique which is

to be used, either in the case of molecules or solids, must differ
fundamentally from traditional perturbation theory. In traditional
theory, we write the Hamiltonian of the system as the sum of an
exactly soluble Hamiltonian and a correction, the latter being
treated as the perturbation. In neither solid-state nor in most
molecular problems does there exist such an exactly soluble Hamiltonian which bears any relation to the true problem. Instead we
seek to define an expansion in terms of the one-particle eigen-
states themselves. We write the true eigenstate $|\psi>$ as a sum
of a guess $|\psi_0>$ and an orbital correction $|\chi>$. We can
always guess an eigenstate so this procedure is always possible.
However, it contains a degree of arbitrariness even once we have
chosen our starting state $|\psi_0>$. This arbitrariness becomes
apparent when we note that there does not ordinarily exist a
natural basis for expanding $|\chi>$. It will be convenient to
expand $|\chi>$ in a complete set such as plane waves. Then, because
we have an additional term $|\psi_0>$, our set is overcomplete; hence
the arbitrariness. There are many ways to make the procedure
unique, but the existence of many ways is of course an arbitrariness in itself. One can also take advantage of this arbitrariness
in formulating the method. For example, one may select orbital
corrections which tend to be small where the potential is largest,
in which case the product of the potential and $|\chi>$ is regarded
as small with respect to $|\chi>$, while of course $|\chi>$ itself is
regarded as small. In this case the perturbation expansion for the
energy becomes

$$E = <\psi_0|H|\psi_0> + \sum_q \frac{<\psi_0|H\text{-}E|q><q|H\text{-}E|\psi_0>}{E - \hbar^2 q^2/2m} + \dots \qquad (1)$$

Here the $|q>$ are plane waves and[1] H is the appropriate one-
electron Hamiltonian. Maarten Heyn[1] has shown that stopping at
second order, as shown above, corresponds to performing the first
step in an iterative procedure which, if carried to all orders,
would lead to an exact eigenvalue of the Hamiltonian. However, of
more interest is the accuracy of the second-order result itself
and that will only be discovered by performing sample calculations.
This approach has proven particularly useful in solids and our preliminary calculations have suggested that it is a very good approach
in molecular systems if the one-electron Hamiltonian in question
does not have deep excited bound states. In fact the one-electron
potential, including free-electron exchange, tends ordinarily not
to have deep excited bound states. In contrast the potential to
be associated with a positive ion, or with hydrogen itself, does
have deep bound states and this particular expansion appears to
lead to appreciable errors.

Of course expansions in overcomplete sets are not new and
many analogous approaches appear in the literature. Probably the
most familiar is that of simple metal pseudopotentials referred to
earlier. In simple metals it is believed that the valence state
can be rather well approximated by single plane waves, orthogona-
lized to every core state in the problem. Thus $|\psi_0 >$ is taken
as such an orthogonalized plane wave, and the matrix element
$< q|H-E|\psi_0 >$ becomes precisely the plane-wave matrix elements of
the non-local pseudopotential, and Equation 1 becomes the basic
equation which enables calculation such as the vibration frequencies in the simple metals.

In order to avoid confusion we should perhaps note that there
is a second method in solids, also called the pseudopotential method, which bears essentially no relation to this approach. The
pseudopotential calculation we have described here, which perhaps
should be called "Pseudopotentials II", can be formulated as the
construction of a Hamiltonian matrix based upon orthogonalized
plane waves and the evaluation of eigenvalues in perturbation
theory, treating the off-diagonal matrix elements as small. The
intent again is to allow the treatment of a wide range of systems
and the study of a wide range of properties. The other approach,
which we might call "Pseudopotentials I", can also be described
in terms of such a Hamiltonian matrix, but the approximation in
this case is an approximate parameterization of the off-diagonal
matrix elements; the motivation is either to reduce the computational effort or to allow the introduction of experimental information in the determination of these matrix elements. In "Pseudopotentials I" the resulting Hamiltonian is diagonalized exactly
just as in a full band calculation and the calculations are limited
primarily to the determination of the electronic structure of perfectly periodic structures.

Returning to the orbital correction method, illustrated by
Equation 1, we consider the treatment of molecules. Certainly then
an orthogonalized plane wave is not an appropriate starting state;
we use something much more closely related to a linear combination
of atomic orbitals. Thus the method becomes a systematic improvement on the LCAO method. In this application to molecules, pseudopotentials do not play an essential role. In addition to being an
improvement on an LCAO, it is also a simplification. Because of
the use of perturbation theory it becomes possible to sum the total
energy of the system without ever performing the diagonalization
which is essential to most LCAO methods. The mathematics of the
step is quite difficult; the essential point is that by systematically discarding higher-order terms the diagonalization can be
avoided. Thus within the context of this approach the size of the
system being treated is not limited. We may work directly with
local orbitals and analyze the interesting portion of the molecule

with the rest being included as a small correction.

Finally, I should like to illustrate the method by giving the results of some calculations on the central hydrides by Richard Meserve[2]. Such simple systems do not require the simplification offered by the orbital correction method. Each property treated could be studied more accurately by another approach. However, this simple system sheds some light on the nature of the method. Starting first with methane, a united atom picture is quite natural and we might for starting states take free carbon orbitals. We find immediately, however, that the first term in Equation 1 gives a positive energy if the methane Hamiltonian is used and the orbitals are taken from the Herman-Skillman tables[3]. Such orbitals are therefore quite inadequate. Meserve therefore re-calculated atomic-like orbitals using the Herman-Skillman program but taking for the potential the spherical average of the methane potential (and accordingly using a neon rather than a carbon configuration). Analogous orbitals may have been used in the past, though perhaps not specifically ones computed numerically with free-electron exchange. We call them "globe orbitals" because the hydrogen nucleii are replaced by globes of charge surrounding the carbon nucleus.

It is of course possible to compute a wide variety of properties of methane using only these zero-order orbitals without the orbital correction. Analogous calculations in the past have given semiquantitatively correct results on a number of properties. With our particular wave functions we can obtain the internuclear spacing to a few percent, the cohesive energy correct on the scale of the best Hartree-Fock calculations, the breathing mode frequency almost exactly correct, torsional frequencies accurate on the scale of perhaps 20%, and of course the correct equilibrium configuration.

The discrepancies with experiments rise partly from inaccuracy in the wave function and partly from an inadequate treatment of the electron-electron interaction. It is not clear that the orbital correction itself will significantly improve such results. Furthermore, this approximate treatment of the molecule, in which the Coulomb interaction between protons plays a central role, appears to give a correct qualitative description of the molecular properties. One might question the necessity of going to higher order. However a moment's thought about the application to ammonia will make it clear that within this approximation ammonia would be found to be a planar molecule and water would be found to be linear. Thus in the context of our calculation the orbital correction is essential to obtaining the correct atomic arrangement in these systems. The next step in the calculation is absolutely essential to obtain even a qualitative understanding of ammonia and water. Preliminary calculations indicate that such distortions will in fact be obtained in the second-order calculation. In

addition there are many other properties, such as the atomic polarizability, which depend directly on an orbital correction and cannot be treated without it.

The orbital correction method itself is intrinsically an
approximate method just as are the pseudopotential methods in
solids. For the simplest systems, such as methane, there will
always be methods which are more accurate. The promise in this
method lies in a conceptual and computational simplicity which may
bring new understanding to molecular problems. On the other hand
some feel that the basic understanding of molecules has been essentially complete for a long time and there is nothing left to do
but larger or more accurate numerical calculations; they may even
argue that the record of theoretical chemistry over the past few
decades supports this view. The orbital correction method was
partially motivated by the belief that this is not the case.

REFERENCES

1. Maarten Heyn, to be published.

2. Richard Meserve, to be published.

3. F. Herman and S. Skillman, Atomic Structure Calculations,
    (Prentice Hall, Englewood Cliffs, N.J., 1963).

PANEL DISCUSSION ON COMPUTATIONAL METHODS

Panel Members:

P. S. Bagus, IBM Research, San Jose (Chairman)
F. E. Harris, University of Utah
W. A. Harrison, Stanford University
K. H. Johnson, Massachusetts Institute of
    Technology
A. D. McLean, IBM Research, San Jose
U. T. Mueller-Westerhoff, IBM Research, San Jose
K. Ruedenberg, Iowa State University
G. A. Segal, University of Southern California
H. S. Taylor, University of Southern California
G. D. Watkins, General Electric Research and
    Development Center, Schenectady, New York
M. C. Zerner, University of Guelph, Ontario

Contributions to the discussion were also made by:

T. L. Gilbert, Argonne National Laboratory
R. K. Nesbet, IBM Research, San Jose
J. A. Pople, Carnegie-Mellon University
L. Salem, Orsay, France and Harvard University

Bagus. Our panel is about to begin. I had originally planned to
organize the panel in the traditional fashion of having the people
stand up and give set responses, but I was overwhelmed by the panel
members who said no, that wasn't a good way to run a panel; that,
in fact, we ought to try to get a general and open discussion going.
This sounds to me like a very good experiment to pursue. What I
plan to do to get things started is to pose four questions to the
members of the panel. The content of some of the questions is related in part to points that were presented in the panel members'
talks and in part to questions that were asked in the discussions
of these talks.

The first question is: For what classes of systems can your own method, within its present limitations, be applied? Second: What extensions, either theoretical or computational, are needed or are foreseeable to extend the method? These two questions ask what sort of things can be done. The remaining two questions are: What properties (in a broad sense, including potential surfaces, energies, transition energies, expectation values) cannot be treated? And, finally: How reliable are values for properties which can be treated? The answer to this last question may go quite far to answer Dr. Marcus' question this morning: How good is Hartree-Fock? I hope that's one specific question that does come up. These last two questions address the value of the results that can be obtained.

I'll start things going with a specific question, and I'll do that by taking two extremes of ab initio calculations that are being carried out at San Jose; extreme in the sense of the size of the systems which are considered. One is the carbon atom in its four lowest states, on which calculations are being carried out by Prof. Sasaki. He has done a rather large CI calculation involving single and double excitations from either the $2s^2 2p^2$ reference state for the ^4S term, which is the fourth term he has considered, the $2s 2p^3$ reference state. He has used approximately 400-500 configurations and he has obtained an absolute error for the non-relativistic energy of 3 milli-Hartrees. That's about 97% of the correlation energy which is in fact better than any calculation to date on the carbon atom. He has in progress calculations involving higher excitations (triple and quadruple replacements). Professor Sasaki expects the error for this calculation to be of order 1 milli-Hartree.

I want to contrast this to a minimum basis set calculation on a system which I believe is the largest system on which an ab initio calculation has been done. The system is 2,4,7-trinitro-9-fluorenone (TNF) with 13 carbon, 3 nitrogen, 7 oxygen, and 5 hydrogen atoms. The work has been carried out jointly by Dr. Batra, Dr. Clementi, Dr. Seki, and myself. The minimum basis set for TNF requires 97 sigma and 23 pi basis functions of the contracted GTO type. About eleven million integrals are generated for this calculation. The contrast, then, is between an extremely accurate calculation of a small system and a minimum basis set calculation for a very large system.

I'd like to address a question to Prof. Zerner. Some six years ago you wrote in a paper that minimum basis set calculations were fairly far off on the horizon. Now that it is possible to do minimum basis set calculations on rather large systems, would you like to have them, or would you prefer semi-empirical calculations? In general, there is the question of what should a semi-empirical calculation model, the minimum basis set calculation or something closer to experiment.

Zerner.   The first question is very much concerned with what you
expect from the calculation. If one is after rather broad concepts
of chemistry dealing with the nature of specific molecular orbitals,
symmetries, ordering, etc., then well-based semi-empirical and approximate methods seem to be successful. I'm not sure that most
people who are interested in chemical problems--as opposed to problems in theory--would have the patience or financing to do exacting
calculations, nor am I sure that much more information of a qualitative nature would be forthcoming from accurate calculations.
The second point, does one get anything more out of a calculation
which stays within an exact theory? What does one have to compare
with? One has a wave function. To my experience, if you can't
play with the wave function, you really can't say much about it.

Taylor.   I don't understand how you can ask the value of something
unless you decide what you want to learn in the first place.   It
depends on what the experimentalist can do with that and what he's
confused about.

Segal.   I'm curious as to why this particular molecule (TNF) was
chosen for study.

Bagus.   It was studied because it is of interest in a charge transfer complex reaction involving photosensitive chemicals.

> (Editors' Note: There was considerable discussion among the
> panel members and among other symposium participants on this
> point. In view of this interest, we have asked Drs. I. P.
> Batra and H. Seki to include in these proceedings a statement
> on the motivation and status of the calculations on carbazole
> and TNF. This statement appears as the following paper.)

Zerner.   Did you get an answer to that problem?

Bagus.   Calculations are still being done.

Harris.   In charge transfer properties, presumably what you want to
know are the gross features of the charge distribution. Something
as crude as a Huckel molecular orbital theory will give the gross
features of the charge distribution if it's been reasonably parametrized.

Zerner.   There has been quite a bit of success using the CNDO/INDO
idea looking exactly at this type of thing. It has worked and the
calculations are very rapid.

Harris.   I think that one could justify, though, calculations of
this sort in terms of learning how to handle and organize very large
amounts of data to get a better idea than you can ahead of time on

how long such calculations take. And there is a place for a very
small number of very crude calculations. I question whether, until
things change very much, this is going to be the preferred way of
getting at routine information about systems of this size.

Mueller-Westerhoff. I would judge it as a pilot for future calculations of the same type. From my experience with semi-empirical
computations I am very sceptical about the immediate value of a
single computation as far as the accuracy of its results is concerned. For CNDO or Hueckel type calculations there exists now a
wealth of correlation between experimental numbers and the data
that you can obtain from calculations, although this correlation
may be limited to certain classes of compounds. For me, an assess-
ment of the absolute value of this specific calculation would require similar computations for a series of related nitro-compounds.

Zerner. I'd just like to point out Dr. Clementi's enormous calculation on the base pairs of DNA in which he addressed himself to
the hydrogen bonding between the pairs. After he had finished his
calculation, he did a very interesting thing. He essentially said
the calculation didn't present him with the answers directly, and
it was too difficult to proceed with more calculations. Arguing by
analogy to what he knew about simple systems, the dimers of formal-
dehyde, he drew conclusions not directly obtainable from the lengthy
calculations. The interesting aspect of this is that argument by
analogy is an essential part of all semi-empirical approximate
theories and is a reason for their success. It has been a far less
common feature in ab initio work.

Segal. With the inclusion of all of the sigma electrons, the fluorenone calculation in the CNDO or INDO approximations would, I would
guess, take a couple of minutes. Just a simple, straightforward
calculation without any geometry optimization. I think that the
ESCA chemical shift results show that the charge densities obtained
would be reasonable. You would gain essentially the information
you can get out of a low level ab initio calculation. Now if you
wanted to try to examine geometry and so forth, that would be possible within the approximate frame of reference, but it certainly
isn't possible here.

Bagus. I believe that you are selling the power of ab initio calculations short. It is true that no geometry optimization has been
performed in the current calculations on TNF. However, Dr. Clementi
is right now in the process of extending his IBMOL program so that
both geometry optimization and extended basis sets will be possible
for the system. In fact, we have plans to perform geometry optimization for a study of the interaction of carbazole and TNF molecules.

Ruedenberg. I think that one does want to calculate certain molecules
as large as the one mentioned. I think it's fairly likely that as

you go to groups of larger molecules you have to establish certain benchmarks by elaborate calculations. Some large calculations will have to be done to set benchmarks for developing approximate calculations. I don't think you can take any of the present approximate methods, which are gauged on small molecules, go to very large systems and believe everything they give. You will have to calculate certain large molecules to gauge again. I've no doubt that large calculations are going to be made for exactly that reason, when the desired information seems worth the expense.

Harrison. Really, the largeness isn't the essential point. You can always do a solid, which is essentially infinite. The question is: How does the nature of the system change when it gets more complex?

Ruedenberg. For certain large systems you can predict exactly what's going to happen; then there is not much point in doing a calculation. But in other large systems, you may have the possibility of rather peculiar bonding situations, which cannot occur in smaller systems, and then you have to do it rather carefully. At every step into a new field, you have to put your foot on solid ground, and solid ground means accurate calculations.

Harrison. There is one aspect of complicated systems that I think needs to be sorted out reasonably clearly. There is a tendency, at least in saturated systems, for interatomic interactions to be local. Sometimes, nevertheless, one part of a very large molecule appears to affect the behavior of some other part which is distant. It would be very nice to have an understanding of the mechanism that communicates across a large molecule. To the extent that there are no such long-range interactions one can study a small part of a molecule by itself. To the extent that there is long-range communication, only if one can isolate that interaction can one deal with the system in a relatively simple way.

Ruedenberg. Chemistry can get complicated. Fluorenone has a six- and a five-membered ring, and unusual things may happen. So you worry about what's going on.

Bagus. Well, you can take as one indication that unusual things are likely to happen with that system the fact that there were severe difficulties in getting the SCF step to converge. The calculations did converge for several states, but with difficulty. This is related to something which had troubled me about the choice, in Prof. Harrison's orbital correction method, of a zeroth order wave function. If the processes which occur are not simple and direct, then you must go to a fairly complicated calculation to see what your zeroth order wave function ought to be.

Harrison. The point is to try a zero-order state. If it converges
badly, you should then question the intuition which led you to
that starting point. To me it is more important to refine one's
understanding than to accurately calculate a set of eigenvalues.
A very powerful computational method might allow one to get good
answers with little understanding but that is not my goal.

Taylor. We don't do large molecule calculations very accurately.
You've got to realize that now. So there should be two lines of
development. One is to attempt to compute accurately. I think
that Frank Harris was talking about that in developing data handling procedures. At the same time, we should say what we can do
for chemistry, because if we go along long enough without ever
turning around and saying we can do something, people are not going
to be particularly interested in this and it's going to become a
very sterile and detached field. I think that there are a great
number of areas in which, working with experimentalists, one can
limit the experimentalists' options. For example: Gerry Segal and
I have worked on electron scattering off CO_2 where all processes
required short-lived states of CO_2^- as intermediates. It was very
important to determine where they would be. We used an inaccurate
theory, that is, INDO. But we did determine, since we knew our
error limits, that there couldn't be a state at one eV above the
ground state of CO_2. This meant that some theories which claimed
that this state was what was causing the cross-section peaks at one
eV were wrong. We suggested angular distribution experiments to
resolve the issue and pin down the effect which we thought was due
to a dipole moment. These are costly experiments--it takes time
to build a machine and to make the measurements--but the experimentalists felt it to be worth doing since they could distinguish between two proposed mechanisms. In the theory we could have been
off even 2, $2\frac{1}{2}$ eV on every point on the potential surface and still
have given the same experimental interpretation. I think this is
useful, and I think that people doing big calculations, whether
they be a little more accurate or less accurate, that their real
service--half of the field--should be going to the experimentalists
and asking for things they can't measure and what they need to know
before they can do a definitive experiment. In some sense, we should
be defining the critical experiment that will pin down the options
among possible physical models. That's where I feel especially the
semi-empirical can make an enormous impact and I think many people
use it this way.

Mueller-Westerhoff. May I make two comments. I believe it to be
generally true that for the "average" experimentalist a Hartree-
Fock calculation is much too complex to handle to allow him to obtain by himself clarification of a question from his main field of
research, so that he has to limit himself to semi-empirical methods.
In this connection, I would find it very interesting, for instance,
to discuss what CNDO or other semi-empirical methods should be tested

against. I think in your comment I noticed a suggestion that you take let's say Hartree-Fock limits as the goal to which to fit CNDO or similar class calculations, rather than purely experimental data. That might be a point worth discussing.

My second comment concerns a different approach that I would like to see discussed. Wouldn't it be a good idea to take something like a CNDO calculation as the first approximation for a Hartree-Fock calculation--just like you take say a Hueckel calculation in a Pariser-Parr-Pople method. I think that this would be quite practical and time-saving especially with respect to the determination of a starting geometry.

Segal. First of all, I think you are quite right. If you're going to construct a semi-empirical theory then it must be well defined. You need rigorous calculations in order to set it up. Otherwise you set it up by going to one experimental property or the other, and you're in a morass of conflicting goals because you tend to bias your theory toward the property you're calculating, usually to the detriment of others. But, on the other hand, there has to be some time when you're going to leave the ab initio calculation and begin looking at experiment. I think that it's true that the minimum basis set calculation on the fluorenone has not much greater probability of being correct, relative to nature, than a semi-empirical calculation. It's a very low level calculation, and therefore is not really something to compare to after a certain point. You want to compare to calculations of this type while building, but you don't need fluorenone to do that.

Ruedenberg. The lesson to be learned is: If even IBM cannot afford hundreds of calculations of this size in one year, then one should think very carefully on which big molecules such a tremendous effort should be expended before going ahead and doing it.

Bagus. It's certainly true that when a calculation is a difficult one to do, one doesn't think in terms of doing a vast number of difficult calculations. One wants to find systems where the answers that a difficult calculation provide can't be obtained otherwise.

I would like to take a different approach to the comparison of the two sets of calculations that I mentioned in my introduction. Professor Sasaki's calculations on the carbon atom are very sophisticated work which aim at obtaining nearly exact solutions of Schroedinger's equation. Our calculations on TNF using a limited basis set are performed using the Hartree-Fock model. Dr. Mueller-Westerhoff pointed out earlier that this model is already very sophisticated, powerful and complex. It is, however, much less sophisticated than the configuration interaction method used by Prof.

Sasaki. With this background, I would like to pose the following
question: What are the practical delimiters between systems for
which very sophisticated quantitatively accurate approaches can be
taken, and systems for which the Hartree-Fock one-electron model
can be used?

McLean. At the small end of the scale, current capability makes it
possible to do calculations on systems of up to three first-row
atoms such as carbon dioxide, which can establish energy differences
between potential surfaces for a given nuclear geometry to within
one or two tenths of an electron-volt. Many properties of these
molecular systems can be evaluated to an accuracy of within a few
percent. These are expensive calculations, and will remain expensive calculations. They can provide data unobtainable from other
sources. The relevant question to ask is: Given the capacity to
produce the data, is it sufficiently valuable, scientifically, that
people will meet the cost of obtaining it?

Segal. I'd suggest that first you do rough calculations which are
not so accurate in order to discover what you really don't know
about the system as against what is already defined by experiment.
If you find information which you need to know and which is difficult to measure, then you spend the money and do the calculation.

McLean. How do you find out what you don't know?

Segal. There's usually a lot of experimental data on any system
you are going to study. You already know a great deal although
you may not have it organized so that you understand the implications of a number of separate experiments taken together. You can
do a rough calculation and you may find after you look at it that
it's not good enough, that the information in the calculation
doesn't define what you want to know about the system when it is
combined with the experimental facts. Then you have to spend the
money and do a better calculation.

Johnson. In the area of small molecules, there is a wealth of
data on excited states of small molecules, multiplets, and so on,
and you would like to do as accurate a calculation as you can on
these.

Segal. But I don't understand. If there is a wealth of data,
then you probably already know the result. Why calculate it? A
great deal of effort presently goes into trying to calculate numbers which are well known. A certain amount of this is necessary
to document the computational methods, but once this is known, calculations should be aimed at filling gaps in the data.

Johnson. First there's the problem of understanding that data. Then, once you've been able to interpret it, you must predict in other systems for which you might not have the data something about the excited states. This is one of the things that worries me about concentrating on one particular large molecule. If you take a method and you apply it to lots of different examples, you really learn what its limitations and advantages are. Then maybe you can use it as a predictive tool for other systems for which you don't have experimental data. That's as important for small molecules as it is for large.

McLean. Recent examples of valuable computations, which pooled the resources of three laboratories, are the determination of the low lying potential curves of CH^+ and CH. The information, unobtainable from any other source, is required to determine the behavior of hydrogen atoms scattering off of either C^+ or C, a process of considerable astrophysical interest. I anticipate that more and more accurate data on small systems will be derived through computation; currently the biggest problem is cost.

Harris. You pointed out something which I think is very important, and that is that one of the places where you can make the main contribution isn't just to produce another number characterizing some excited state, but to get information about repulsive potential curves for systems in chemical reactions. When we do that, I think it's also very important to recognize that most of the time you won't be able to do it successfully in the Hartree-Fock approximation because the processes you're studying involve changes in the number of electron pairs, changes in the organization of these pairs. You probably should make such calculations using configuration interaction. You should project out proper spin states or at any rate, have the different kinds of spin couplings present in the problem. This means that you are talking about something which is comparable to taking a modest sized basis and doing a full CI. It is probably much more important to have the freedom in the N-electron basis so that you get the electrons to couple in all the possible ways for a modest basis set rather than to have a large basis set and artificially restrict yourself not to have all the coupling. When you've done that then you can get very nice potential curves and surfaces for two and three heavy-atom systems. That's well within economic practicality.

Bagus. Directed to this point, there are many properties that are extremely difficult to obtain by experiment. To take the example of the system that Dr. McLean just mentioned, CH, transition probabilities among electronic states are extremely difficult quantities to measure experimentally. Yet, as soon as wave functions are available, they are not terribly difficult quantities to compute.

Mueller-Westerhoff. That's not really the kind of argument that we are having here. We are concerned with large systems. I would not argue with you on a small system. That is one area in which you should apply the best possible computation.

Johnson. I'd like to put in a word about large versus complex systems. There are small systems involving few atoms (e.g., diatomic molecules) which contain heavy atoms. I understand, for example, there's an increasing amount of interest in diatomic transition metal oxides and rare earth compounds. What does one do in these cases to achieve the most accurate calculation? I understand that the complexities of ab initio LCAO techniques-- the same ones which limit you in large molecules--also limit you in dealing with large numbers of electrons per atom. Is that not true?

Bagus. One of the great hopes for ab initio calculations is that the complexity of the calculation is a function of the number of valence shell electrons. That is, that it is a function of the number of electrons for which one has to provide a sophisticated wave function. Thus, a large atom does not present problems of size because it contains a large core; there's no real need to treat the core electrons in a very sophisticated way. Presumably, what is happening that's important to the system is happening in the outer shells, i.e., is happening in the valence shells. Now large atoms may pose problems because the bonding for d and f electrons may be more complicated than for s and p electrons, but this is a different question. In any case, one is only obliged to have a sophisticated treatment for the valence shell electrons and not for the core electrons. Even on a level of a Hartree-Fock calculation where we talk about the importance of a double-zeta representation and polarization functions, it's a relatively common practice to treat the core electrons in a crude way with a very limited basis set and to use a flexible basis only for the valence shell electrons. The recent work of Prof. Moscowitz and his collaborators on compounds of Xe with F is an excellent example of this sort of treatment. When one introduces correlation effects it's again rather common practice to treat the correlation effects just in the valence shell orbits. There have been calculations to justify the accuracy of this practice. I would say the difficulty of the calculation doesn't increase in any sort of linear or quadratic or n^4 way with the number of electrons in the system.

Mueller-Westerhoff. As long as you neglect the core.

Bagus. No, not neglect the core, but treat it in a less sophisticated way. Treat it at the level of a Hartree-Fock approximation.

Zerner. What do you do with systems which take you away from reality when you don't include spin-orbit, spin-spin couplings, and so forth?

Bagus. First, that is a tough one for me to answer because I haven't done calculations that involve spin-orbit couplings and magnetic effects, except for the fermi contact term. My answer to that would be again I'd expect that spin-orbit coupling is important for the open shells; it is not important for the closed shells in the core. So again the problem is complex because of the complexity of the open shell structure rather than the complexity of the closed shell structure. These effects are normally treated using perturbation theory. If we have included electron correlation properly in the solution for the electrostatic Hamiltonian, this should be a satisfactory procedure at least for systems containing relatively light atoms.

Harrison. There's another question I think we might discuss. If you have a well-defined problem, you just need to have a big enough machine or a clever enough method, and that's one thing. The more interesting situation is when you do not have a well-defined problem. If it's true, for example, that a five-membered ring makes a molecule behave peculiarly, it is important to understand why. It seems unlikely, however, that I will get that understanding by calculating all the eigenvalues for a molecule with a five-membered ring. We need to know the right question to ask, and then maybe we can see which calculations will be useful. We should be putting more effort into seeing which questions should be asked, rather than proceeding blindly to larger molecules with CNDO.

Ruedenberg. Science is a bootstrap operation. You have some idea about the right questions; you try to answer them; then you find you have asked the wrong question; that gives you a better idea of what the right question is. It's an iterative process.

I feel that the discussion of small molecules had a point: It raised the question whether the money for heavy ab initio calculations is more wisely spent on small systems, to get more insight, than on large systems. A case can be made for that.

There is one other thing for which ab initio calculations are indispensible, namely when you calculate systems, small or large or whatever, away from their "normal" conformations. I am thinking of the transition states which are of interest for reaction kinetics. Some transition states have strange bonding situations. Are present approximate methods properly gauges for these? In some of these intermediate states, which have the atoms in rather peculiar positions, rather accurate calculations will probably have to be used to regauge the approximate methods.

<u>Segal</u>. Well, one doesn't know, does one? One has to document what the method can do for the problem. First of all, as you said, transition states are rather odd bonding situations; there are other odd bonding situations you can look at. And secondly, you can do a few ab initio calculations of surfaces so you can look at this. But after a while, if things work out, then you can proceed. You can't believe what any semi-empirical calculations predict without corroborative evidence of some sort.

<u>Ruedenberg</u>. And with many states, the only corroborative evidence that one can get is from an accurate calculation.

<u>Segal</u>. And you do know some things which tell you in advance that a given method is ill-suited to a particular problem. For instance, you know that CNDO gives very poor stretching force constants and, therefore, poor potential energy curve shapes along stretching coordinates. You know the force constants come out very much too large. Therefore, you know that if you do a surface, this method probably isn't going to be much help in treating motions of atoms away from each other.

<u>Ruedenberg</u>. What do you do then?

<u>Segal</u>. You don't use INDO.

<u>Ruedenberg</u>. What do you use?

<u>Segal</u>. Perhaps you do a small ab initio Hartree-Fock calculation within the limits that you know it has when applied to that sort of surface. Or perhaps you go to a higher formalism, if you want to, which will take care of the part of the correlation energy which is important to the effect one is studying.

<u>Taylor</u>. There is another alternative. Realizing the limits of accuracy of your curves, you can define certain experiments which can be done; very specific experiments which will distinguish between a limited set of physical model possibilities. In other words, in planning experiments, the experimentalist starts out in a vacuum; he can do many, many things. In the field I'm working in, for example, there are all sorts of cross sections that the experimentalist can measure at all energy ranges and at all angles. It's a lot of work. But it would be better to tell him: "I can understand the mechanism of certain reactive processes if you look at the angular distribution at 50° or 90°. If the cross section is big at 50°, I know it's one way; if it's big at 90°, I know it's

the other." Sometimes semi-empirical potential surfaces will actually distinguish between these two things. I would not feel ashamed
telling the experimentalist I can't answer your question; what I
can do is limit, though, the things that you have to do and tell
you a range of interesting problems. It would be nice if programs
for semi-empirical calculations were more generally available so
that experimentalists could use them in designing experiments.
This would be a great contribution that one can make without being
highly accurate. Right now, in the present state of the art, this
is where we can make a contribution to the study of big molecules.
I think we ought to continue trying to make computation more and
more accurate. But I think we really should realize that there is
definitely something we can do now and that we don't have to be
ashamed to say we can't get the final answer; often the experimentalist can't either. He may have to come back to us and ask us to
do another calculation. But, hand in hand, I think, in this way,
a great deal of light can be shed on the chemistry of big molecules.

Bagus. Is that loop going to be convergent?

Taylor. It can be. You have no guarantees in science at all that
it's going to be, but what you're trying to do--mainly in chemistry
and I think in solid state physics, also--is to get an understanding. Your systems are so complicated that you feel that you understand them when you can predict experiments, magnitudes of measured
numbers, and phenomena. And I do think that with semi-empirical
theories, used with proper experience, that you can begin to predict
phenomena. You may not be able to predict it to one-tenth of an
eV; Dr. Segal and I have predicted the production of certain negative ions in certain reactions. Now that's not big systems; but
we have predicted the energies at which certain negative ions wanted
by experimentalists could be produced. Most of the time we turned
out luckier than we should be and they were found experimentally.
It's worth it. This is the kind of thing you can do today, and I
think it's important.

Harris. If you read into what you said, what I want to hear there,
part of what you said, I think, involves us getting to some extent
out of the business of calculating some of the things for the experimentalist and putting into his hands what he needs to calculate it
for himself. These people are very clever at knowing what questions
to ask about the things they're trying to do; probably much better
than we are. I think one very important direction that quantum
chemistry should take is to make it easier for the standard kinds
of calculations to be done by more people. That will also sell
computers!

Bagus. It'd be nice if it got more people to do computations.
That leads to an interesting question. When Dr. Marcus this

morning asked "How good is a Hartree-Fock calculation?" one of the answers he got was that you have to tell case by case. You have to look at the chemistry and see whether you're going to be able to get out the answers you want. To what extent can an experiment-alist be handed a black box for which there are input parameters, a button is pressed to say start, and after some time some numbers come out.

Harris. But that's his game. He'll put in some numbers and see what he gets out. After a few days and a few thousand dollars or whatever, he'll know.

Bagus. I'd be tempted to say there's one class of calculation for which this can be done quite regularly now, and those are Hartree-Fock calculations. Those are calculations where there's been enough experience built up to define quite well the procedure for doing a calculation. But we must go one step further, and define those conditions under which a Hartree-Fock calculation or a scattered wave SCF calculation will be useful.

Ruedenberg. I think you can trust the experimentalists, at least those of the future, to be able to use a tool like that wisely. It is true that you find many calculations today where the money shouldn't have been spent, because the method really didn't apply to the case under study. But I think that's only a temporary situ-ation. The experimentalist can get smart and learn to distinguish the case where he can get useful answers from the case where he would be wasting his time and money using an approximate method.

Johnson. My comment to that is that you don't have to wait for the future. To use our own scattered-wave $X\alpha$ applications as examples, many of them have actually been calculated by experimentalists, both experienced ones and recent Ph.D.'s. Most of the calculations relating to the interpretation of ESCA data were actually carried out with the help of experimentalists from the ESCA group of Uppsala University who visited the University of Florida recently. The cal-culations on phosphorus and on guanidinium phosphate were also car-ried out by an experimentalist, a former student of F. A. Cotton.

Ruedenberg. I am quite aware that there are experimentalists now who are able to do this. But you also do find misuse of simple programs.

Johnson. They have to be tutored for a while. You can't just mail out the programs indiscriminately. I don't think it's possible to really make a program completely into a black box.

Segal. Yes, it is! You can send a semiautomatic program out and everybody will be able to use it. The result will be that people will constantly tend to push a method further and further into new areas until they finally attack problems that are beyond its capabilities.

Johnson. That's better than not giving it to them at all.

Bagus. I'm not sure there's a real choice there. It's a very difficult thing to say no to someone who makes a reasonable request of you. What we must do is provide extensive documentation, not only of our programs but also of the capabilities and limitations of the models on which they are based. This is hard work but it must be done before we can distribute programs to people who are not computational experts in quantum chemistry.

Taylor. But I'm saying something more. We're asking ourselves at this meeting what type of work we are supposed to be doing on large molecules. My answer is to go out and find out what the experimentalists want to know and to work with them as at least half our program. The other half is learning how to do it more accurately. I'm sure that twenty years from now with the more accurate methods there will be other questions for which we'll be doing the same thing we're doing now but at a higher level. I see it time and time again--photo-ionization, photo-dissociation, all these things-- you get a bunch of potential surfaces, they're not even accurate to ± 2 eV. But just their general shapes suggest a whole slew of experiments, explain a whole bunch of things, and proceed to actually lead to what chemists would call an understanding. I think that we ought to actively see that these programs get out, and to their misuse, I think we ought to make sure we have darn good refereeing on some of the cheaper journals which tend to publish the bad INDO and CNDO calculations without refereeing. I think there should be a stop to that.

Bagus. Does the audience have questions they would like to address to the panel--specific members or the panel in general?

T. L. Gilbert, Argonne National Laboratory. In respect to going to large molecules, I've been waiting to hear something about the building block approach and some of the problems involved in that. I haven't heard anything yet. Are there any comments on that particular path of dealing with large molecules?

Johnson. I can make a comment on that. That's exactly what we're trying to do in our X_α scattered wave applications to biological macromolecules, for example, guanidinium phosphate and hemoglobin. Moreover, I think the localized defect or impurity problem in solids is again another area where you have to build from the small to the large groups of atoms. In our own case, what we're simply doing is to start with the self-consistent charge distributions for component

clusters of atoms, and then proceeding from there. I think that's
something you can do with other theoretical methods also, at least
in principle. I don't know if anyone's done it in practice.

Bagus. In the LCAO approach, we regularly use basis set information
taken from parts of the system, particularly from the separated
atoms. This is a very great help in molecular calculations. However, I believe that Dr. Gilbert was referring to the transfer of
information with more physical content. Professor Ruedenberg, do
you have a comment there based on your work on transferable orbitals?

Ruedenberg. Transferable orbitals _are_ useful as building blocks.
But even if you use a building block approach, so that you have a
prediction for the building blocks of your wave function, you still
are faced with practical problems when you come down to make a calculation. What kind of calculation will you do? Scattered wave
X_α or INDO or a more rigorous one? The approach alleviates some of
the problems, but you still have with you the problem of the level
of the basis function, etc., etc. which you have to make up your
mind on.

Johnson. I'm not so sure that what you want to transfer is wave
functions but rather charge densities, because they are more convenient.

Ruedenberg. I feel that you're going to end up wanting to do more.

Bagus. Well, you may want to know about the distortions.

Johnson. In our calculations we divide space up into regions of
local potential energy and charge density. It's easier, in fact,
for us to work with these quantities than it is to work with numerical functions that constitute our wave functions.

Ruedenberg. You are saying that you adopt your kind of computational
mechanics and in that case you can proceed with it. But suppose
your kind of computational mechanics isn't adequate for a certain
purpose. There are two things: One is the building block approach
and the other is the mechanics appropriate for the desired accuracy.
You have to make a decision on the latter. You are not relieved
from that decision by the building block approach.

L. Salem, Orsay, France and Harvard University. Would the panel
care to comment on what seems to be one of the major difficulties
in ab initio calculations, and that is the difficulty in extracting
qualitative information or explanations for what is calculated.
Just to give an example, the barrier to internal rotation in ethane
was calculated with a minimal basis set by Pitzer at least ten years

ago, and although we have the number and the calculation has been done, I don't think that one can seriously say that one understands where this energy comes from. As Professor Zerner mentioned this morning, extended Hueckel theory had a great success with the organic chemists. I think there are good reasons for this. Would the panel care to comment on it?

Harrison. I would like to comment. The scale that I drew earlier with experiment at one end and intuition at the other, has some relevance. A complete calculation is a little like an experiment. Pitzer's calculation was like an experiment in that when he was through he had little more than a number. In a more intuitive calculation we focus on one aspect of the problem and see how well it fits. It is ordinarily less accurate but it yields a larger message.

Ruedenberg. I agree that quantum chemical interpretations are important. The extraction of such interpretations from rigorous wavefunctions presents, however, separate problems with non-trivial difficulties. I have been much interested in this in the past. But I have found that there seem to be two kinds of quantum chemists: those who appreciate interpretations and find them useful, and those who find interpretations useless and consider the time spent on them wasted. They raise two objections: (1) To find interpretations takes considerable theoretical and computational effort which does not result in any new concrete numbers that can be compared with experiment; therefore, no new "hard information" is generated; and (2) In a specific case, several different interpretations may be possible and equally valid; thus, no unique answer is found. I consider both objections shortsighted and invalid and feel that more needs to be done on the problem of interpreting rigorous electronic wavefunctions.

Bagus. A comment might be that a lot of our interpretation is in terms of the kind of calculations we do. It's a respectable answer to say that correlation effects are important, but really what that is saying is that one particular model hasn't worked. The situation for ethane would have been complicated had there been calculations of the rotation barrier that had given very bad results. It seems a fortunate property to calculate, in that most calculations give rather nice results for that particular property. So there's no comparison that says one model works, one model doesn't work. We can't provide an understanding in terms of what model works and what model doesn't work.

Ruedenberg. Some wavefunctions are better than others. If you have a wavefunction which has a relatively simple interpretation and which gives a pretty good result, then that interpretation is a very useful interpretation. This remains true even if you can get a better result using a wave function which is completely unintelligible.

That's why SCF is useful; it's sometimes very good and in those cases
its interpretation is illuminating, even if you have to correct it
to get higher accuracy.

Harris. Some questions are harder to answer than others, and he's
picked a question (on the rotation barrier of ethane) which proved
to be very difficult to interpret in a simple way. There have
been lots of other things which the chemists have been quite successful at interpreting in terms of the gross features of the charge
distributions that come out of calculations and things of this sort.
This just happens to be too subtle for us right now.

R. K. Nesbet, IBM, San Jose. I'd like to ask a major general question
of the panel before we all have to go home. What technological niches
are going to be filled by the three major classes of methods we have
under consideration? At one extreme are the ab initio first principles calculations aimed at quantitative results. At the other
extreme we have semi-empirical methods which drift off into things
like Hueckel models which you can apply to certain enormous systems
and get away with something. In the middle we have the new MS-X_α
method based on solid state statistical exchange approximations.
What are the boundaries and limitations of these methods? Where
are they going to be most useful?

Johnson. I feel that the areas where the scattered wave method is
most useful in large systems, and in small systems containing heavy
atoms. And I include very heavy atoms. We plan to treat systems
as heavy as trans-uranium and super-heavy element complexes. We've
already made some progress in putting the relativistic corrections
into the program. I didn't have a chance to show some of the results that we've gotten with those relativistic programs. But
that's the sort of area for which we feel the scattered wave method
will make valuable contributions. We're also focusing on those
problems where there is an interface between solid state physics
and molecular physics. We feel that there are certain aspects of
reaction chemistry which are accessible with this technique. I'll
mention one in particular, namely, ligand-exchange reactions in
transition metal complexes. There's a wealth of literature on this
subject, but the basic features of the electronic structure which
you need to make transition metal reaction chemistry truly a quantitative theoretical subject have not been tractable until the development of the X_α scattered wave method. We're not interested in
doing very much with small molecules involving lighter atoms, except as a test of the method to find out what its limitations are.
But we've already, I think, accomplished a good part of that goal.

Zerner. It's been really my experience in looking through the literature that if you give a reasonable wave function and a reasonable
method to a clever person, you'll get an awful lot of information

out of that. Much more information than if you give a less clever
person an ab initio method, where you will get only wave functions
and eigenvalues. If you give a clever person a method which does
not quite explain his phenomenon, he goes out and he fixes it up
because of some physical intuition. I think a perfect example of
that was Professor Karplus' next nearest neighbor betas, for example,
when he found something was wrong. Or to calculate force constants
in a method which as far as I know hasn't been done before. So I'm
saying simply that the success of any molecular method depends on
the persons doing the investigation.

Johnson. It also depends on where one's research funding comes
from. That's not a scientific issue; it's a practical one. There's
the problem of relevance, and at least in some laboratories that's
a very important criterion in choosing the problems that you work on.

Taylor. I've been told by my organic chemist friends that the
Woodward-Hoffman rules are very important, and I gather that an
important step in discovering that insight were Hoffman's semi-
empirical calculations. Now this mass of calculations, each one
of which was not that accurate, has therefore played a very important tole in chemistry. I think that is something that still can
be done for other fields. I think that one could hopefully look
forward to those type of insights by doing large numbers of calculations in line with experiment; some of them will be wrong, but on
the whole from the mass of this data you begin to get understanding
and a feeling for potential surfaces and reaction coordinates and
things like this. There's a big role to be played with methods
from Professor Johnson's which is semi-accurate down to the semi-
empirical. I think large quantities of data in the hands of clever
experimentalists, with theoreticians working with them, will produce
big insights into chemistry and probably solid state physics.

Bagus. I will try to provide an answer to Dr. Nesbet's question
for ab initio methods. I would say, for ab initio methods that go
reasonably well beyond the Hartree-Fock approximation, the limit
to the size systems that can be treated are systems that have got
four, five or six atoms other than hydrogen. That this, with
present-day techniques, with the ideas that we presently have, is
a limit which I think is placed on calculations that can go in a
rather regular way beyond Hartree-Fock calculations.

Segal. The point is that's already a reasonably expensive calculation at just one geometric configuration. To do chemistry, in the
end, one wants to move those nuclei around; it becomes quite expensive.

Bagus. But at that limit, variations of geometry are still possible.

Taylor. But I think it's still a question of cost. I think you
should really ask what the chemist wants to know. Cost is important,
so I don't think you should go running off doing ab initio.

Ruedenberg. I would be interested in Dr. Pople's opinion because
he has worked successfully with semi-empirical methods as well as
with ab initio methods.

J. A. Pople, Carnegie-Mellon University. I think that for any level
of scientific problem one goes after, one should use the simplest
technique for which there is reasonable evidence that it is going
to work. This is a difficult thing to get at and one may have to
go a long way before you get satisfaction. I think we are, however,
as technology improves, moving to a stage where many more problems
can be studied at some simple ab initio levels than were possible
some years ago. So I think to some extent there's a movement from
the semi-empirical methods to rather better methods. The question
of geometries has been raised, and we have devoted a lot of time,
as I indicated in my talk, to studying the geometries of structures
of small organic molecules in recent years. I feel, at the present
time, that although the semi-empirical methods work very rapidly,
they're not as good as the simplest minimal basis ab initio methods.
We have done statistics on that; we have showed that. So for this
range of molecules, and I was perhaps thinking more of three heavy
atoms and some hydrogens, I think now the correct method to go at,
extensively and widely, is minimal basis ab initio. I think we're
beginning to get a real feeling for the deficiencies of that method
and a real feeling for where our predictions may be right. Of
course, one can probably still find cases where it is necessary to
go further. But let us first fully explore the simplest methods.

AB INITIO SCF CALCULATIONS OF THE CARBAZOLE AND

2,4,7 TRINITRO-9-FLUORENONE (TNF) MOLECULES

I. P. Batra and H. Seki

IBM Research Laboratory

San Jose, California 95114

Our interest in these calculations is based on our desire to
understand the properties of a photosensitive polymer. The organic
polymer film consisting of a 1:1 monomer molar ratio mixture of
poly-n-vinylcarbazole (PVC_z) and 2,4,7-trinitrofluorenone (TNF) is
a sensitive organic photoconductor currently being used in the IBM
electrophotographic copier. In this laboratory we are presently
investigating the electrical and optical properties of this and
related materials. The carbazole monomer unit and the TNF molecule
(see Figs. 1 and 2) are known to form charge transfer complexes
about which relatively little is known. But these molecules are
mainly responsible for the photosensitivity in the visible region.
Our experimental results indicate that optical absorption, photo-
generation efficiency of free carriers, and the very small charge
carrier mobilities are all strongly field dependent. These results
support the description of these materials as disordered molecular
solids and we expect that the properties of the basic molecular
constituents can be expected to have a relatively direct correlation to the properties of the solid as a whole.

The general SCF program developed by E. Clementi opened up
the possibility of investigating these molecular properties using
a minimum basis set. Initially, we considered the carbazole and TNF
molecules. Details of these calculations will be published elsewhere. These molecules are planar in structure and in the ground
state the orbits are closed. The existing program could be expected
to handle these molecules individually on the IBM 360/195 computer
in a reasonable time. We also have expectation that further modification and development of the program would soon enable us to compute molecules two to three times that size.

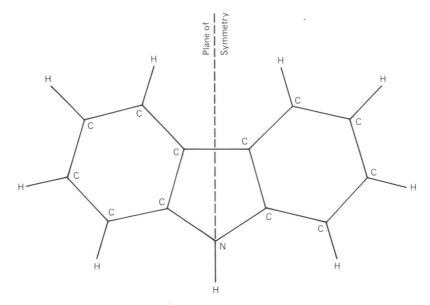

Figure 1. Carbazole

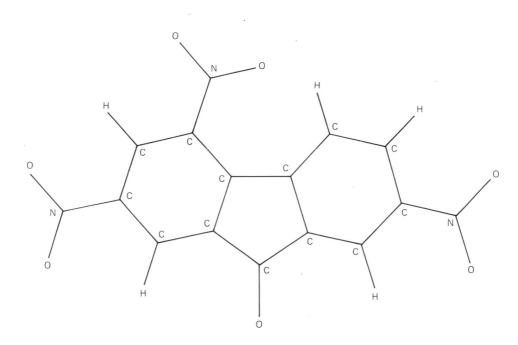

Figure 2. 2, 4, 7 Trinitro − 9 − Fluorenone (TNF)

It is hoped that the results of these computations will provide us with a molecular basis to support some of the physical models that are being invoked to explain the properties we observe.

Specifically:

1.   We would like to obtain a molecular basis for the intermolecular hopping model for the charge transport process. In effect, we want to obtain the Hartree-Fock potential for the outermost electron of the molecule or ion and how it is perturbed due to external fields. To begin with, we would consider a single molecule or ion and then continue to the case of two adjacent molecules and ions. Needless to say, such Hartree-Fock potentials will be the basis for further one-electron band calculations for crystals consisting of such molecules.

2.   We want to gain further insight into the properties of the charge transfer complex. Initially from the static charge density distribution of the individual molecules we can determine the optimum relative orientation that the two molecules are likely to find themselves in as they approach each other to form the complex. When it becomes possible to treat the two molecules as one, the intermolecular interactions are accounted for and we can then hope to determine the geometry, energy, degree of charge transfer, and Hartree-Fock potential for the ground and various excited states of the complex. We would, of course, be concerned about how these are perturbed by external fields.

It is important to stress that one of the primary concerns here in these computations is that we want to know how the molecule is perturbed due to the environment, i.e., external fields, dipole distribution, etc. This is partly due to the fact that we are interested not so much in the isolated molecule but as it exists in the solid. Certainly one of the important reasons for choosing the Hartree-Fock approach rather than some of the more approximate methods is the flexibility which this method provides. The environmental conditions can easily be added to the computation in terms of distributed point charges and the modified Hartree-Fock potential results in a natural way. Furthermore, the nature of the approximation involved is at least theoretically clear. Once the isolated molecule is computed it is only necessary to compute the one-electron integrals which involve the point charges which define the environment for the perturbed case.

At this stage it is difficult to predict how much agreement we can expect between computed results and measured characteristics. We are confident that, at the least, considerable molecular insight into our efforts to interpret our experimental observation will be obtained from the computations.

Note Added in Proof

The electronic structure of these molecules is now also being studied by the scattered wave method (1) using a statistical exchange potential. This work is being done by one of us (I. P. Batra) in collaboration with D. A. Liberman.

(1)   K. H. Johnson, J. G. Norman, Jr., and J. W. D. Connolly, this volume, p. 161.

5. Localized States
and Disordered Solids I

SURFACE STATES AND LEED

P. M. Marcus and D. W. Jepsen

IBM Research Center

Yorktown Heights, N. Y., 10598

INTRODUCTION

The following remarks, which introduce the session on localized states in solids, concern the simplest kind of localized
state and its close relationship to the scattering states used in
low-energy electron diffraction (LEED). These simple localized
states are the so-called intrinsic surface states, which can exist
on the surface of an ideal crystal and are localized in the vicinity of that surface; they are described by a wave function which
has the usual Bloch periodicity parallel to the surface (with real
wave numbers k_x and k_y), but attenuates exponentially both going
into the crystal and into the vacuum outside the crystal, hence
these states are localized in one direction - the z direction.
The idea of such states goes back a long time, to Tamm[1] (1932),
and has been discussed by many workers since.[2] In fact such states
appeared in the calculations of George Watkins, on the energy level
of a finite cluster, described at this Conference. In Watkins'
calculations they were a nuisance, since he was trying to simulate
the behavior of a bulk crystal, and he tried to get rid of them,
whereas now we are specifically interested in these states and
concerned with getting an accurate description for the case of the
semi-infinite crystal. The point of the present remarks is that the
same formulation and computational method which gives accurate
LEED states applies also to these surface states, so that we can
obtain their energies and wave functions accurately for a specified potential of a semi-infinite crystal.

We first describe a model of the crystal potential at a
surface and then the formulation and solution of the LEED problem.
The solution is put in a form which reveals the common aspects of

the LEED and intrinsic surface-state problem - both follow from the
properties of a scattering matrix M which relates the wave function
in vacuum to the wave function in the crystal. We then illustrate
the relationship between the LEED reflection and transmission
spectra and the extended band structure - extended into complex k_z
space. Finally we discuss the conditions for the occurrence of a
surface state, first of the stationary kind which occurs below the
vacuum level, and then of the quasi-stationary kind at positive
energies with respect to vacuum.

THE LEED PROBLEM

The LEED problem is described schematically in Fig. 1, which
shows one simple model of a semi-infinite crystal that has been
fairly successful in accounting for observed LEED spectra. In this
model the potential is constant in vacuum, the real part of the
potential drops abruptly by the amount $|\Delta|$ at a surface plane to
the level designated the muffin-tin zero, and then continues periodically into the crystal; the imaginary part drops abruptly from
zero in vacuum by amount β at the surface plane and continues

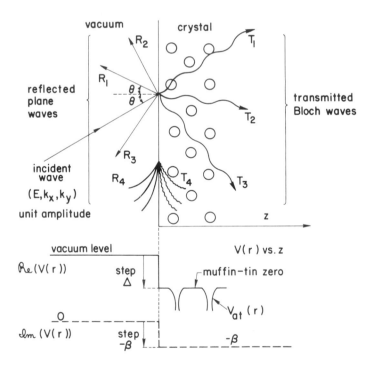

Fig. 1. Muffin tin - abrupt step model of potential of semi-
infinite crystal.

constant in the crystal. An incident plane wave of unit amplitude,[3]
energy E, and wave number components parallel to the surface k_x
and k_y, falls on the surface and gives rise to reflected plane
waves in vacuum with amplitudes R_i and transmitted Bloch waves in
the crystal with amplitudes T_i, all with energy E and the same reduced k_x and k_y values, designated $[k_x]$ and $[k_y]$. These waves
include an infinite number which attenuate on both sides of the
surface. The imaginary part $-i\beta$ of the potential in the crystal
describes the effects of incoherent scattering processes; when β
is finite, all the Bloch waves attenuate in the + or $-z$ direction.

We solve the LEED problem by a method conveniently called the
layer-KKR method[5] which applies multiple-scattering theory in the
KKR form to a single atomic layer, periodic in x and y. A detailed
exposition of the method has been published,[5] and we just mention
here some features relevant to the surface-state problem. The
method uses a convenient discrete representation for the wave
function, the beam representation, which specifies the amplitudes
of the set of plane waves of given E, $[k_x]$, $[k_y]$ mentioned above, including waves propagating or attenuating toward both $+z$ and $-z$. This
description is a natural one in vacuum, where these plane waves
are the eigenmodes, but it can also be used in the crystal with
amplitudes which are functions of z. However now the eigenmodes
of the crystal, the Bloch waves, must be found by solving the
eigenvalue problem for a matrix Q, the transfer matrix, which
describes the scattering properties of the single layer which, on
repetition, makes up the whole crystal. The eigenvalues of Q, for
the given E, $[k_x]$, $[k_y]$, are simply related to the k_z values which
belong to the corresponding eigenvectors; these eigenvectors include the attenuating Bloch waves, even when V(r) is real. The
matrix Q is found within the KKR procedure by introducing transformations between the spherical representation needed to calculate
individual atomic scatterings and the beam representation needed to
find the Bloch waves.

Now the R's and T's of the LEED problem, or its generalization
to any set of incident plane waves or Bloch waves (at the given E,
$[k_x]$, $[k_y]$),are found by solving a set of linear matching equations
which match a linear superposition of these Bloch functions to a
superposition of plane waves - both the wave function and its z
derivative are matched at the surface plane of the model in Fig. 1.
The final form of these matching equations in an Nth order approximation which uses N plane waves and N Bloch waves (in both $+z$ and
$-z$ directions) in the representation is

$$
M \begin{pmatrix} R_1 \\ \vdots \\ R_N \\ T_1 \\ \vdots \\ T_N \end{pmatrix} = \begin{pmatrix} \alpha_1 \\ \vdots \\ \alpha_N \\ \gamma_1 \\ \vdots \\ \gamma_N \end{pmatrix} \tag{1}
$$

where M is a 2N by 2N matrix whose elements depend on E, $[k_x]$, $[k_y]$,[6] and are calculated by the layer-KKR method from the model potential; it is essentially a scattering matrix for the semi-infinite problem relating incoming to outgoing wave amplitudes in both media. In (1) is the amplitude of the ith incident plane wave, and γ_j is the amplitude of the jth incident Bloch wave.

For the LEED problem, which has one incident beam from vacuum, a solution to (1) is desired when $\alpha_1 = 0$, $\alpha_i = 0$ for $i \neq 1$, $\gamma_j = 0$ for all j. However for the surface-state problem, which has no inci-dent waves, a solution to (1) is sought when all α_i's vanish, as well as all γ_i's, as will be discussed later. Many examples of[5,8] LEED calculations with the layer-KKR method have been published in which realistic values of β, Δ are chosen, various angles of incidence, θ, ϕ are used, and the spectra in the interesting energy range up to 100 eV or greater are given and compared with experiment. For the discussion here, we shall use the Cu(001) surface in the energy range of the d bands, and of the energy gap in the s bands produced by interaction with the d bands, between about 2 and 5 eV above the muffin-tin zero (MTZ), where we ex-pect an intrinsic surface state to occur. Hence we have calculated reflection and transmission spectra for Cu from the MTZ to the Fermi level (0 to 8 eV), choosing $\theta = 0$ (normal incidence) to bring out an interesting consequence of symmetry, $\beta = 0$ since absorption is small in the vicinity of the Fermi level, and $\Delta = 0$, needed to obtain a LEED spectrum in this energy range, i.e., to yield propagating waves in vacuum. However for intrinsic surface states, Δ will have to be finite, and will have some value that puts vacuum above the Fermi level.

In Fig. 2 we show three plots on a common energy scale. The top plot is the reflected flux in the 00 beam (specularly reflected beam) relative to the incident beam flux, i.e., a plot of $|R_1|^2$, which shows the total reflection ($|R_1|^2 = 1$) in the gap between 2.3 and 4.5 eV. At 2.3 eV $|R_1|^2 = 1$ is expected   since there are no propagating Bloch waves, as we see from the band structure along [001] given in the middle plot, and only the 00 beam propagates (the next beam, the 11 beam, starts propagating above 20 eV).

However, the d bands provide propagating Bloch functions above 2.7
eV, but are not excited because at normal incidence only waves of
Δ_1 symmetry are excited. This absence of coupling is again exhibited in the bottom plot of relative transmitted flux (propor-
tional to $|T|^2$ and to the velocity of the Bloch wave in the z
direction), which shows no transmission into any d band, including
the band from 5.0 to 5.4 eV. However Fig. 2 shows the transmission
into the a band and the split transmission into the b_1 and b_2 bands
on the two sides of the band edge at 4.5 eV.

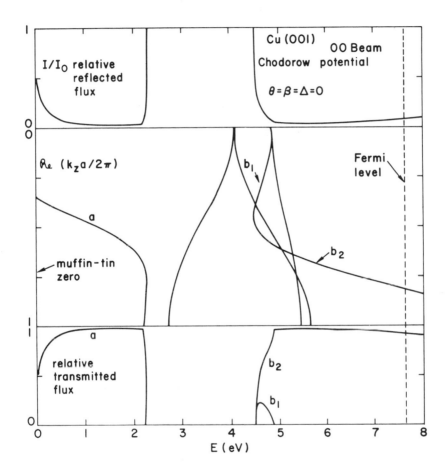

Fig. 2.   LEED specular beam reflection spectrum (top plot) and
transmission spectra (bottom plot) and band structure along [001]
(middle plot) for Cu(001) on common energy scale between muffin-
tin zero (MTZ) and Fermi level. Calculation by KKR-layer method
used Chodorow potential, zero absorption, zero discontinuity between MTZ and vacuum and normal incidence.

The continuation of the bands at the band edges along "real
lines" (lines on which the energy remains real) into complex k_z
space is shown in Fig. 3.   The first and third plots show the
imaginary part of k_z for band edges at the center and edge of the
zone, continued along the real line.   The second and fourth plots
show the real and imaginary part of k_z [10] along a real line between
band edges in the interior of the zone.

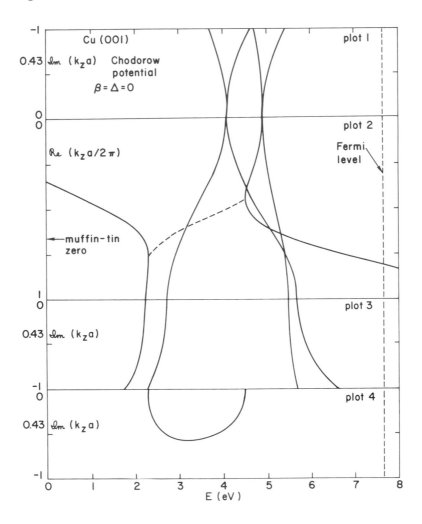

Fig. 3.   Cu band structure along [001] extended into complex k_z
plane for the same potential as in Fig. 2.   Real lines at center
of zone (plot 1), band structure plus interior real line, dashed
(plot 2), real line at X point of Brillouin zone surface (plot 3),
imaginary part of the interior real line of plot 2 (plot 4).

SURFACE STATES

The preceding discussion shows that the LEED problem is solved
by solving for the amplitudes of outgoing waves produced by "driving" incident amplitudes inserted in Eq. (1); such solutions exist
at all E (and $[k_x]$, $[k_y]$). However the surface-state problem requires a set of finite amplitudes for outgoing waves in the absence
of driving incident amplitudes, hence requires a solution of the
homogeneous set of 2N equations in 2N unknowns obtained from (1)
with right-hand-side zero. Such solutions can be expected only at
special values of E, for given $[k_x]$, $[k_y]$, and these special values
 must be in the energy gaps and below the vacuum energy, otherwise
charge will leak out of the state through propagating Bloch waves or
plane waves. Accordingly one would look for zeros of the determinant
of M between band edges, such as the band edges in Cu between 2.3 and
4.5 eV for $[k_x]=[k_y]=0$ (normal incidence). A study by Forstmann
and Pendry of surface states in d-band metals,[11] based on a
simplified model of the potential (only the d phase shift is retained and its energy dependence is approximated by a simple
resonance formula) led to a surface state approximately 0.01 Ry
(0.14 eV) above the lower edge (2.3 eV) of the above energy gap in
Cu. We have not as yet confirmed this result with the complete Cu
potential used in the above calculations.[12] In any case, we note
that the exact location and structure of the surface state will be
sensitive to the details of the potential in the transition region
between vacuum and crystal interior, since most of the surface
state exists in the region. Accordingly, the truncated potential
model in Fig. 1 is probably too crude for good surface state
analysis. We are now modifying this potential by using a transi-
tion-region potential obtained from self-consistent surface charge
calculations,[13] and will return to the surface state calculation
with the modified potential.

The surface state described above is a true intrinsic surface
state, which can exist only at energies below the vacuum threshold
because otherwise it would leak charge into propagating plane waves.
However it has been pointed out[14] that there exists an analogue of
such surface states at energies above the vacuum energy, which
leaks charge but is relatively stable if that leakage is small.
Such states have been called quasi-stationary surface states by
McRae,[14] and they are possibly even more interesting than the true
stable surface states because they can produce observable effects
in LEED. A description of these states within the framework of the
present analysis is easily obtained as follows. Suppose that at a
particular E, $[k_x]$, $[k_y]$ of interest, I plane waves and J Bloch waves
out of the N used in (1) propagate (when the potential is real).
Then consider the submatrix of M obtained by leaving out the I+J
rows and columns referring to these propagating waves, and look for
zeros of this (N-I-J) by (N-I-J) sub-matrix. Such zeros will be

analogous to the true surface state discussed above. Now the interactions with the propagating waves are described by the complete M
matrix, but if these couplings are weak enough, the surface state
will retain some physical identity. Thus it could become only
slightly leaky and might be properly described as quasi-stationary.
In particular, one can expect to find sharp resonant features in beam
reflection spectra at that energy, so that the leakiness provides
a mechanism for observation. McRae suggests that such resonances
occur just before the emergence of new beams (the grazing emergence
condition) when parts of the wave function have strong intensities
in the surface region.

In summary, it appears that we have the tools to find both the
true surface states and these quasi-stationary surface states with
a realistic three-dimensional crystal potential with the same high
accuracy as the LEED calculation, and may be able to explore the
nature of the potential at the surface if the corresponding effects
can be identified in observed LEED spectra.

REFERENCES AND FOOTNOTES

1.   I. Tamm, Z. Physik $\underline{76}$ 849 (1932),
              Phys. Z. Sowjet. $\underline{1}$, 733 (1932).

2.   For recent reviews see, for example,
      S. G. Davison and J. D. Levine "Surface States" in Solid State
      Physics, Vol. 25, p. 2. (H. Ehrenreich, F. Seitz, D. Turnbull,
      eds.) Academic Press, 1970,
      Peter Mark, W. R. Bottoms "Electronic States of Free Surfaces"
      in Progress in Solid State Chemistry, Vol. 6, p. 17 (H. Reiss,
      J. O. McCaldin, eds.), Pergamon Press 1972.

3.   For convenience, all plane waves and Bloch waves are normal-
      ized to unit flux at the surface plane.

4.   The discrete set of diffracted plane waves or beams has k_x
      and k_y values differing from the incident plane wave by re-
      ciprocal lattice vectors of the surface net.

5.   D. W. Jepsen, P. M. Marcus, F. Jona Phys. Rev. B$\underline{5}$, 3933 (1972).

6.   An equation equivalent to (1) is the matrix equation (10) of
      reference 5, on introducing the general amplitudes of α_i and γ_j
      and bringing the T's and R's to the same side.

7.   There is no difficulty in generalizing the model and Eq. (1) to
      the case of any finite number of layers differing from bulk
      between vacuum and bulk, including the transition layer as one

leaves vacuum.   Essentially a product of different transfer
matrices is used to carry the wave function from vacuum to the
beginning of bulk crystal, where the matching to a superposition of Bloch waves is made.

8.   P. M. Marcus, D. W. Jepsen, F. Jona, Surface Sci. 31, 180 (1972).

9.   For LEED spectra calculations, we actually use an artificial
      boundary condition instead of exact matching of solutions at
      a surface plane. This boundary condition, called the no-
      reflection boundary condition, suppresses all reflections at
      the surface plane, hence is more realistic than matching wave
      functions at a discontinuity of potential since the true, more
      gradual, change of potential between vacuum and crystal will
      give small reflection. We find that the no-reflection bound-
      ary condition makes little difference in LEED spectra, but it
      could have a drastic effect on surface states.

10.   The calculations for Figs. 2 and 3 required $N \geqslant 21$ beams to
      achieve an accuracy of better than 0.01 eV, hence the
      representation required 18 or more attenuating waves to describe
      the region of the d bands accurately; at N=17, splittings of
      order 0.06 eV appeared in the energies of states that should
      have been degenerate at the center of the Brillouin zone,
      indicating the accuracy of the representation at that N (note
      that the calculation of the energy bands is not variational).

11.   F. Forstmann, J. B. Pendry, Z. Physik 235, 75 (1970).

12.   The determinant of M becomes very large ($\sim 10^{150}$) when N=21 and
      many attenuating waves are included in the representation,
      which inclusion, as footnote 10 points out, is needed for
      adequate accuracy. This large magnitude complicates the
      determination of the zeros.

13.   N. D. Lang, W. Kohn, Phys. Rev. B1, 4555 (1970).

14.   E. G. McRae, Surface Sci. 25, 491 (1971).

WANNIER FUNCTIONS

Walter Kohn

University of California, San Diego

La Jolla, California 92037

I. INTRODUCTION

During the last four decades the electron theory of solids has been formulated almost entirely in the language of Bloch waves and energy bands. Of course, this representation highlights the itinerant character of the electrons.

In several projects in which our group is currently engaged, which deal with rather tightly bound crystal electrons, we have felt it to be most desirable to emphasize the similarity of the electronic structure to that of isolated atoms. For this reason, we have reviewed the current status of the theory of Wannier functions, which because of their equivalence to Bloch waves and their resemblance to atomic states offer a promising avenue of approach. We found that although the general theory of Wannier functions for periodic structures is rather well developed[1-10], they have not been actually calculated for realistic models, let alone been used in quantitative solid state calculations.

We have, therefore, addressed ourselves to the problem of calculating quantitatively Wannier functions ab initio (that is without prior calculation of Bloch waves) for typical solid state situations, in particular for cases of degenerate or hybridized band structures. Our work is described in detail in a paper currently in press[11]. Here we shall highlight some of our results and conclusions.

II. VARIATIONAL PROCEDURE

We propose to use variational methods for the calculation of Wannier functions. This possibility was originally pointed out in refs. 4 and 5.

Let $\varphi_{m,k}(r)$ ($m=1,\dots\bar{m}$) be a set of Bloch-waves belonging to a composite band consisting of $\bar{m}$ connected (or disconnected) single bands. Denote the corresponding Wannier functions, centered on the lattice sites n, by $a_m(r-n)$. Their symmetry is known[9]. To calculate the a_m variationally, we begin with a set of trial functions, $f_m(r-n;\ \gamma_1,\gamma_2,\dots)$, of the correct symmetry but in general not orthogonal. The quantities $\gamma_1,\gamma_2,\dots$ are variational parameters. The f_m themselves cannot yet be used as variational trial functions because of the so-called non-ortho-gonality catastrophe[12]. One must first construct from the f_m orthogonal trial functions of the correct symmetry[2,9],

$$a_m^t(r-n;\ \gamma_1,\gamma_2,\dots).$$

Now consider, for mathematical convenience, a set of spin-less fermions whose number is precisely right to just fill all the Bloch states (m,k) of the bands in question. The exact ground state ψ of this system is given by either of the following two expressions

$$\psi = (\bar{m}N!)^{-\frac{1}{2}}\ \text{Det}\ |\varphi_{m,k}(r)|$$

$$= (\bar{m}N!)^{-\frac{1}{2}}\ \text{Det}\ |a_m(r-n)| \tag{2.1}$$

and the corresponding total energy per particle,

$$\mathcal{E} = N^{-1}\ \frac{(\psi,H\psi)}{(\psi,\psi)}\ , \tag{2.2}$$

where H is the Hamiltonian. The expression (2.2) is well-known to be stationary with respect to small variations of ψ. If the second form of (2.1) is used with the trial Wannier functions $a_m^t(r-n)$ one finds a trial energy

$$\mathcal{E}^t = \sum_{m=1}^{\bar{m}} (a_m^t(r;\gamma_1,\gamma_2,\dots),\ H\ a_m^t(r;\gamma_1,\dots)), \tag{2.3}$$

whose stationary property can now be used to determine the parameters $\gamma_1, \gamma_2, \ldots$.

Non-uniqueness Problems

In constructing the Wannier functions a_m there are certain non-uniqueness problems not encountered in other applications of the variational method. Consider, as a simple example, a hybridized s + p band in a SC lattice of lattice constant d. Let a_s; a_x, a_y, a_z be a set of well localized orthonormal Wannier functions of the correct symmetry. Then the functions

$$\bar{a}_s(r) = a_s(r) + \epsilon\{[a_x(x-d,y,z)-a_x(x+d,y,z)]$$

$$+ \quad [a_y(x,y-d,z)-a_y(x,y+d,z)]$$

$$+ \quad [a_z(x,y,z-d)-a_z(x,y,z+d)]\} \, ,$$

$$\bar{a}_x(\underset{\sim}{r}) = a_x(\underset{\sim}{r}) + \epsilon[a_s(x-d,y,z)-a_s(x+d,y,z)] \quad , \tag{2.4}$$

etc.

where ϵ is a small quantity, are also orthonormal and well-localized and hence also acceptable as Wannier functions.

There are more serious non-uniqueness problems when the bands m=1, ...$\bar{m}$ are attached to and overlapped by other bands. These are discussed in detail in ref. 11.

III.   ENERGY BAND PROPERTIES

Although our primary purpose has been to construct Wannier functions and work with them directly, we would like to remark here that once they are constructed, properties of the corresponding energy bands are easily obtained.

For a composite band with Wannier functions, a_m, define the matrix elements

$$E(\underset{\sim}{n})_{m'm''} \equiv (a_{m'}(\underset{\sim}{r}), \, H \, a_{m''}(\underset{\sim}{r}-\underset{\sim}{n})), \, (m',m'' =1,2,..\bar{m}) \tag{3.1}$$

and the derived quantities

$$\mathcal{E}(\underset{\sim}{k})_{m'm''} \equiv \underset{\underset{\sim}{n}}{\Sigma} E(\underset{\sim}{n})_{m'm''} \, e^{i \, \underset{\sim}{k} \cdot \underset{\sim}{n}} \tag{3.2}$$

Then the eigenvalues $E = E_{m,k}$ of the composite band structure are given by

6)

$$\text{Det } | \, \mathcal{E}(k)_{m'm''} - E \, \delta_{m'm''} \, | = 0 \qquad\qquad (3.3)$$

This is a determinental equation of order $\bar{m}$.

Alternatively one can also easily obtain the energy bands from the non-orthogonal and, in general, better localized functions $f_m(r)$.

It is also straightforward to construct the Bloch waves $\varphi_{m,k}$, from the Wannier functions a_m (or from the f_m), even if there is degeneracy or hybridization.

One can calculate the moments of the density of states $n(E)$ directly from the Wannier functions a_m. Thus,

$$M_s \equiv \int n(E) \, E^S \, dE = \sum_m (a_m(r), \, H^S \, a_m(r))$$

$$= \sum_{n_1+n_2+\ldots n_s=0} \quad \sum_{m,m_1,m_2,\ldots} E(n_1)_{mm_1} \, E(n_2)_{m_1 m_2} \, E(n_s)_{m_{s-1}m}$$

(3.4)

Finally matrix elements of an operator Q, between Bloch states $\varphi_{m,k}$ and $\varphi_{m',k'}$ can all be simply expressed in terms of a few numbers,

$$Q(n)_{m_1,m_2} \equiv (a_{m_1}(r), \, Q \, a_{m_2}(r-n)) \, . \qquad\qquad (3.5)$$

or the corresponding integrals involving the $f_m(r-n)$. Because of the exponential localization of the $a_m(r)$ and $f_m(r)$ these quantities decrease very rapidly with n.

IV. CONCLUDING REMARKS

We believe we have now a better grasp of the various issues which arise when one wishes to actually calculate realistic Wannier functions. These issues include (1) methods for ab initio calculations, (2) problems associated with composite bands, (3) localization, (4) orthogonality, (5) non-uniqueness. As a result we believe it is now possible to make explorations in the following directions.

(I) Numerical ab initio calculations of Wannier functions as an alternative to numerical calculations of energy bands and Bloch waves. There would seem to be real value in having all the informa-tion about an $\bar{m}$-fold composite band contained in $\bar{m}$ localized functions $a_m(\underset{\sim}{r})$ (or $f_m(\underset{\sim}{r})$). It remains to be seen whether in some cases this approach to electronic band structure also can offer some economy of computing time.

(II) Once Wannier functions are actually constructed, one can explore their quantitative use in solid state applications, e.g., in the theory of superconductivity of narrow-band metals [13].

(III) Based on better understanding of Wannier functions in periodic lattices it has been possible to make some progress defining and constructing Wannier functions for non-periodic systems. This work will be reported in the near future.

REFERENCES

1. G. Wannier, Phys. Rev. 52, 191 (1937).

2. P. O. Löwdin, J. Chem. Phys. 18, 365 (1950).

3. G. G. Hall, Phil. Mag. 43, 338 (1952).

4. G. F. Koster, Phys. Rev. 89, 67 (1953).

5. G. Parzen, Phys. Rev. 89, 237 (1953).

6. J. C. Slater and G. F. Koster, Phys. Rev. 94, 1498 (1954).

7. W. Kohn, Phys. Rev. 115, 809 (1959).

8. R. I. Blount in Solid State Physics, 13, 305, F. Seitz and D. Turnbull, eds. (Acad. Press, New York and London, 1962).

9. J. Des Cloizeaux, Phys. Rev. 129, 554 (1963),
                       Phys. Rev. 135, A685 (1964),
                       Phys. Rev. 135, A698 (1964).

10. P. W. Anderson, Phys. Rev. Letters 21, 13 (1968).

11. W. Kohn, Phys. Rev. B, in press.

12. T. Arai, Phys. Rev. 126, 471 (1962).

13. J. Appel and W. Kohn, Phys. Rev. B, 4, 2162 (1971).

DEVELOPMENTS IN LOCALIZED PSEUDOPOTENTIAL METHODS

P. W. Anderson
Bell Laboratories, Murray Hill, New Jersey, and
Cavendish Laboratory, Cambridge, England

J. D. Weeks
Cavendish Laboratory, Cambridge, England

In his recent Buckley Prize lecture, J. C. Phillips showed a slide (Fig. 1) in which he compared the actual usage of, as opposed to the purely theoretical interest in, various electronic structure methods in the band theory of solids. In this slide pseudopotentials and the APW method heavily outweighed all others, and in the upper right-hand corner was a tiny square representing LCAO. If Jim had, however, been interested in comparing the total number of electronic structure calculations without limiting himself to solids, all other methods would have paled in comparison to LCAO. At least half of all LCAO calculations are Huckel theory ones. In terms of the practical application of quantum theory to real electronic structures, then, one must submit that the Huckel theory and other rather crude LCAO concepts are by far the most successful. It is reasonable to suppose that most of this success must be accounted for by the fact that these methods really work.

From any realistic point of view it is rather a surprise to find these methods working for chemistry as well as they do. For example, the successful methods in band theory, to which I just referred, are all based on plane wave functions, not atomic orbitals. Again, the overlap integral between Huckel orbitals, which is often neglected in the standard Huckel theory, is .25. Yet again, the use of hybridized sp^3 bond orbitals for saturated carbon structures is highly satisfactory but the same concepts applied to the neighboring element B, or without severe modification to N, or even to unsaturated C, leads to nonsense; yet, other ways of using LCAO work just fine in all these cases. The reasonable success of LCAO methods in the ionic alkali halides in terms of energies is well known; band structures also are now shown,

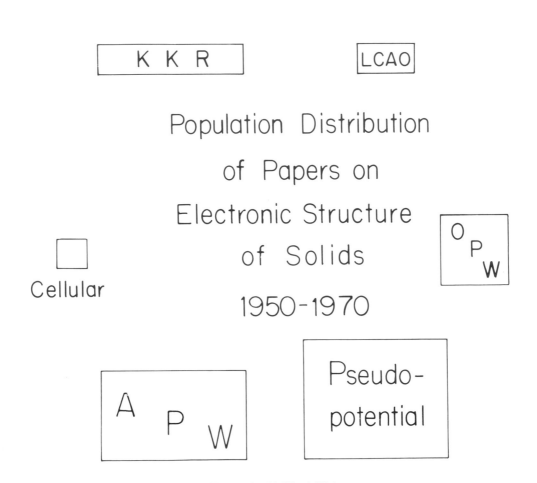

Figure 1. Phillips' Slide

by A. B. Kunz, to be well described. But other properties - e.g. dielectric constants, don't work - why?

To my mind the most urgent task of quantum chemistry is to provide a foundation for understanding and helping to systematize all the empirical information chemists have accumulated over the years, preferably using the conceptual structures they have developed and found useful: such as the pair bond, hybridization, and the Huckel theory. Next to this in importance is the attempt to develop new concepts and new ways of looking at the facts of chemistry. After all, in almost all cases it is easier to measure the strength of a chemical bond or the configuration of a molecule than it is, or ever will be, to calculate it; it is theory as a way of systematizing and understanding empirical facts, and as a way of suggesting new relationships, which is really valuable to the chemist. It is also very important to point out when not to use a certain simple conceptual set, as well as when to use it: we will see an example of this shortly.

It seems to me that one of the most important steps in the direction of providing a real theoretical background for chemistry - or, more properly, a theory for real chemistry - was taken only about a decade ago with the papers of W. H. Adams (1) and of Gilbert (2). With these papers one could for the first time understand why and in what circumstances the use of a localized, atomic-like basis set to describe the low-lying electronic states was justified. They showed that a defining equation for such a basis set could be written down by the use of pseudizing techniques, in which the influence of the distant environment is indeed a weak perturbation. More recently I suggested a simplified version of this theory (3), which is particularly adapted to the goal I have set out above: deriving and supporting such empirical parameter-izations as the simple Huckel theory; and I later showed (4) how the Huckel theory in its simplest form follows from these equations, and that the appropriate localized orbitals for the Huckel theory are almost exactly the simple atomic orbitals.

In this paper I will bring out some further developments which have taken place in this general area, three of them by me and my associates at Cambridge.

Actually, from the point of view of this conference, and perhaps from any point of view, the most promising development in the field is the highly successful computation by Barry Kunz (5) of alkali halide band structures, using a very similar methodology. I think Kunz has done a most important job in demonstrating the power of our method, but since I am not closely associated with that work I can't say much about it, and will stick to our simpler or more formal approaches.

The first two of these developments begin with one very
practical, but much simpler, calculation which has been carried
out by my student A. G. H. Davidson. His choice of a problem
was to extend my derivation of the Huckel theory to systems containing heteroatoms, the simplest of these, of course, being
pyridine. ·

In doing this we ran into certain difficulties, the resolution of
which is the subject of the formal problems with which John Weeks,
Allan Davidson, and I have been wrestling recently and which have
led to a rather surprising resolution (6).

There are a number of somewhat inequivalent versions of the
localized pseudopotential equations. The most direct is the
original scheme of Adams. We suppose we have a good one-electron
Hamiltonian H (say using a Slater xα potential, for instance).
The orbitals we wish to find are eigenstates

$$H \left| \Psi_i \right> = E_i \left| \Psi_i \right> .$$

We choose to work at first only within a certain "band" of these
states, to which certainly all the occupied orbitals belong, but
as in the Huckel theory we often want to include some of the
relevant unoccupied states also: the whole π "band." The success
of our localization program will depend specifically, as shown in
my paper, on this "band" being well separated in energy from all
excited states (but later we will show how to evade that difficulty).

Adams defines a projection operator on this band

$$P = \sum_{i \text{ in band}} \left| \Psi_i \right> \left< \Psi_i \right|$$

and asks what equations one would wish a best set of localized
orbitals φ_a, which one wants to span this band, to obey. The
first says that it belongs to the "band"

$$\left| \varphi_a \right> = P \left| \varphi_a \right> \quad \text{or} \quad H \left| \varphi_a \right> - PHP \left| \varphi_a \right> = 0$$

and the second involves defining an "atomic" Hamiltonian H_a for
each state which is the part of the Hamiltonian one wants φ_a to
belong to. For instance, one might use a single atom's potential

$$H_a = T + V_a$$

or a bond potential

$$H_{ab} = T + V_a + V_b$$

or whatever one likes. Adams suggests that we define $|\varphi_a\rangle$ by satisfying H_a as best we can in the band:

$$PH_a P|\varphi_a\rangle = \varepsilon_a |\varphi_a\rangle .$$

We define $U_a = H - H_a$, the "rest" of the interactions. Then this is just the same as

$$H_a \varphi_a + (U_a - PU_a P)|\varphi_a\rangle = \varepsilon_a |\varphi_a\rangle .$$

This is the basic equation for the φ_a and demonstrates clearly that the effect of the "rest of the system" U_a is strongly damped out by subtracting all of that which acts within the band: $PU_a P$. This is often enough that the residual effect of the rest of the system may be treated by perturbation theory or ignored; thus the qualitative behavior, at least, is completely explained by localized, atomic-like wave-functions. It is also clear that the variational theorem applied to $(PH_a P)\varphi = \varepsilon\varphi$ gives us a unique set of φ's; this is very important, because in this respect these localized orbitals are far superior to Wannier functions - the chemist's symmetry orbitals - in that the latter can, in principle, not be defined uniquely, at least by a Schrödinger equation.

We pay something for this: every φ_a is an eigenfunction of a different Hamiltonian, and so there is no orthogonality relationship between them. But if the H_a are correctly chosen, the $|\varphi_a\rangle$ will be a complete set, and it is possible, therefore, to write down the secular equation which determines the energies of the actual eigenstates of H.

The simplest way to choose, but not the only one or the best, is just to write down

$$|H_{ab} - ES_{ab}| = 0$$

which is indeed reasonably localized, but is relatively untransparent. A second, more complicated way is to orthogonalize the functions φ_a, thereby creating a set of Wannier functions:

$$W_a = S^{-1/2}|\varphi_a\rangle$$

and solve the "Wannier" secular equation

$$\left| S^{-1/2} H_{ab} S^{-1/2} - E\delta_{ab} \right| = 0$$

where the easiest way to do $S^{-1/2}$ is the Lowdin expression in terms of powers of the overlap. But this is not at all guaranteed to take ideal advantage of the locality of our wave functions, though as we shall see it actually does in many cases.

This is where we get to the simplifications which come about from my way of doing it. First we make a simplification of Adams' equation: since $P|\varphi_a\rangle = |\varphi_a\rangle$, we leave that out and have

$$[H_a + (U_a - PU_a)]\varphi_a = \varepsilon_a \varphi_a .$$

From this equation we can directly deduce the effect of $H = H_a + U_a$ on φ_a:

$$H\varphi_a = \varepsilon_a \varphi_a + PU_a \varphi_a .$$

If all the φ's were orthogonal, we could just write $P = \sum_b P_b$, the

sum of the projections on all the local states, and the equation would simplify into a form similar to that we will soon demonstrate. Instead, my equation is equivalent to writing $U_a = \sum_{b \neq a} P_b V_b$ (a

perfectly allowable choice, if we don't insist on the absolute optimum H_a!) where V_b is some potential chosen to be the best approximation to what we want the wave function φ_b to satisfy. Then since $PP_b = P_b$, we have simply

$$H_a |\varphi_a\rangle + \sum_{b \neq a} \{ V_b |\varphi_a\rangle - \langle \varphi_b | V_b | \varphi_a \rangle | \varphi_b\rangle \} = \varepsilon_a |\varphi_a\rangle$$

or

$$H|\varphi_a\rangle = \varepsilon_a |\varphi_a\rangle + \sum_{b \neq a} \langle \varphi_a | V_b | \varphi_a \rangle | \varphi_b\rangle .$$

In this form of the equation it is particularly easy to see how to do an actual calculation of the φ_a, as I did in my Huckel paper, starting from a good estimate of all φ_a. The equation for $|\varphi_a\rangle$ involves all other φ_b's; clearly we choose a reasonable starting set and iterate, using, at each stage, the "other" atomic functions from the previous one. Rapid convergence has been demonstrated in several cases. This is the source of another, slightly misleading, name for my method: "self-consistent pseudopotentials."

By simply expanding the exact eigenfunction Ψ_i in terms of φ_b's:

$$\Psi_i = \sum_a c_{ia} \varphi_a$$

we can see that the exact secular equation is simply

$$\left| (\varepsilon_a - E)\delta_{ab} + (\varphi_a | V_b | \varphi_a) \right| = 0 .$$

Since V_b is more local than H, this is likely to be even more local than $|\overline{H - ES}| = 0$. If the system is perfectly periodic or cyclic, it is the same as the Wannier secular equation: one can show that this corresponds to

$$\left| S^{-1}H - E \right| = 0$$

rather than

$$\left| S^{-1/2}HS^{-1/2} - E \right| = 0$$

which is the same if $[S,H] = 0$. That was the case with the problem I treated before, so I didn't notice the difference: but there is a very real difference if $V_b \neq V_a$, i.e. if the system is not periodic. In that case the ideal local secular equation is, regrettably but perfectly manageably, non-Hermitian.

This was the point to which Allan's attempt to do pyridine led us - when he tried to derive an appropriate set of α's and β's for it, he found that either he had to accept a non-Hermitian secular equation: $\beta_{N-C} \neq \beta_{C-N}$ or he had to transform back to Wannier functions, in which case he found that the Hermitian Huckel type secular equation was not quite accurate: there is a small but finite α on the neighboring carbons to the nitrogen. It is your choice which you prefer, and in any case experiment is not entirely accurate enough really to pick the effect up.

The next thing about pyridine is the dipole moment. It has long been known that with reasonable values of the α's and β's and the standard apparatus of the Huckel theory, one does not get enough dipole moment from the π electrons, and it is necessary to postulate that a fair amount is contributed by the σ-bonds. For quite a while Alan worked on the σ bonds and it began to appear that it was not particularly reasonable to expect them to be polarized as much as is normally assumed.

Now the standard apparatus of Huckel theory puts the charge in a given Wannier function on the center of the appropriate atom.

That leads to a beautiful mathematical theory but we began at
this point to realize that it wasn't right. The wave-functions
are either truly nonorthogonal and local (our functions) or
they are orthogonal and rather widely spread out (Wannier) and in
either case the dipole moment as opposed to the secular equation
is <u>strongly</u> affected by the overlap terms. In terms of the local
orbitals, overlap leads to a normalization difference in the φ_i's,
which moves charge a long distance. In terms of Wannier functions,
$\int \varphi_a x \varphi_b \neq 0$ even if $\int \varphi_a \varphi_b$ does. When he calculates using our best
estimates of α's and β's, and correct φ_C's and φ_N's which are for
all practical purposes atomic orbitals, he finds that there is a
much larger π contribution. His work is still in process and
quantitative results will be published later.

The overlap contribution to μ is not unknown, of course, but
may not be adequately appreciated. Unpublished work of mine in
1950, and some nice work by Overhauser (7), showed long ago that
it contributes strongly to alkali halide dipole moments. Another
interesting case is the large moment carried by adsorbed rare gas
atoms such as Xenon on metal surfaces (8); or at least so, without calculation, I would assume. I expect this to play a role
in H bond dipole moments as well.

I want finally to mention two uses for localized, self-
consistent pseudopotentials which as yet we haven't got under
way from a numerical point of view, but which I think are very
interesting. The first makes use of the fact that in my own,
non-Hermitian version, the eigenvalues ε_a are the diagonal
elements of the secular equation and so by the well-known trace
theorem the total energy of our "band" is given by the sum of the
ε_a's. After correction for many-electron effects and direct core-
core interactions, this then is the basic binding or repulsive
energy of any saturated system of electrons. A particularly
straightforward application should be to intermolecular repulsive
forces.

More directly in the main stream of the subject-matter of
this meeting, I should like to project the possibilities of this
method into the future and discuss how one might use it to understand amorphous materials. Since after all it is set up to deal
with complex molecules, which are also irregular systems, it should
be ideal for such problems.

The first straightforward point one should make is that the
total energy is, as I have just intimated, the sum of the ε_a's
and that these depend only on local surroundings. Thus this
method is directly adapted to allowing the local bonding to adjust
itself to the local configuration energetically. The concept of
the random covalent network with local bonding orbitals adjusted
to the local bonding situation follows very directly.

Taken simply, then, this would lead to some such description as Weaire's (9) for amorphous covalent structures, based on the Hall (10) type of LCAO theory for semiconductors. I would like to do better than this for two reasons.

1. The Hall theory does not give a good description even of the valence bands of crystalline semiconductors: the errors are larger than the band gap itself.

2. LCAO is particularly bad for the conduction bands, even of very wide band gap materials such as alkali halides or rare gases. As Barry Kunz has shown, for such a problem one needs a mixed representation (11) theory: one must include both atomic orbitals and plane waves. This is the attempt I am undertaking, with a student, Mr. Bullett, to put computational flesh upon. Briefly, however, it is very easy to describe the essential idea, since it fits perfectly within the above scheme, particularly if we start from the "Anderson" version and not the Adams one.

We expand our basis set to include a certain number of plane waves by letting one of the H_a's simply be the kinetic energy T, so that the unperturbed orbitals φ_{00} are plane waves e^{ikr}. These will be modified by the projection process in some minor way, to represent the nearest thing to plane waves which are available within a basis set made up of the lowest N+N' wave functions. For this eigenequation

$$T\varphi_k + \sum_{alla} V_a \varphi_k - \sum (\varphi_a|V_a|\varphi_k)\varphi_a = \varepsilon_k \varphi_k$$

we keep, not just the lowest wave function as for the local functions but the lowest N' solutions, N' being the number of plane waves we wish to keep in our basis.

With this simple expansion the scheme becomes so general as to include all the known pseudopotential methods under one roof. If the φ_a's are deep in energy and can be treated as "core" functions, not to be appreciably varied, this is standard pseudopotential theory. As we have seen, if we leave out $(H_a)_0 = T$, the scheme is the localized pseudopotential method. In general, it is a mixed-set scheme essentially equivalent to the Mueller-Ehrenreich-Pettifor scheme for d bands except for two points: (1) it shows why in those schemes the use of simple atomic orbitals for the tight-binding part of the d bands is a success empirically; (2) the resulting Hamiltonian is non-Hermitian, if one uses the "Anderson" scheme (III) of deriving it. Of course, one may use schemes I (equivalent to Mueller) (12) or II (Pettifor) (13) but it may well be that an asymmetrical secular equation is as good as or better than these.

To get back to the question of insulating or semiconducting amorphous materials, it seems to me that there is a very fundamental point to be made here. Molecular liquids, polymers, and glasses represent the "type" or "model" cases for such materials, in the sense, for instance, that one can think of Al as the most metallic metal or diamond as the extreme covalent semiconductor. As I have said, in these "type" cases it is quite clear that the valence band is a collection of deep, bound atomic or bonding orbitals, while the conduction band must be based on plane waves. As we narrow the gap it is clear we must keep both of these orbital types; but it is just possible we can get by without any further complications, e.g. perhaps we need no antibonding orbitals in amorphous semiconductors.

We would like to acknowledge discussions with V. Heine, and kind permission by J. C. Phillips to use his Fig. 1.

REFERENCES

1. W. H. Adams, Jr., J. Chem. Phys. $\underline{37}$, 2009 (1962).

2. T. L. Gilbert, in <u>Molecular Orbitals, a Tribute to Mulliken,</u> (Pullman and Lowdin, eds.) Academic Press, N.Y., (1964).

3. P. W. Anderson, Phys. Rev. Letters $\underline{21}$, 13 (1968).

4. P. W. Anderson, Phys. Rev. $\underline{181}$, 25 (1969).

5. A. B. Kunz, Jr., Phys. Stat. Sol. $\underline{36}$, 301 (1969); Phys. Rev. $\underline{132}$, 2224; Phys. Rev. $\underline{134}$, 609, 1374, 4639 (1971).

6. J. D. Weeks, P. W. Anderson and A. G. H. Davidson, to be published.

7. B. G. Dick and A. Overhauser, Phys. Rev. $\underline{112}$, 90 (1958).

8. T. Engel and R. Gomer, J. Chem. Phys. $\underline{52}$, 5572 (1970).

9. D. Weaire, Phys. Rev. Letters $\underline{26}$, 1541 (1971).

10. G. G. Hall, Phil. Mag. $\underline{43}$, 338 (1952).

11. A. B. Kunz, Jr., J. Phys. C. $\underline{3}$, 1542 (1970); Phys. Rev. $\underline{132}$, 5015 (1970); and Ref. (5).

12. F. M. Mueller, Phys. Rev. $\underline{151}$, 557 (1956).

13. D. Pettifor, J. Phys. C. $\underline{3}$, 366 (1970); J. Phys. C. $\underline{2}$, 1051 (1969).

LOCALIZED DEFECTS IN SEMICONDUCTORS[*]

F. L. Vook and K. L. Brower

Sandia Laboratories

Albuquerque, New Mexico 87115

ABSTRACT

Recent progress in experimental and theoretical studies of localized defects in semiconductors is reviewed. Most of the present information on localized defects has been obtained from electron paramagnetic resonance and optical absorption measurements. In addition, new innovative experimental approaches, such as isotopic ion implantation, ion implantation damage production, depth distribution measurements of specific defects, and channeling techniques, are contributing greatly to current progress in understanding defects. Computational studies which complement the experimental studies are also discussed.

INTRODUCTION

The purpose of the present paper is to review recent progress in experimental and theoretical studies of localized defects in semiconductors.[1] The importance of localized defects in semiconductors is that small concentrations are found to completely dominate many of the solid state properties. We will first briefly discuss the experimental approaches that have proven useful in the past and then review some of the information that has been obtained

---

[*]This work was supported by the U. S. Atomic Energy Commission.

on localized defects in silicon, the best understood semiconductor
material. We then center our discussion on recent innovative experimental approaches that we believe will have the greatest impact
on current and future progress. The implications of these experimental approaches on calculations are then discussed.

REVIEW OF DEFECTS IN SEMICONDUCTORS

Most of the information on localized defects in semiconductors
has been obtained by combining radiation damage experiments (particularly electron and neutron irradiation) with optical absorption or
electron paramagnetic resonance (EPR) measurements. These measurements have provided the basis for our current understanding of
irradiation produced defects in silicon.[2-7] Included in the
detailed information that is now available on defects in Si are:
annealing kinetics and energetics, stress response, impurity
dependence, electronic energy levels, and incident particle energy
dependence for specific defect production. All of these characteristics are useful for correlation of the effects of irradiation on electrical, photoconductive, luminescent, thermal, and
other physical properties of Si. Defects identified in EPR studies
in irradiated Si are sketched in Fig. 1 along with information on
the annealing energetics, energy levels, and approximate center
annealing temperatures observed in isochronal (approximately 10-60
minute periods) experiments.[8-10]

Optical absorption bands produced by high energy particle
irradiation of Si fall into two general classes: (1) localized
vibrational bands and (2) electronic excitation bands. Both of
these classes are included in the irradiation produced bands shown
in Fig. 2.[11] Prominent bands in the spectra at 1.8, 3.3, 3.9, and
5.5 microns are associated with electronic transitions involving
the divacancy defect in Si. The 12-micron band is a vibrational
band associated with vacancy-oxygen (substitutional oxygen) A-center
defect in irradiated Si. The 9-micron band is a vibrational band
associated with interstitial oxygen and can be present in
unirradiated Si. The 8-micron band is an electronic excitation
associated with arsenic; for phosphorus doped samples, the comparable peak is at 8.7 microns.[11]

The association of electronic transition bands at 3.3, 3.9, 1.8,
and 5.5 microns with the divacancy was accomplished by correlating
the stress response and the annealing behavior of these bands with
those for the divacancy as determined in EPR measurements. In fact
uniaxial stress has been one of the important tools which has been
used to obtain information on the symmetry properties of defect

centers both in optical absorption and EPR measurements. Another
identification technique has been the use of isotopes, such as the
use of ^{18}O in the identification of the vibrational absorption
bands for the interstitial and substitutional oxygen centers in
silicon^{12,13} and germanium.^{14} Isotopes with nonzero nuclear spins
give rise to EPR hyperfine spectra which are indispensible for
establishing the structure of defects. For example, the ^{29}Si
hyperfine interaction was used to establish the structure of the
defects shown in Fig. 1.

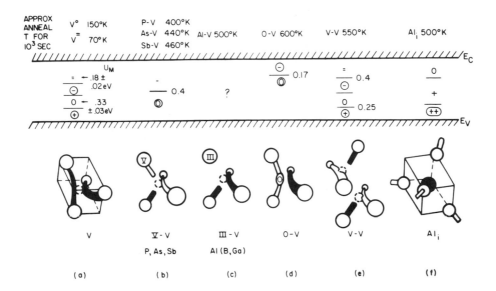

Fig. 1.   Models of defects in Si as deduced from EPR studies.
          Where known, the electrical level structure is indicated,
          and the charge state shown in the circle is that observed
          by EPR. Also shown are the approximate isochronal
          annealing temperatures for ~ 10³ seconds. (a) Configura-
          tion of the isolated vacancy. The electrical level
          structure is shown along with the activation energies for
          motion determined from annealing studies in n- and p-type
          material; (b) through (e) single vacancy - defect pairs
          including interaction with: (b) substitutional Group V
          donors, (c) substitutional Group III acceptors, (d)
          interstitial oxygen, (e) other vacancies to form diva-
          cancies; and (f) aluminum interstitial. The solid bonds
          indicate the predominant localization of the paramagnetic
          electrons. (Refs. 8-10,41)

Schematic Diagram of Infrared Absorption Bands in Irradiated Silicon

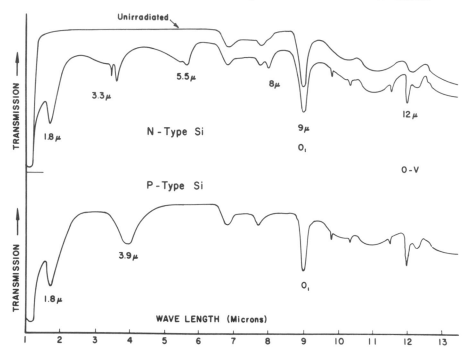

Fig. 2.   Schematic diagram of infrared absorption bands in irra-
          diated silicon (both n- and p-type) in the wavelength
          region 1 to 13 μ at room temperature (Ref. 11).

NEW EXPERIMENTAL AND THEORETICAL TOOLS TO STUDY DEFECTS

A.   Ion Implantation and EPR

        Recent innovative experimental approaches, such as ion implan-
tation,[15-17] have been coupled with the older experimental techniques
of EPR and optical absorption to give new information on defects in
semiconductors. The neutral vacancy-oxygen center in Si (Fig. 1d,
Si-SL1) has been the focal point of two innovations: ion implantation[18] and spin-spin interactions.[19,20] Although there are many
facets to ion implantation, the first idea we wish to pursue here
is the use of ion implantation in isotopic doping. The ability to
dope a crystal with a particular isotopic impurity offers distinct
advantages to many spin resonance studies, as we shall demonstrate
in the case of the neutral vacancy-oxygen center. Furthermore, ion
implantation is not a thermal equilibrium process. Consequently,
the doping of a crystal by ion implantation tends to be limited by

technological techniques rather than by high temperature thermal
equilibrium processes. Also, the efficiency for incorporating into
a sample an isotope, which is usually available only in limited
quantities and chemical forms, is very high by ion implantation.[18]

The molecular structure of the vacancy-oxygen center is illustrated in Fig. 3. Although the existence of oxygen in the
negative vacancy-oxygen center (Si-Bl) was established by Watkins
and Corbett[9,13] using stress and infrared techniques, they were
unable to detect the ^{17}O hyperfine spectrum associated with this
particular charge state using EPR because the paramagnetic electron
is in an antibonding molecular orbital, and the oxygen atom is located in the nodal plane. However, in the case of the neutral
vacancy-oxygen center, which can be excited by light into a triplet
spin state, one of the paramagnetic molecular orbitals transforms
as Γ_1 so that a sizeable ^{17}O hyperfine interaction is expected.

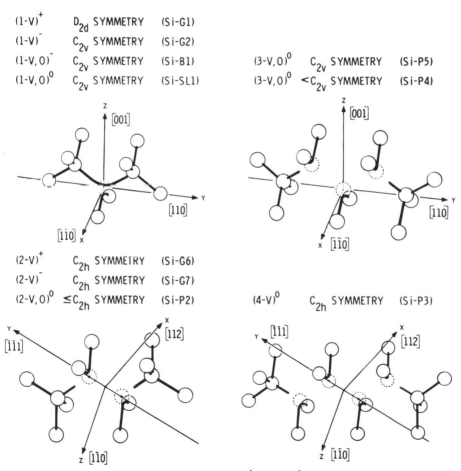

$(1-V)^+$ D_{2d} SYMMETRY   (Si-G1)
$(1-V)^-$ C_{2v} SYMMETRY   (Si-G2)
$(1-V,0)^-$ C_{2v} SYMMETRY   (Si-B1) $(3-V,0)^0$ C_{2v} SYMMETRY   (Si-P5)
$(1-V,0)^0$ C_{2v} SYMMETRY   (Si-SL1) $(3-V,0)^0$ $<C_{2v}$ SYMMETRY   (Si-P4)

$(2-V)^+$ C_{2h} SYMMETRY   (Si-G6)
$(2-V)^-$ C_{2h} SYMMETRY   (Si-G7)
$(2-V,0)^0$ $\leq C_{2h}$ SYMMETRY   (Si-P2) $(4-V)^0$ C_{2h} SYMMETRY   (Si-P3)

Fig. 3. Family of multiple vacancy (oxygen) defects.

The only stable oxygen isotope with a nonzero nuclear spin is ^{17}O ($I = 5/2$), and it is only 0.037% naturally abundant. Under these circumstances, the ^{17}O hyperfine spectrum is ordinarily very weak and unresolved from the fine structure and ^{29}Si hyperfine spectra.[19] This difficulty was overcome by implanting ^{17}O into silicon samples and thereby significantly enhancing the intensity of the ^{17}O hyperfine spectrum relative to the fine structure spectrum as illustrated in Fig. 4. The concentration of ^{17}O in the implanted layer relative to the ^{16}O was enhanced by $\approx 10^4$.[18]

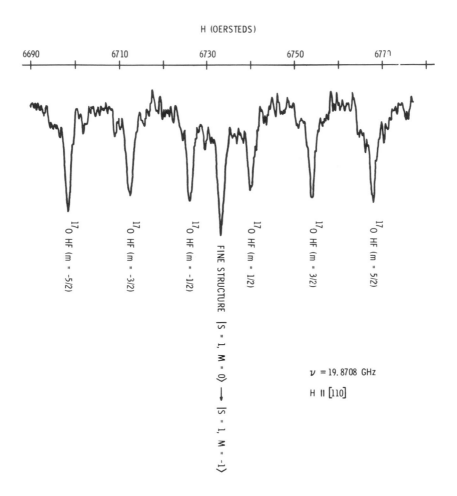

Fig. 4. Part of the Si-SL1 EPR spectrum observed in Si implanted with ^{17}O. Because the ^{17}O hyperfine and the nuclear Zeeman interactions lift the 6-fold degeneracy of the ^{17}O nuclear spin states, the ^{17}O hyperfine spectrum is characterized by six nearly evenly spaced resonances (Ref. 18).

The results of the analysis on the Si-SL1 [17]O hyperfine spectrum indicate that only 4% of $\langle \chi(\Gamma_1)|\chi(\Gamma_1)\rangle$ and 1.8% of $\langle \chi(\Gamma_4)|\chi(\Gamma_4)\rangle$ are localized on the oxygen atom in this defect. $\chi(\Gamma_1)$ and $\chi(\Gamma_4)$ are the two one-electron molecular orbitals which were examined by virtue of the spin resonance arising from their unpaired electrons. Furthermore, ≈ 40 and 60% of $\chi(\Gamma_1)$ localized on the oxygen atom arise from ψ_{2s} and ψ_{2p_z}, respectively.[18]

As these and many other spin resonance studies indicate, the hyperfine interaction is analogous to a projection operator which extracts from the paramagnetic molecular orbitals the admixture of s- and p-like orbitals localized on each site from which a resolved hyperfine spectrum originates. It is through measurements and experimental techniques such as these that we are able to get a glimpse of the actual electronic structure of defects.

From a semiquantitative description of the wave functions for the paramagnetic electrons it is possible to calculate in certain instances the spin-spin interaction between the unpaired electrons associated with defects in the triplet spin state. In the case of the neutral vacancy-oxygen (Si-SL1) center, the spin-spin interaction between the paramagnetic electrons is dominated by the magnetic dipole-dipole interaction. This result suggested that the weaker spin-spin interactions associated with the Si-P2, -P3, -P4, and -P5 centers are also dominated by the magnetic dipole-dipole interaction. In the case of the Si-P4 and -P5 centers, the g tensor is remarkably similar to that of the Si-SL1 center; however, the spin-spin interactions differ by about an order of magnitude. These factors provided the key leading to the models for the Si-P2, -P3, -P4, and -P5 centers illustrated in Fig. 3.[20]

The centers in Fig. 3 belong to a family in the sense that they are characterized by one or more vacancies in a single (110) plane. Furthermore, oxygen can be bonded between a pair of silicon atoms adjacent to a vacancy.

Previous studies of the [29]Si hyperfine interactions indicate that the unpaired electrons tend to be localized mostly in s,p hybrid orbitals directed along $\langle 111\rangle$ directions in the plane containing the vacancies. For even vacancy defects, the dangling bonds are antiparallel; whereas for odd vacancy defects, the dangling bonds are canted with respect to each other. This feature is reflected in the anisotropy of the g tensor and allows one to discriminate between even and odd vacancy (oxygen) defects.[19,20]

Since even- and odd-vacancy (oxygen) defects can be distinguished by the symmetry and anisotropy in their g and A tensors, the number of intervening (oxygen-) vacancies can be estimated

from the spin-spin interaction. Figure 5 shows the results of a
calculation of the principal values of the D tensor as a function
of the separation between the end silicon atoms, assuming a magnetic
dipole-dipole interaction. In these calculations, the one-electron
orbitals for the paramagnetic electrons were represented by 3s, 3p
hybrid orbitals whose localization and 3s and 3p character agreed
with that deduced from the [29]Si hyperfine interactions. The smooth
progression in the experimental values for the D tensor as well as
the agreement with the calculated D tensor made it possible to
assign the Si-SL1, Si-P2, -P3, -P4, and -P5 centers in terms of
particular multiple-vacancy (oxygen) structures.[19,20] It appears
that this procedure is now beginning to contribute new insights
into the structure of spin 1 centers in diamond.[21]

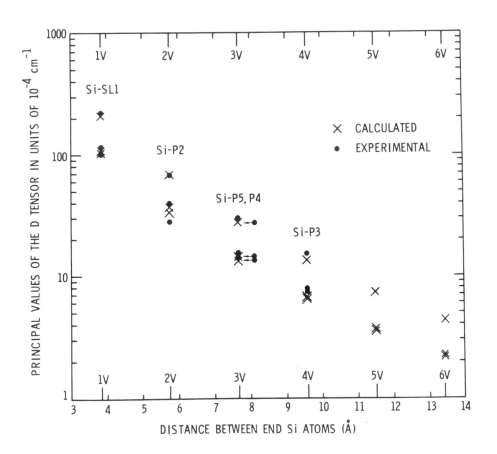

Fig. 5. Calculated and experimental values of $-D_{11}$, D_{22}, and D_{33}
         as a function of the separation between the end silicon
         atoms in the $|S = 1, M\rangle$ spin formalism (Refs. 19,20).

Recently, another example has been given of the use of ion implantation for the isotopic doping of a solid to produce new localized defects. An optical absorption band has been observed at 473 cm^{-1} in nitrogen implanted GaAs and is attributed to the localized vibrational mode of N substitutional on the As sites.[22] Previously, it was not possible to incorporate nitrogen in GaAs by other techniques.

Another important use of ion implantation to study localized defects is in radiation damage experiments. Stein[23] has recently used low temperature ion implantation in germanium together with optical absorption measurements to obtain a band which he has tentatively identified with the Ge divacancy. This is an important defect in a material where very few defects have been identified by EPR measurements.

Ion implantation can also be used to study high concentrations of localized defects. The crystallinity of ion-implanted silicon has been investigated as a function of ion mass and ion fluence by infrared measurements of divacancy absorption.[24,25] In a parallel study recent EPR measurements of ion-implanted silicon, as a function of ion fluence,[26] have shown how increasing concentrations of crystalline defects result in an accumulation of defect complexes which are characterized by electrons which tend to be delocalized among the defects in the complex. At even higher concentrations the crystalline silicon is converted to the amorphous state with the observation of a characteristic isotropic EPR resonance. The results of the optical and EPR experiments are consistent with localized defect domination of the lattice disorder when the energy deposited into atomic processes is $\leq 10^{20}$ keV/cm^3 and with amorphous layer formation when the energy deposition is $\sim 10^{21}$ keV/cm^3.[24,26]

B. Measurements of the Depth Distribution of Defects

Another new experimental approach involves the use of depth distribution measurements of specific defect centers on a few hundred Å scale. These results have stimulated calculations of the spatial deposition of energy into atomic and electronic processes by radiation damage cascades and other higher energy events in solids. The depth distribution of tetravacancies (Si-P3) in 400 keV O$^+$ implanted silicon has been determined using EPR measurements in conjunction with anodization and stripping of the implanted layer.[27] Similar measurements were made of the depth distribution of divacancies using infrared measurements of the characteristic 1.8 μ absorption band.[28] These distributions of localized defects differ from the distribution of implanted ions and compare favorably to theoretical calculations by Brice for the depth distribution of the energy deposited into atomic processes.[29]

In order to obtain the depth distribution of energy into atomic processes, Brice derived integro-differential equations governing the spatial distributions of ions of intermediate energy E' moving in a solid target after the ions were incident on the target surface with energy E. The solution of the spatial distribution of ions incident on a solid target as a function of instantaneous energy E' allows the solution of the spatial distribution of reaction products of any energy dependent interaction of the incident ions with the target atoms. One such reaction is the energy into atomic processes.

Recently, defects produced by electronic processes have been used to make the first measurements[30] of the depth distribution of energy deposited into electronic processes by ion implantation. Measurements were made of the radiophotoluminescence generated in silver-activated phosphate glass by the electronic energy losses of 200-400 keV ions (H^+, He^+, O^+) incident on the glass. In this glass ionizing radiation forms an absorption band (probably from Ag^{++}) centered around 3600 Å. A broad luminescence band at about 6200 Å can be excited from the irradiated glass (radiophotoluminescence) by light pumped into this absorption maximum. The intensity of the luminescence is proportional to the deposited energy into electronic processes per unit volume below 10^{18} keV/cm^3. For lower deposition values detailed measurements of the integral depth distribution of luminescence intensity were made as material was removed by vibratory polishing (~ 1000 Å steps). The measured depth distributions compared favorably to the integral depth distributions of electronic energy losses predicted by LSS[31] theory (He^+ and O^+) and experimental values of thin film losses (H^+). These measurements indicate that the luminescent centers are formed in the glass close to ($\leq 10^3$ Å) the point of the initial deposition of electronic energy.

C. Ion Backscattering-Channeling Study of Defect Lattice Location

Another new experimental technique is the ion backscattering channeling-effect technique[15,16,32] which gives information on the lattice location of impurities in crystalline solids. When the angular dependences of the reductions in the backscattering yield along various crystal directions are the same for the impurity as for the lattice atoms, then the impurity atoms are determined to be on substitutional sites. Such measurements have been widely used to determine the lattice location of heavy impurities implanted into silicon and germanium.[15] When the impurity atoms occupy nonsubstitutional sites, interpretations of the impurity lattice location are aided by combining experimental measurements (of angular scans near various channeling directions for both the impurity and the lattice atoms) with channeling calculations as a function of lattice position. Recently, such measurements and calculations have been made for Bi implanted in silicon.[33] Best agreement with the single and double alignment data was obtained for the case of $\approx 50\%$ of the Bi displaced 0.45 Å from the Si lattice sites (approximately along the $\langle 110 \rangle$ direction) and the remaining Bi atoms located substitutionally on Si lattice sites.

Much information on the lattice location of impurity interstitials in silicon has been obtained from channeling effect measurements. The lattice location of implanted Yb atoms in Si was determined to be slightly displaced from the tetrahedral position, and the authors suggest that a displacement along the $\langle 100 \rangle$ direction is consistent with their data.[34] The interstitial positions of Zr, Hf, Tℓ, and Hg atoms implanted in Si were measured by the channeling effect technique and interpreted to be near the tetrahedral interstitial position.[35] On the basis of previous Rutherford backscattering experiments without detailed angular scans,[36,37] In, Ga, Tℓ, Zn, Cd, and Hg implanted into Si all exhibited interstitial positions.

The channeling technique has also been used to determine the lattice location of interstitial B in B ion implanted Si by use of the B^{11} (p, α) nuclear reaction.[38] The interstitial B atoms do not occupy the tetrahedral interstitial sites but appear to lie along the $\langle 110 \rangle$ atomic rows, but not midway between the $\langle 110 \rangle$ row lattice sites. Watkins[39] has also suggested that his EPR measurements indicate that the B interstitial is nestled asymmetrically between two substitutional lattice sites, in agreement with the channeling results. In addition, EPR measurements indicate that the Aℓ interstitial occupies the tetrahedral interstitial sites in Si.[40-41] Extensive EPR measurements on transition metal elements in silicon, many of which occupy tetrahedral interstitial sites, have also been made.[42]

The importance of these results is seen by the fact that there is essentially no experimental information on the configuration of the Si interstitial in Si or the interstitial-Si-vacancy close pair in Si. As previously discussed, there is much more EPR information on the defect configurations of vacancy-type defects in Si, and thus the backscattering channeling effect measurements of interstitial defects complement the EPR studies.

REFERENCES

1. Recent reviews of defects in semiconductors are given in
    Refs. 2-7.

2. Radiation Damage in Semiconductors, edited by P. Baruch
    (Dunod, Paris, 1965).

3. J. W. Corbett, Electron Radiation Damage in Semiconductors
    and Metals (Academic Press, New York, 1966).

4. Radiation Effects in Semiconductors, edited by F. L. Vook
    (Plenum Press, New York, 1968).

5. Radiation Effects in Semiconductors, edited by J. W. Corbett
    and G. D. Watkins (Gordon and Breach, New York, 1971).

6. H. J. Stein, Rad. Effects $\underline{9}$, 195 (1971).

7. J. W. Corbett and J. C. Bourgoin, IEEE Trans. Nucl. Sci.
    $\underline{\text{NS-18}}$, 11 (1971).

8. G. D. Watkins, p. 67 of Ref. 4.

9. G. D. Watkins and J. W. Corbett, Phys. Rev. $\underline{121}$, 1001 (1961);
    $\underline{134}$, A1359 (1964); $\underline{138}$, A543 (1965); $\underline{138}$, A555 (1965).

10. G. D. Watkins, J. Phys. Soc. Japan $\underline{18}$, Suppl. II, p. 22
    (1963); p. 97, Ref. 2; Phys. Rev. $\underline{155}$, 802 (1967).

11. L. J. Cheng, J. C. Corelli, J. W. Corbett, and G. D. Watkins,
    Phys. Rev. $\underline{152}$, 761 (1966); C. S. Chen, J. C. Corelli, and
    G. D. Watkins, Bull. Am. Phys. Soc. $\underline{14}$, 395 (1969); C. S. Chen
    and J. C. Corelli, Phys. Rev. B $\underline{5}$, 1505 (1972); Rad. Effects $\underline{9}$,
    75 (1971); L. J. Cheng and P. Vajda, Phys. Rev. $\underline{186}$, 816 (1969).

12. H. J. Hrostowski and R. H. Kaiser, Phys. Rev. $\underline{107}$, 966 (1957);
    J. Phys. Chem. Solids $\underline{9}$, 214 (1959).

13. J. W. Corbett, G. D. Watkins, R. M. Chrenko, and R. S. McDonald,
    Phys. Rev. $\underline{121}$, 1015 (1961).

14. R. E. Whan, Phys. Rev. $\underline{140}$, A690 (1965).

15. J. W. Mayer, L. Eriksson, and J. A. Davies, Ion Implantation in
    Semiconductors (Academic Press, New York, 1970).

16. Ion Implantation, edited by F. H. Eisen and L. T. Chadderton
    (Gordon and Breach, New York, 1971).

17. *Ion Implantation in Semiconductors*, edited by I. Ruge and J. Graul (Springer-Verlag, Berlin, 1971).

18. K. L. Brower, Phys. Rev. B **5**, 4274 (1972).

19. K. L. Brower, Phys. Rev. B **4**, 1968 (1971).

20. K. L. Brower, Rad. Effects **8**, 213 (1971).

21. Y. M. Kim and J. W. Corbett (to be published).

22. A. Kahan, F. Euler, T. Whatley, and W. G. Spitzer, Bull. Am. Phys. Soc. **17**, 27 (1972).

23. H. J. Stein, Bull. Am. Phys. Soc. **17**, 154 (1972).

24. H. J. Stein, F. L. Vook, D. K. Brice, J. A. Borders, and S. T. Picraux, Rad. Effects **6**, 19 (1970).

25. H. J. Stein and W. Beezhold, Appl. Phys. Lett. **17**, 442 (1970).

26. K. L. Brower and W. Beezhold, J. Appl. Phys. **43**, 3499 (1972).

27. K. L. Brower, F. L. Vook, and J. A. Borders, Appl. Phys. Lett. **16**, 108 (1970).

28. H. J. Stein, F. L. Vook, and J. A. Borders, Appl. Phys. Lett. **16**, 106 (1970).

29. D. K. Brice, Appl. Phys. Lett. **16**, 103 (1970); Rad. Effects **6**, 77 (1970); Rad. Effects **11**, 227 (1971).

30. G. W. Arnold and F. L. Vook, Rad. Effects **14**, 157 (1972).

31. J. Lindhard, M. Scharff, and H. E. Schiott, Kgl. Danske Videnskab Selskab, Mat. Fys. Medd. **33**, No. 14 (1963).

32. *Proc. Intl. Conf. on Atomic Collisions in Solids*, Gausdal, Norway, Sept. 1971 (Gordon and Breach, to be published).

33. S. T. Picraux, W. L. Brown, and W. M. Gibson, Phys. Rev. B **6**, 1382 (1972).

34. J. U. Andersen, O. Andreasen, J. A. Davies, and E. Uggerhøj, Rad. Effects **7**, 25 (1971); or Ref. 16, p. 315.

35. B. Domeij, G. Fladda, and N. G. E. Johansson, Rad. Effects **6**, 155 (1970); or Ref. 16, p. 425.

36. L. Eriksson, J. A. Davies, N. G. E. Johansson, and J. W.
     Mayer, J. Appl. Phys. 40, 842 (1969).

37. O. Meyer, N. G. E. Johansson, S. T. Picraux, and J. W. Mayer,
     Solid State Comm. 8, 529 (1970).

38. J. C. North and W. M. Gibson, Appl. Phys. Lett. 16, 126 (1970).

39. J. D. Watkins, Radiation Effects on Semiconductor Components
     (Journies D'Electronique, Toulouse, France, 1967), Vol. I, Al.

40. G. D. Watkins, IEEE Trans. Nucl. Sci. NS-16, 13 (Dec. 1969);
     also p. 97 of Ref. 2.

41. K. L. Brower, Phys. Rev. B 1, 1908 (1970).

42. G. W. Ludwig and H. H. Woodbury, Solid State Physics, edited
     by F. Seitz and D. Turnbull (Academic Press, New York, 1962),
     Vol. 13, p. 223.

TRANSMISSION OF ELECTRONS THROUGH A DISORDERED ARRAY OF

POTENTIALS*

Paul Erdös
Department of Physics, The Florida State University
Tallahassee, Florida   32306
and
R. C. Herndon,
Nova University, Fort Lauderdale, Florida   33314

I.   INTRODUCTION

There is a close relationship between localized states and
the transmission properties of solids.

By localized states[1] we mean, for instance, localized
electronic states in a metal or semiconductor, or localized spin
deviation states as found in magnetically ordered materials.
Also, we think about localization as arising from certain types of
disorder, rather than being the result of intrinsic properties of
the atomic or crystal structure, such as the localization of
tightly bound electrons or excitons.

An example is shown in Fig. 1. Here, a linear chain of
spins is coupled by Heisenberg-type exchange interactions. Were
it not for the "impurity" spin S_0, at a given time any deviation from the fully aligned state of all spins would have equal
probability to be anywhere on the chain. The impurity captures
the spin deviation and converts it to a localized eigenstate.
In the example shown, the parameters of the spin system are such,
that the localized state has the lowest energy of all eigenstates.

Generally speaking, an eigenstate of a quantum mechanical
system is obtained by solving an eigenvalue problem, defined by
a differential equation and boundary conditions such as the
"particle in the box" problem. For crystals, electronic eigenstates can be found which, when occupied, carry a current.
Transport theory is formulated in terms of these states.

275

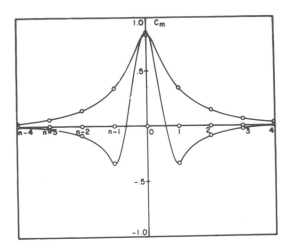

Fig. 1.   Example[2] of a localized spin deviation state due to
an impurity spin S_0, with Landé factor g_0 which replaces the
spin S with Landé factor g at the origin in a linear chain of
spins coupled by Heisenberg-type exchange interactions.   The
spin sites are numbered on the abscissa, and on the ordinate
is plotted c_m, the deviation of the spin from its value when
fully aligned along the z-axis.   The two curves result for
different values of the spin and of the external magnetic
field.

For disordered systems the studies show that all electronic
eigenstates are localized,[3] in the sense that there is a non-
zero probability for the electron to be in a finite volume of
an infinitely large sample.   An electron in a localized eigenstate does not move (hence carries no current) unless an
external force induces it to transfer from one localized state
to another.   One approach to transport phenomena, such as
electrical conduction in disordered solids is, therefore, to
study how the localized eigenstates change into current carrying
states.   Mathematically, this is accomplished by finding
approximations to the two-particle Green function of the system.
Since every disordered system has a different configuration of
atoms, the average Green function is the aim of the calculations.

Theory has been moderately successful in calculating average
two particle Green functions, which in the one-electron approximation reduce to the average of the product of two one-particle

Green functions (Kubo-formula).[4] This should be contrasted to the simpler calculation of the density of states, which requires only the averaging of one-particle Green functions.

Experience shows that the conductivities of different samples of the same disordered material are only infinitesimally different, provided the samples are sufficiently large. Therefore a suitably calculated average conductivity will be representative of the conductivity of any sample. This fact is, by no means, self-evident, and does not need to hold for all theoretical models.

Another approach to the solution of the problem of conductivity is to study, how a particle, originally localized, will become delocalized in the course of time. Naturally, the particle is now not meant to be in the localized eigenstate defined above, because it would, of course, always stay there. Suppose, however, that the electron is placed in a δ-function type state, or Wannier orbital. Then it may or may not "float" away gradually. This problem has been studied by many workers,[5] but results of convincing mathematical rigor are difficult to obtain.

In the typical conductivity experiment the electrons entering the sample are neither in a localized eigenstate, nor in any other localized state. Rather, they are injected into the disordered system with a certain momentum. This point of view was first emphasized by R. Landauer.[6] Thus we are not dealing with an eigenvalue problem, but a transmission problem ("particle in a box with open ends").

Possibly the simplest, but still instructive example is the transmission of electrons through a linear array of N potentials. In the following, we describe results obtained for such a model. It should be noted, that this is not a model for conductivity, because the electrons are not distributed in energy according to a Fermi-function. We consider monoenergetic electrons. It is felt, that some insight can be obtained into the transmission properties of disordered systems by studying this model. Besides its theoretical interest, there are various applications of the model to real systems· These are

1. the transmission of electrical signals through a chain of four-terminal networks whose parameters are randomly distributed within certain tolerance limits. Of particular interest is the case of electrical cables, where there is a continuous distribution of parameters, such as inductance, leakage, etc. The knowledge of the properties of imperfect cables (e.g., telephone lines) is of great importance for high-speed data transmission.

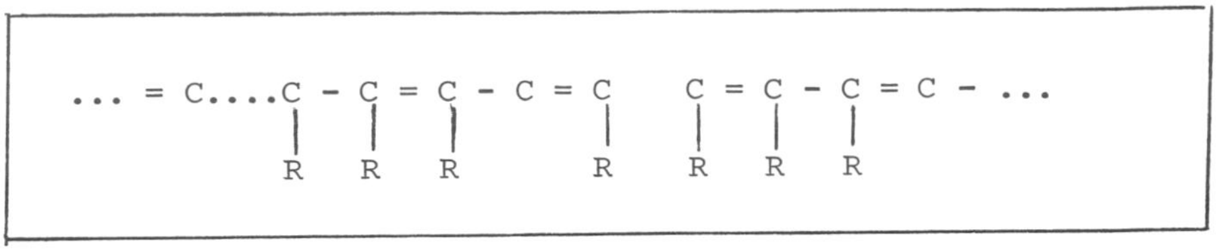

$$... = C....C - C = C - C = C \qquad C = C - C = C - ...$$

with substituent groups R below.

Fig. 2.   Polymer chain with conjugation defects.   The
defects are, e.g., broken bonds.

2.   the transmission of monochromatic light through a medium
whose refractive index is a random function of position.   The
equations governing light propagation and electron propagation
have a one-to-one correspondance; therefore, the results of the
electron problem can be transferred to the light propagation
problem.

3.   the electrical conductivity of polymer chains with conjugation defects.[7] An example of such a chain is shown in Fig. 2.
The defects are, for instance, broken bonds.   As shown in Ref. 7,
the electron propagation in this system can be described by the
simple model introduced in this paper.

4.   the transmission of electrons through thin metallic foils,
described by the results obtained for the special case of an
ordered system of finite length.

II.   DESCRIPTION OF THE MODELS

In this section we discuss the calculation of the transmission
of electrons through a linear array of N non-overlapping potentials
at arbitrary positions and of arbitrary sign and strength.   Three
models called A, B and C will be introduced: they differ in the
type of correlation which exists between the positions of adjacent
potentials.   The following assumptions are common to all models:
(a)   Each potential is of the form $V_n \delta(x-x_n)$.   (b)   The strength
parameters

$$g_n = 2mh^{-2}V_n \tag{1}$$

are statistically independent of the location of the potentials.
(c)   All potentials have the same probability distribution
function of their strength parameters.

In Model A we consider a line segment of length L and think
of the N potentials as being thrown down successively, in a
random fashion such that the probability density $f(x_k)$ of the
position of the kth potential thrown is L^{-1} for $0 < x_k < L$ and zero
elsewhere.   We will call this the model of complete spatial
disorder.   It may be considered as representative of substitutional disorder.   In a host crystal, impurity atoms are embedded
at random, and the difference between the impurity and host
potential is given by $V_n(x)$.

In Model B the location x_n of each potential is a random
variable, and the probability distribution $f(x_i - <x_i>)$ of the
positions of all potentials around their mean positions $<x_i>$ is
the same.   This model is representative of thermal disorder,
with long range order preserved.   The average interpotential
distance is ℓ; hence,

$$<x_n> = n\ell = nLN^{-1} . \tag{2}$$

In Model C the distance between adjacent potentials is a
random variable and the probability distribution $f(x_i - x_{i-1})$ of all
position differences is the same.   (See Fig. 3.)
This could be representative of an amorphous material.

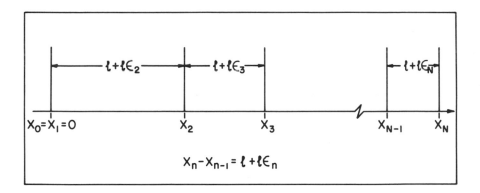

Fig. 3.   Model C with the interpotential distances being the
spatial random variables.   The potentials are located at
$x_1, \ldots, x_N$.   The average interpotential distance is ℓ.   The
deviation of the distance between the nth and (n-1)st potential from its average is $\ell\varepsilon_n$.   The first potential is always
at the origin $(\varepsilon_1 = -1)$.

Cases with short-range order, but no long-range order (C) or neither (A) or both (B) type of order are included in the three models.

In all cases considered, a monochromatic electron beam of wave number k and of unit intensity impinges (from the left) on the first of N potentials on a line segment. To study the reflected and transmitted beams, the transmission matrix $\mathcal{M}^{(N)}$ for the array of N potentials is defined through

$$\mathcal{M}^{(N)} = \prod_{n=1}^{N} M_n = \mathcal{P} \prod_{n=1}^{N} U_n^+ R_n U_n. \qquad (3)$$

The matrix $M_n = U_n^+ R_n U_n$ connects the coefficients $a_{n-1}^{(\alpha)}$ and $a_n^{(\alpha)}$ ($\alpha=1,2$) of the wave function (exp(ikx),exp(-ikx)) on the two sides of the potential at x_n where

$$U_n(1,1) = U_n^+(2,2) = \exp(ikx_n),$$
$$\qquad (4)$$
$$U_n(1,2) = U_n(2,1) = 0,$$

and R_n depends on g_n and k but not on x_n. The elements of R_n are subject to the relations

$$R_n(1,1) = R_n^*(2,2); \quad R_n(1,2) = R_n^*(2,1), \qquad (5)$$

$$\det R_n = 1,$$

and $\mathcal{P}$ orders the matrices with increasing n from right to left. The elements of R_n may be related to the strength parameter of the potential V_n through

$$R_n(1,1) = 1 - ig_n(2k)^{-1}, \quad R_n(1,2) = -ig_n(2k)^{-1}. \qquad (6)$$

The transmission properties of the set of N potentials can now be represented by this two-by-two complex matrix $\mathcal{M}^{(N)}$ given by Eq. (3) which expresses the wave function to the right of the Nth potential in terms of the wave function to the left of the first potential such that

$$a_N = \mathcal{M}^{(N)} a_0. \qquad (7)$$

For an incident beam of unit intensity $a_0^{(1)}=1$, and for the case of no beam entering through the last potential $a_N^{(2)}=0$, it follows then from the properties of $m^{(N)}$, (which are the same as those of R_n, Eq. (5)) that the ratio of the reflected-to-transmitted beam intensities, S_N, is given by

$$S_N=|a_0^{(2)}/a_N^{(1)}|^2=|m_{12}^{(N)}|^2 . \qquad (8)$$

Certain transmission properties of the system are calculated by ensemble averaging. This process denoted by $<...>$, consists of integrating the product of the bracketed quantity with the joint probability distribution of the position and strength parameters of the N potentials. The ratio S_N of reflected to transmitted beam intensity can be averaged exactly for distributions which do not overlap. It should be noted from Eq. (3) that an ordering problem arises for distribution functions $f(x_i)$ which do overlap.

For brevity, we will present here only the results for Model C in which we have used Gaussian distributions of the inter-potential spacings with small standard deviation ($\sigma<0.3$). Calculations (not presented here) with non-overlapping distributions indicate that for small standard deviations the overlap is, indeed, not important for the results shown in this paper.

It should be noted that in the averages of S_N the square of $m^{(N)}$ occurs. Since each random variable occurs in more than one matrix, and the matrices M_n do not commute for different n, one must group together those matrix elements which depend on the same random variable for the averaging. This is accomplished by the introduction of 4 x 4 matrices for S_N acting in the product space spanned by the basis $a_n^\alpha a_n^\beta$, ($\alpha,\beta=1,2$). By group-theoretical arguments, these matrices may be block-diagonalized with respect to the invariant subspaces of the product spaces. The quantity $<S_N>$ may then be expressed in terms of the characteristic roots E_i of the secular equations of the irreducible blocks.

An expression is found which gives the average reflection-to-transmission ratio $<S_N>$ as a function of the number of potentials, which is the sum of exponentially increasing terms and an oscillatory term. The latter is absent for certain ranges of the wave number k. These ranges, in which the oscillatory term is absent, are denoted as "overdamped" regions and coincide, for zero disorder, with the well-known forbidden regions of the Kronig-Penney model.[8] Explicitly, $<S_N>$ is a sum of three real exponentials E_i^N (i=1,2,3) for the overdamped wave-number regions.

The ranges of wave number for which the expression $<S_N>$ contains an oscillatory term as well as an exponentially increasing term are denoted as "underdamped" regions. In the underdamped regions, one has a pair of complex conjugate roots E_1, E_2 and a real root $E_3 > 1$, and in this case

$$<S_N> = -\frac{1}{2} + aE_3^N + b|E_1|^N \cos(N\phi + \psi). \qquad (9)$$

The parameters a, b, ϕ and ψ which appear in Eq. (9) are all real. These parameters, and the roots E_i ($i=1,2,3$) depend on $k\ell$, $<g>\ell$, $<(g-<g>)^2>\ell^2$ and $\chi(2k\ell)$ where χ is the Fourier transform of the spatial distribution function of potentials. It should be noted that in the expressions for $<S_N>$, N appears only as an exponent of the roots E_i ($1=1,2,3$), and in the underdamped regions, in the argument of the cosine provided that ℓ is kept constant when varying N.

III. DISCUSSION OF THE RESULTS

In this section we discuss a few representative results for Model C. In Fig. 4 we show $<S_N>$ as a function of the number of potentials N. This figure is for two values of the spatial disorder σ in an underdamped band. There is no strength disorder of the potentials and for this case the equation which determines the three roots which appear in $<S_N>$ (see Eq. (9)) simplifies to

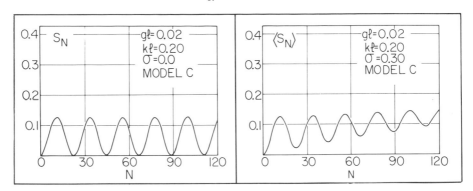

Fig. 4. Average reflection-to-transmission ratio $<S_N>$ as a function of the number N of potentials for an average interpotential distance ℓ, and an incident electron beam of wave number k in an underdamped region. The strength parameter for the δ-function potentials is g. The model of Gaussian random interpotential distances of standard deviation σ is used and there is no strength disorder. The curves are defined only for integral N.

$$E^3 - E^2[|\chi|\beta + (1-|\chi|)(1+\tfrac{1}{2}g^2k^{-2})]$$

$$+E[|\chi|\beta - |\chi|(1-|\chi|)(1+\tfrac{1}{2}g^2k^{-2})] - |\chi|^2 = 0 . \qquad (10)$$

Here,

$$\beta = 4[\cos(k\ell) + \tfrac{1}{2}gk^{-1}\sin(k\ell)]^2 - 1 , \qquad (11)$$

and χ is the Fourier transform of the Gaussian distribution function for the interpotential distances given by

$$\chi = \chi(2k\ell) = \exp[-2(k\ell\sigma)^2] . \qquad (12)$$

In the left figure $\sigma=0$; therefore, $\chi=1$ and S_N varies between zero and a maximum. The periodicity p of this oscillation is determined from the preceding equations and is given by

$$p = 2\pi\{\arctan[(3+2\beta-\beta^2)^{\frac{1}{2}}(\beta-1)^{-1}]\}^{-1} . \qquad (13)$$

For the curve in which $\sigma=0.3$, we see the effect of $\chi\neq1$. Since k is in an underdamped region, the real root in Eq. (9) is greater than one; therefore, we see very clearly the superposition of the sinusoidal effect on the exponentially increasing term.

Fig. 5 shows an example of the overdamped and underdamped regions in the $(k\ell,\sigma)$-diagram as obtained from Eqs. (10), (11) and (12) for Model C for no strength disorder. For $\sigma=0$, the forbidden and allowed bands of the usual Kronig-Penney model appear. For $\sigma>0$, however, note the appearance of additional overdamped regions between the ones obtained for the ordered lattice. These new regions, for $\sigma=0$, degenerate to points; however, since the solutions of Eq. (10) are the roots of unity, no damping (attenuation) occurs. It should be noted that the first band in this figure is an overdamped one. It is easy to see why this should be so by considering the case when $\sigma=0$. In this event one root of Eq. (10) is unity and the other roots are given by

$$E_{1,2} = \tfrac{1}{2}(\beta-1)\pm[(1-\beta)^2-4]^{\frac{1}{2}} . \qquad (14)$$

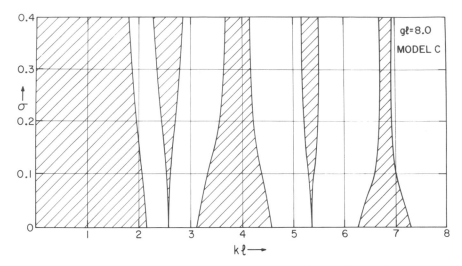

Fig. 5.   Dependence of underdamped and overdamped (cross-hatched) regions on $k\ell$ and σ. The electron wave number is k, the average interpotential distance is ℓ and the standard deviation of the Gaussian distribution of interpotential distances is σ. The strength parameter for the δ-function potentials is g.

It is clear from Eqs. (11) and (13) that $E_{1,2}$ are complex roots of unity as long as

$$0<\left|\cos(k\ell)+\frac{1}{2}\,gk^{-1}\sin(k\ell)\right|<1. \tag{15}$$

This, of course, is just the well-known Kronig–Penney condition[8] for the allowed bands of an electron in a one-dimensional array of δ-function potentials. Eq. (15) shows, that for small $k\ell$, the first region will always be forbidden for positive g (potential barriers). For negative g (potential wells) and

$$|g|\ell<4 \ , \tag{16}$$

the first region will be allowed,[9] while for $|g|\ell>4$ the first region again becomes forbidden. For $\sigma>0$ this feature persists, although the edge of the first region moves toward smaller values of $k\ell$ as the disorder increases.

To study the attenuation of the electron beam through the lattice, we plot in Fig. 6 a quantity y, called the transmittivity, defined as

$$y= (1+<S_N>)^{-1} .$$

(17)

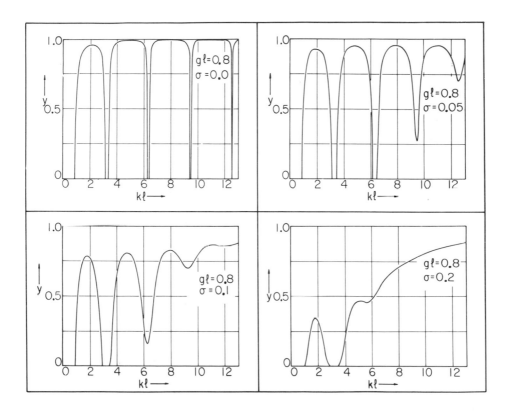

Fig. 6.   Transmittivity $y = (1+<S_N>)^{-1}$ vs. $k\ell$ for an array of N = 120 potentials for increasing spatial disorder in Model C. The standard deviation of the Gaussian distribution of interpotential distances is σ, the average interpotential distance is ℓ, and the strength parameter for the δ-function potentials is g. The electron wave number is k. The fine structure of the curves has been omitted.

This quantity can vary between 0 for complete reflection and unity
for complete transmission. When the transmitted and reflected
intensities are equal, y=0.5. The curves show y as a function of
the electron wave number for 120 potentials and $g\ell$=0.8 and no
strength disorder. Not shown is a fine structure which appears
at the shoulders of the maxima. The effect of increasing disorder can be clearly seen from this figure. The bands for small k
are least affected by the disorder since the electron wavelength
is large compared to the width of the distribution of potential
positions and cannot detect a degree of disorder whose scale is
much less than its wavelength. The regions where y appears to be
zero do not represent exactly zero transmission, since a finite
number of potentials is always slightly transparent. Conversely
the transmission is not complete in the high transmission regions
which is also due to the finite length of the array.

In order to illustrate the fine structure of y($k\ell$), we have
included Fig. 7 which is a computer plot of the transmittivity
vs. $k\ell$. This figure is for the case of an ordered array of
potentials which shows the fine structure most emphatically.
It arises from the interference of the waves scattered from the
ends of the array. As the disorder increases, the fine structure
diminishes more quickly in the higher allowed regions than in the
lower.

We have included in Fig. 8 an example of what happens to
the transmittivity y as a function of $k\ell$ in an array in which
there is no spatial disorder, but only a strength disorder of
the δ-function potentials.

The transmittivity of the perfect lattice is shown on the
left for positive and negative δ-function potentials, in the top
and bottom diagrams, respectively. These two diagrams show, that
for positive potentials of any strength there is a forbidden band
at k→0. For negative potentials, and for the value of $g\ell$=-0.8,
(see Eq. (16)) the band at k→0 is allowed. As a strength disorder is introduced (diagrams on the right), the transmittivity
in the underdamped regions is reduced. The least reduction
occurs at the top of the underdamped band for $<g>\ell>0$, and at
the bottom of the underdamped band for $<g>\ell<0$. This effect is
peculiar to arrays of δ-function potentials and would, in general,
not occur for other types of potentials. It is due to the fact,
that for $<g>\ell>0$ the top of the allowed band for an array without
disorder always occurs at multiples of $k\ell$=π, independent of the
strengths of the potentials. Therefore, the disorder can only
smear out the bottom edges of allowed bands, but not their top
edges. The reverse holds for $<g>\ell<0$, with the exception of the
first allowed band, which derives from the bound state of the
δ-function potentials.[9]

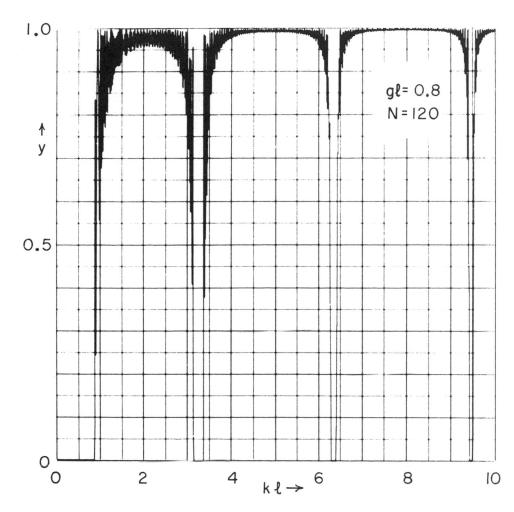

Fig. 7.   Computer plot of the transmittivity $y = (1+<s_N>)^{-1}$
vs. $k\ell$ for an array of $N=120$ equally spaced δ-function potentials of strengths g (Kronig-Penney model of finite length).
In the overdamped (forbidden) regions the transmittivity in-
creases with $k\ell$. At $k\ell \simeq 9.5$ it has reached a value $y = 3\cdot10^{-4}$.
The fine structure at the transmittivity shoulder is due to
the interference of the waves scattered from the two ends of
the array.

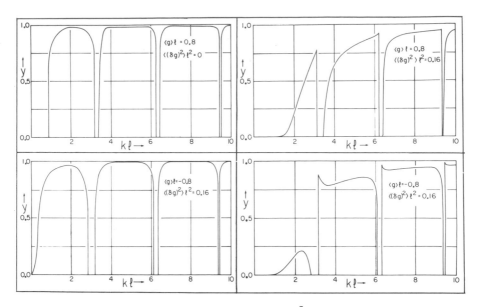

Fig. 8.   Transmittivity y = $(1+\langle S_N \rangle)^{-1}$ vs. kℓ for an array of
N=120 equidistant potentials for increasing strength disorder
$\langle (\delta g)^2 \rangle$. There is no spatial disorder (σ=0). The average
strength of the δ-function potentials is $\langle g \rangle$, and the interpotential
distance is ℓ, and the electron wave number is k. The two top
curves are for potential barriers of increasing strength disorder and the two bottom curves are for potential wells of
increasing strength disorder. We define g=$\langle g \rangle$+δg.

        An interesting feature of the array of δ-function potentials
is, that the transmittivity depends only on the first two moments,
$\langle g \rangle \ell$ and $\langle (\delta g)^2 \rangle \ell^2$ of the probability distribution of the strengths.
The reason for this is, that the elements of the transfer matrix
(Eq. (6)) depend on the product of a function of the strength of
the potential with a function of the electron wave number k. For
general potentials, such a separation into a product is not
possible, and the transmittivity will depend on all moments of the
probability distribution functions of the shape parameters of the
potential.

        In Fig. 9 we present a semilogarithimic plot of the results
for Model C for $\langle S_N \rangle$ vs. N in an underdamped region and σ=0.1
(open circles) and no strength disorder. Note that N increases
up to 1000 potentials in this case and that for large N the
behavior of $\langle S_N \rangle$ is essentially exponential. This is a result of
Eq. (9), in which for large N the predominant term is the

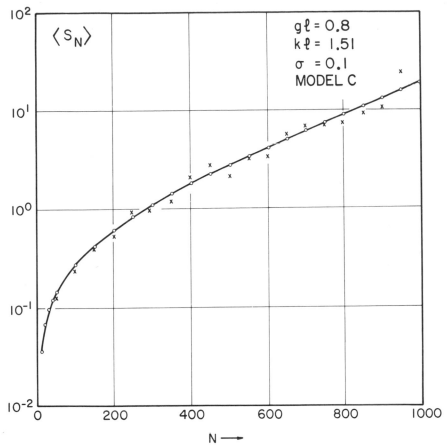

Fig. 9. Semilogarithmic plot of the average reflection-to-transmission ratio $\langle S_N \rangle$ vs. the number N of potentials in an underdamped region of the wave number k. The circles are analytic results, and the crosses are the results of a numerical Monte Carlo calculation. The standard deviation of the Gaussian distribution of interpotential distances is σ, the average interpotential distance is ℓ, and the strength parameter for the δ-function potentials is g.

exponentially increasing one, and the modulation of this exponential behavior by the oscillatory term is completely obscured except when the number of potentials is small. It is also possible to calculate $\langle S_N \rangle$ using a completely different technique of a Monte-Carlo procedure (Fig. 9, crosses). This method consists of the numerical evaluation of the average of Eq. (8), using random numbers appropriate to the distribution for the independent random

variables which occur in the particular model considered. It
does not afford the physical insight which the analytical
calculation provides, and requires a considerable amount of
computing time to get the statistical uncertainty small
enough to give useful results. Nonetheless, it does permit a
check on the analytical method. In Fig. 9, 85% of the
numerically computed points lie within one statistical uncertainty of the analytical curve.

REFERENCES

*This research was sponsored by the Air Force Office of Scienti-
fic Research, Office of Aerospace Research, United States Air
Force, under grant numbers AFOSR-69-1745/70-1940.

1. See overview by D. Adler, Critical Reviews in Solid State
    Sciences, 2, 317 (1971).
2. P. Erdös and E. Rhodes, Helv. Phys. Acta 41, 785 (1968).
3. N. F. Mott and E. A. Davis, Electronic Processes in Non-
    Crystalline Materials, Clarendon Press, Oxford, 1971.
4. B. Velický, Phys. Rev. 184, 614 (1969), and references quoted
    in his paper. See also H. J. Fischbeck, Phys. Stat. Sol. (b)
    49, 829 (1972).
5. E. N. Economou and M. H. Cohen, Phys. Rev. B5, 2931 (1972),
    and references quoted in their paper.
6. R. Landauer, I. B. M. J. Res. Dev. 1, 223 (1958); Phil. Mag.
    21, 863 (1970), and R. Landauer and W. F. Woo, Phys. Rev. B5,
    1189 (1972).
7. I. D. Mikhailov and A. P. Kolesnikov, Soviet Phys. Solid State
    13, 791 (1971).
8. R. de L. Kronig and W. G. Penney, Proc. Roy. Soc. (London)
    130, 499 (1931).
9. R. A. Smith, Wave Mechanics of Crystalline Solids, 2nd Ed.,
    Chapman and Hall, London, 1969.

6. Localized States
and Disordered Solids II

INTRODUCTORY REMARKS

Douglas Henderson

IBM Research Laboratory

San Jose, California 95114

In the past, the development of solid state physics has been based very strongly, perhaps excessively so, on the concept of the ideal crystal and the resulting periodicity and symmetry. One of the key areas for future development is the study of amorphous or disordered solids. This study is of interest not only because of what it will tell us about amorphous solids but also because of what it will tell us about the basic concepts of the solid state physics of crystals. For example, until the recent work of Weaire and his collaborators, which is ably summarized by Weaire and Thorpe, most of us believed that the existence of a gap was a result of the periodicity and symmetry of the lattice structure. Weaire has shown that, at least for his simple Hamiltonian, the gap results from the short range order of the atoms. Bloch's theorem is sufficient but not necessary! The study of amorphous solids is of interest not only because of what we can learn. In addition to these friendly intellectual pursuits we have the exciting prospects for technological advances. All in all I feel that the study of amorphous solids is one of the exciting frontiers of physics. Thus, it is highly appropriate that this should be the concluding session of this conference. The dessert should be served last.

The rapid growth of an area of physics usually follows from the development of a canonical model. For example, most work on phase transitions has been based on the Ising model or on one of its refinements. One canonical model which has been useful in the theory of amorphous solids is the Anderson model of a compositionally disordered solid. But what of topological disorder where there is no lattice? There now seems to be considerable evidence that the random tetrahedral network (RTN) may be useful as a

293

canonical model. In this model each atom is bonded to four
neighbors. It appears to be possible to build RTN's which are as
large as one wants without dangling bonds and without large distortions in the bond lengths or angles. Thus, the RTN seems an
appropriate canonical model for such systems as amorphous
germanium or silicon.

However, there are some questions to be resolved before one
can be sure that this is the case. Those RTN's which have been
constructed so far seem to be more dense than the densest known
samples of amorphous germanium or silicon. Also, the RTN is not a
well-defined system. Several RTN's have been built which differ
somewhat in such structural features as the number of 5- and 6-atom
rings which they contain. The relationship of these structural
features to the electronic and optical properties of these networks needs to be more fully explored.

In their paper Weaire and Thorpe review their work. In particular, they show that, within the limitations of their Hamiltonian,
a gap can exist in the energy spectrum even in the absence of a
crystal lattice. In addition, they draw attention to the Si 3 and
Ge 3 structures. These structures share many of the features of
the RTN. They have distorted tetrahedral bonds and, in the case
of Ge 3, 5-atom rings. These structures, especially the Ge 3
structure, are promising simplified models of amorphous Si and Ge.

Weaire and Thorpe suggest the energy gap is larger for systems
with 5-atom rings than for those with 6-atom rings. However,
Henderson and Ortenburger in a comment at the symposium which has
been expanded into a paper show that for a slightly more elaborate
LCAO model, the gap in Ge 3 is smaller than in cubic Ge at the density of Ge 3. This suggests that the role of 5-atom rings, although important, may not be as clear as originally supposed.

Keller discusses the cluster scattering techniques which have
been applied by him and his colleagues at Bristol to amorphous
solids. For small clusters they find a minimum in the energy which
becomes more pronounced as the size of the cluster is increased.
If periodic boundary conditions are applied, the region of low
density of states is transformed into a perfect gap.

Kirkpatrick and Eggarter examine a simple model of a substitutional alloy using the LCAO approximation and find a new class of
localized electronic states, not observed before, in the middle of
the band. They show that this surprising result is in accord with
our prejudice that localized and extended states do not coexist.

Freed examines the transition between localized and extended
electronic states and finds it can be characterized by exponents in
analogy with the critical exponents which characterize phase transitions.

ELECTRONIC STRUCTURE OF AMORPHOUS SEMICONDUCTORS

D. Weaire and M. F. Thorpe

Becton Center, Yale University

New Haven, Conn. 06520

I.   INTRODUCTION

Until very recently, amorphous solids were a good deal more
common in nature than in solid state theory textbooks. This can
be only partially excused on the grounds that they are embarrass-
ingly difficult substances to analyse theoretically in any fundamental way. The lack of periodicity (and hence Bloch's Theorem)
presents severe obstacles to the formulation of a theory that is
at once fundamental, rigorous, realistic and tractable. One must
always hope for the last of these qualities but the other three
can hardly be attained simultaneously, so the problem must necessarily be investigated from different (complementary rather than
competing) points of view, each of which has its own particular
merits. We shall review these briefly in the next section, before
embarking on an exposition of our own recent efforts. However it
is first necessary to say something about structure. This obvious
first step in the description of a solid is, in the case of amorphous solids, still a rather hesitant and uncertain one. While
many of the remarks which we shall make are of wide validity, it is
appropriate at this point to define our field of immediate interest
as being confined to the very simplest amorphous semiconductors,
namely Si, Ge and related compounds.

Let us begin by explaining what we mean by amorphous. The
word is perhaps best defined in a negative way. By an amorphous
solid, we mean one which is not crystalline; nor may it reasonably
be regarded as made up of microcrystallites, that is, small regions
of perfect crystalline order separated by disordered boundaries.
In the past, experimentalists have often not bothered to worry
about the distinction between an amorphous and a fine polycrystal-

line structure, and amorphous samples have been called polycrystalline and vice-versa. However, today the distinction is regarded
as an important one, and in the case of non-crystalline films of
Si and Ge, the weight of evidence has fallen on the side of their
being genuinely amorphous in many cases, although they can, of
course, be made in polycrystalline form as well. In the case of
these amorphous semiconductors and amorphous covalently bonded
systems in general there is strong <u>short</u> range order because of
the local bonding requirements of the atoms, in this case the tetrahedral bond. Thus we have short range order in the form of tetrahedral bonding, but no long range order. If we assume that there
are no dangling bonds, we have the <u>random</u> <u>network</u> model for such a
system. Polk (1) has shown that this is quite a satisfactory
model for Si and Ge as far as fitting the observed radial distribution function is concerned. That is not to say that the alternative microcrystallite idea is quite dead--it appears to be alive
and well and living in Cambridge (2).

The striking thing about the random network model is the pres-
ence of <u>five</u> <u>fold</u> <u>rings</u> <u>of</u> <u>bonds</u>. These are absent in the diamond
cubic structure, which has only <u>six</u>-fold rings. If we accept this
model, our problems are still not over, as far as the definition
of the structure is concerned. Although it is often tacitly assumed that this random network is unique, there seems no reason why
it should not be possible to build different networks with different statistical properties and if Nature indeed builds such networks, why she should not do so. (Remember that these are only
metastable phases.) Even if we confine our attention to the <u>crys-
talline</u> phases of Si and Ge, it turns out that there are at least
four different tetrahedrally bonded structures observed, namely,
the usual diamond cubic structure, the wurtzite structure (3), the
Si III structure (4) and the Ge III structure (4). The polymor-
phism of other tetrahedrally bonded systems, such as SiO_2 and H_2O
is perhaps better known.

Figs. 1 and 2 illustrate the ring statistics of the above
crystal structures and Polk's random network model. The full lines
indicate what appears to be bounds on the number of five and sixfold rings which can pass through a given site without unreasonable
distortion of the bonds. Given that such widely differing <u>crystal</u>
structures are possible, it would appear that Polk's construction
is far from unique. Even if the random network model is accepted,
the question then remains--what kind of random network? The de-
bate over the structure of such amorphous systems is therefore far
from over. In what follows we implicitly accept the random network
model, but our studies have not yet progressed to a sufficient level
of detail to be much concerned with the finer details of the structure.

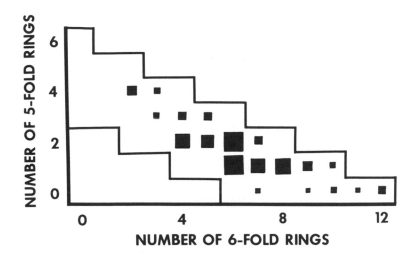

Fig. 1   The size of each square represents the proportion of atoms
having a particular number of five- and sixfold rings passing
through it, in the model constructed by D. Polk (1).

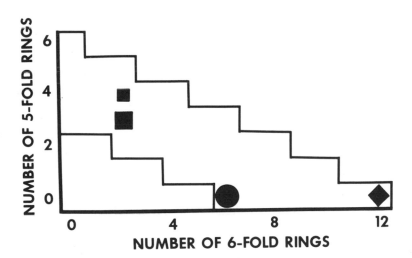

Fig. 2   A similar plot to that of Fig. 1, for the diamond [◆],
Si III [●] and Ge III [■] structures.

II.   A REVIEW OF RECENT THEORETICAL WORK

There have, as we remarked in the introduction, been quite a
few different approaches to the problem of the electronic properties of amorphous semiconductors. This is not merely because of
the predilection of theorists to attack a given problem from every
conceivable angle, but rather stems from the difficulty of the
problem. Some aspect or other must be satisfied in order to make
progress. In some cases all contact with reality may have been
sacrificed, which would be regrettable, but it is not yet easy to
judge.

At the outset there is a dichotomy between those who wish to
use as realistic a Hamiltonian as possible (e.g. a pseudopotential
Hamiltonian) and those who use less realistic model Hamiltonians.
Those in the former category (5-9) must indulge in questionable
approximations at the point at which the structure enters the theory. There are two obvious choices. One may take the diamond
cubic structure and try to incorporate the effects of disorder in
some ad hoc way (5-7), or one can treat finite clusters of N atoms
(8,9). The cluster approach is perhaps less arbitrary, but has
the disadvantage of uncertain boundary effects, although the recent calculations of Keller (9) have extended the method to sufficiently large clusters (N $\sim$ 50) that the neglect of boundary effects becomes more than just a pious hope.

The various studies which have used model Hamiltonians have
generally been based on a tight-binding or LCAO formalism. (A recent exception is Hulin's proposal of a model (10), similar to the
"network model" of solids and molecules, in which the electrons
are treated as if they moved in pipes along the bonds). In a

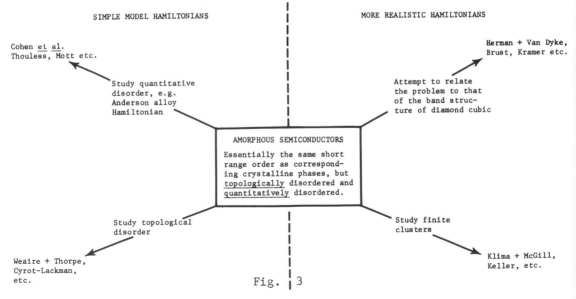

Fig. 3

tight binding theory there is a natural distinction between quantitative and topological disorder. The former refers to the random
variation of the matrix elements in the Hamiltonian from one atom
to the next. However if we neglect interactions beyond nearest
neighbors and bear in mind that there is strong short range order
(bond lengths and angles are fairly constant), we might choose to
neglect this variation. In that case we are left with only topological disorder, meaning that the topological network defined by
the structure is non-periodic. The secular matrix would have only
a few different values recurring throughout it but they crop up in
unpredictable places. The study of topological disorder has been
our own particular speciality. On the other hand one may study
quantitative disorder in isolation, by setting up a Hamiltonian
whose matrix elements vary but which is defined on a periodic
structure (11). This is, of course, the natural description of an
alloy and such studies therefore have a history which predates the
current interest in amorphous semiconductors.

Such are the various approaches from which one would hope
eventually to synthesise a decent understanding of amorphous Si
and Ge. We shall attempt no such synthesis here, for lack of
space and temerity, but these remarks may serve to place the later
sections properly in context.

III. THE HAMILTONIAN

The natural choice for a simple Hamiltonian appropriate to Si
or Ge is that in which we allow four basis functions per atom
("sp^3 orbitals"), one for each bond, and include in the Hamiltonian only matrix elements V_1 between basis functions associated
with the same atom and V_2 between basis functions associated with
the same bond. Formally we write

$$H = V_1 \sum_{i,j,j'} |\phi_{ij}><\phi_{ij'}| + V_2 \sum_{i \neq i',j} |\phi_{ij}><\phi_{i'j}| \qquad (1)$$

where subscripts i and j label respectively all the atoms and bonds
of the system. (The basis functions are taken to be orthogonal.
The summation in the first term includes $j = j'$, which turns out
to be mathematically convenient).

Clearly the V_2 term tends to split the band structure into
bonding and antibonding bands, and the V_1 term tends to split it
into s-like and p-like bands.

Some of the consequences of such a Hamiltonian for the case
of the diamond cubic structure were discussed a long time ago, by
Leman and Friedel (12) especially. We now know that it is not very
realistic, although it does give quite a good qualitative descrip-

tion of the valence band of Si and Ge. However we shall postpone
the discussion of the spectrum produced by (1) for the moment, in
order to present the "one band--two band transformation" which reduces the problem of the determination of the spectrum of (1), for
an arbitrary structure, to a somewhat simpler one.

IV. THE ONE BAND–TWO BAND TRANSFORMATION

It is the object of this section to show that the eigenvectors and eigenvalues of the Hamiltonian (1) are closely related to
those of a much simpler Hamiltonian. In this we associate only
one basis function ϕ_i with each atom and a single interaction V
between basis functions associated with neighbors

$$H^{(1)} = V \sum_{\substack{i,i' \\ \text{neighbors}}} |\phi_i> <\phi_{i'}| \tag{2}$$

It is perhaps obvious that there ought to be some relation between
the properties of $H^{(1)}$ and H since both depend critically on the
same topological properties of the structure. That there is an
analytic transformation relating the two spectra is not so obvious.
Having noticed that it was so for the case of diamond cubic, we
developed a rather complicated proof of this, based on the use of
diagrammatic expansions and Green's functions (13). That proof
relates only to the density of states n(E), but we will show below
that there is a close relationship between the individual eigenfunctions. Specifically, if H has an eigenfunction

$$|\psi> = \sum_{ij} a_{ij} |\phi_{ij}> \tag{3}$$

corresponding to energy E, then $H^{(1)}$ has an eigenfunction $|\psi^{(1)}>$
defined by

$$|\psi^{(1)}> = \sum_{i} b_i |\phi_i> \tag{4}$$

where

$$b_i = \sum_{j} a_{ij} \tag{5}$$

at an energy ε

$$E = 2V_1 \pm \sqrt{4V_1^2 + V_2^2 + V_1 V_2 \varepsilon/V} \tag{6}$$

The coefficient associated with $\psi^{(1)}$ at a given site is just the
sum of the four coefficients associated with that site, in ψ.

We focus attention on a given atom i and simplify the notation

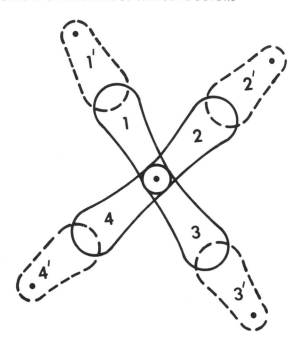

Fig. 4   The coefficients of the orbitals 1-4 are here called u_j, while those of 1'-4' are called v_j.

by replacing a_{ij} by u_j. The coefficients $a_{i'j}$ associated with the other orbitals on the four bonds j are now called v_j (See Fig. 4).   For convenience, let j take values 1 to 4 for the four bonds and we will further use it to define b_j to be the sum of coefficients over all the orbitals associated with the corresponding neighbor, and $b_o = \sum_j u_j$, the same sum for the central atom.   Then the secular equation immediately gives

$$V_1 b_o + V_2 v_j = E\ u_j \qquad (7)$$

and

$$V_1 b_j + V_2 u_j = E\ v_j \qquad (8)$$

and the elimination of v_j, together with summation over j immediately yields

$$4V_1 b_o + \frac{V_2}{E} (V_1 \sum_j b_j + V_2 b_o) = E\ b_o \qquad (9)$$

or

$$(E^2 - 4V_1E - V_2^2)b_o = V_1V_2 \sum_j b_j \tag{10}$$

We now identify this as simply

$$V \sum_j b_j = \epsilon b_o \tag{11}$$

where the correspondence between ϵ and E is as given in Eq. (6).
Equation (11) is, of course, just the secular equation associated
with $H^{(1)}$ for the given site and we have proved the required relationship. What this kind of analysis adds to the previous proof
(13) is the statement that, in the case of a periodic structure,
there is a 1-1 correspondence between the states in any band generated by H (for which the quantities b_j do not simply vanish
everywhere) and eigenstates of $H^{(1)}$, with the same k-vector. The
parenthetical proviso is an important one--it turns out that half
of the eigenfunctions of H are in flat, purely p-like bands (i.e.
delta functions in the density of states) for which b_j does indeed
vanish while the other half map into corresponding bands of $H^{(1)}$
in the manner described. Straley (14) and Hulin (15) have also
advanced alternative derivations of the transformation.

Armed with this theorem, we need not study the properties of
H directly for any given structure, but can rather sidestep a good
deal of complexity by studying $H^{(1)}$, and transforming the results.
This statement applies even to more complicated generalisations of
H, as discussed in §§ VI, VII and IX.

We might note in passing that for even-ringed structures, a
further transformation is possible. Such structures are divisible
into two interpenetrating sublattices. If we consider $(H^{(1)})^2$ the
two sublattices are decoupled and the coefficients of an eigenfunction of $H^{(1)}$, of energy ϵ, which are associated with one sublattice
give an eigenstate of $(H^{(1)})^2$ of energy ϵ^2. The resultant analytic
transformation, for the case of the diamond cubic structure, relates the density of states of $H^{(1)}$ to that of the same kind of
Hamiltonian for fcc. We used this relationship in our original
work (16), without realising the generality of the relationship.
It is amusing that, confronted with a problem involving four states/
atom on a diamond cubic lattice, one can reduce it to one which
involves only one state/atom on a fcc lattice, by analytic transformations! (It is also very convenient since the latter problem
has been the subject of careful numerical work). Trickery of this
kind is the justification for the use of such a crude Hamiltonian.

V. SOME PROPERTIES OF THE HAMILTONIAN

We have shown that for many purposes, including the determination of the density of states n(E), it is sufficient to consider

the properties of the simpler Hamiltonian $H^{(1)}$. The spectrum of
this Hamiltonian for the special case of the diamond cubic lattice
is shown in Fig. 5. The spectrum extends from -4 to $+4$ if we set
$V = 1$. In fact, it follows from a celebrated theorem of Perron
that these values are bounds on the spectrum whatever the structure. When this statement is transformed, by use of Eq. (6), into a statement about H, the corresponding part of its spectrum
has two allowed ranges, with a gap between them, regardless of
the structure (17). When the delta functions are added, the resulting spectrum for the diamond cubic case is shown in Fig. 6.
All the band edges in the figure are bounds on the spectrum for
any structure. There are two cases depending on $\left| V_1/V_2 \right| \lessgtr \frac{1}{2}$. The
upper sign places the gap between "valence" and "conduction" bands
with four electrons/atom each (counting spin), giving the kind of
band structure which is characteristic of Si and Ge. With the
lower sign, the delta function (flat band) is shifted to the other
side of the gap and we get, for the diamond cubic structure, the
kind of semimetallic "inverted" band structure found in grey Sn.
We will return to this interesting case later, but in what follows
immediately we assume $\left| V_1/V_2 \right| < \frac{1}{2}$. We have shown that the gap be-
tween valence and conduction bands must be at least $\left| 2 \left| V_2 \right| - 4 \left| V_1 \right| \right|$
for any structure. There are a number of other structure inde-
pendent properties of H. One which we have already mentioned is
the existence of the two delta functions in the spectrum. In addition, the average s-like (p-like) and bonding (antibonding)
character of a wave function anywhere in the allowed bands is completely specified at a given energy. At the four bounds this re-

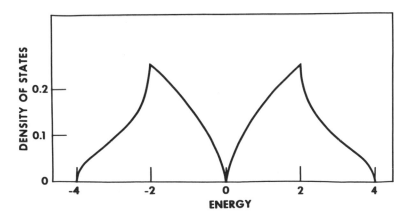

Fig. 5 Spectrum of the Hamiltonian $H^{(1)}$, for the diamond cubic
structure (as a function of E/V).

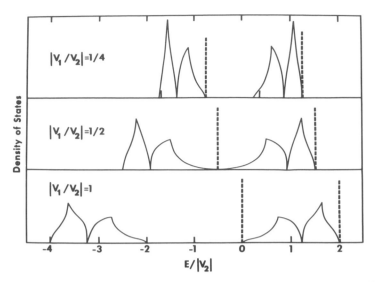

Fig. 6   Spectrum of H, for the diamond cubic structure, illus-
trating the two cases $\left|V_1/V_2\right| \lessgtr \frac{1}{2}$.

sult demands pure s-like or p-like and pure bonding or antibonding
functions, as the case may be. This brings us to the question of
the band edges, a subject dear to the hearts of experimentalists
and theorists alike in the field of amorphous semiconductors. The
edges of the continuous bands in Fig. 6 are of two sorts associated respectively with the top or bottom of the spectrum in Fig. 5.
The limiting state at the top edge is the Γ_1 state in the diamond
cubic structure.   This is defined by setting $b_i = +1$ for all i
in Eq. (4) and it is obvious that this can be defined, and is an
eigenfunction of $H^{(1)}$, for _any_ structure.   One suspects, therefore,
that the edge is attained for any reasonable structure although
one must confess that the mathematical argument is incomplete, as
we have demonstrated only the existence of a single state, giving
a contribution of order 1/N to the density of states.   For _perio-
dic_ structures, of course, there is no problem--indeed k · p theory can be used to derive an interesting formula for the associated square root edge (21).   However, for topologically disordered
structures it is not easy to proceed and, although Cyrot-Lackman
has suggested that there is always a square root edge (22), it remains to be proved.   We believe such a result would depend on a
subtle restriction of "topological homogeneity".

The situation at the lower edge of the spectrum of $H^{(1)}$ is
quite different and, since this corresponds to the bottom of the
conduction band, it is of some importance.   The limiting state has

$b_i = \pm 1$ with differing signs on the two sublattices and clearly
can only be constructed if the structure has such interpenetrating
sublattices, i.e. has only even rings. For structures with odd
rings, two things can happen. If there are enough of such rings,
distributed fairly homogeneously throughout the structure, the band
edge shifts inwards. (A specific example is given in § X). For
instance, one can show that if a lattice is dissectable into five
fold rings of bonds, this band edge is bounded by $-4V \cos \frac{\pi}{5}$ instead
of $-4V$. On the other hand, if one has only a small number of odd
rings, randomly distributed, then the band edge would not be
shifted but the lower part of the band would presumably consist of
a tail of localised states. The situation here is closely analogous to the more familiar tails which result from quantitative disorder, and one might study mobility edges, and so forth, just as
is currently fashionable for quantitative disorder. It would be
even more difficult, however, so it may not be pursued with the
same vigor.

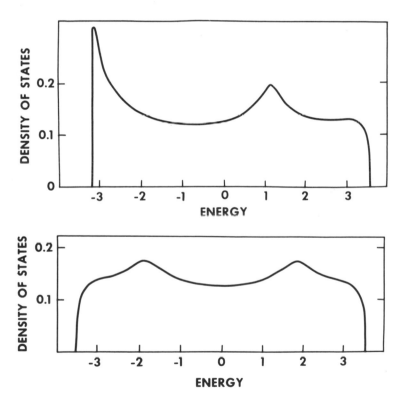

Fig. 7 Spectrum of $H^{(1)}$ for Husumi cactus structures made up of
fivefold rings (a) and sixfold rings (b).

What about the overall shape of the spectrum? Of course, the
singularities in the spectrum shown in Fig. 5 are connected with
the long range order and cannot be present for a topologically
disordered structure. However we are here asking about the broad
features of the spectrum. What, in general terms, gives rise to
its having two peaks with a valley between them? Again, we believe this to be caused by the ring structure. One can solve exactly the case of a fourfold coordinated "Bethe lattice", which is
an infinitely branching tree-like structure, with no closed rings
(13). The result is a flat featureless spectrum. However if we
build up such a branching structure out of rings of $\underline{n}$ bonds (a
"Husumi cactus") then peaks appear in the spectrum at approximately
twice the eigenvalues of a single ring (23). The results for five-
and sixfold rings are shown in Fig. 7. This suggests, although it
does not exactly prove, that the two peaks in Fig. 5 are indeed
attributable to the sixfold rings and that a random structure with
five-, six-, and sevenfold rings would have a washed-out spectrum,
since the different rings tend to enhance the spectrum in different places.

Finally we should mention that the sequence of transformations
mentioned at the end of the last section can be used to show that
all $\underline{polytype}$ structures (diamond, wurtzite....) have the $\underline{same}$ density of states for the Hamiltonian H, since it can be shown that
the simple Hamiltonian given by $(H^{(1)})^2$ generates the same density
of states for fcc, hcp, etc. (24). There have been studies of the
density of states generated by the pseudopotential for different
polytype structures of Si, which are remarkably similar in many
respects (25). While there is as yet no convenient mathematical
bridge between our Hamiltonian and the more realistic pseudopotential Hamiltonian, the above theorem should help in the understanding of the pseudopotential results.

VI.   EFFECT OF MORE DISTANT OVERLAP

The Hamiltonian defined in Eq. (1) contains two parameters V_1
and V_2. There are two other overlap parameters that one might
reasonably expect to be relatively independent of structure. In
Fig. 4 these would be V_3, V_4 the overlap between, for example, the
states marked 1 and 2', 1' and 2'. Only states on adjacent bonds
are involved and so we would not expect much variation in such
quantities if the bond angles do not vary much. All other overlaps
will depend on such considerations as the dihedral angle, making it
unreasonable to neglect quantitative disorder.

The method of §IV may be readily adapted to give the energy
spectrum in terms of the one band energy spectrum. We define b,

as in §IV (i.e. the sum of the amplitudes associated with the
states 1,2,3,4 in Fig. 4, and also c_j which is the sum of the amplitudes associated with 1',2',3',4'. The leads to the following
matrix equation that is a generalization of (10)

$$
\begin{bmatrix} E - 3V_1 + V_4 & -V_2 - 2V_3 \\ -V_2 - 2V_3 & E + V_1 - 3V_4 \end{bmatrix} \begin{bmatrix} b_o \\ c_o \end{bmatrix} = \begin{bmatrix} V_3 & V_4 \\ V_1 & V_3 \end{bmatrix} \begin{bmatrix} \sum_\delta b_\delta \\ \sum_\delta c_\delta \end{bmatrix}
\tag{12}
$$

Taking appropriate linear combinations of b and c to give normal
modes, we find that the energy eigenvalues can be written in terms
of ε

$$
E=2V_1+V_3\varepsilon/V+V_4\pm\sqrt{4(V_1-V_4)^2+(V_2+2V_3)^2+(V_1+V_4)(V_2+2V_3)\varepsilon/V+V_1V_4\varepsilon^2/V^2}
\tag{13}
$$

As before, this leads to two bands and four band edges at the
energies appropriate to the bonding s, antibonding s, bonding p
and antibonding p; the latter two edges having delta functions associated with them as before. The energy gap between the two
bands is

$$
|2|V_2 + 2V_3|-4|V_1 + V_4||
\tag{14}
$$

It is surprising that the eigenvalues of this rather complex
Hamiltonian with four parameters can be related to the eigenvalues
of the simple hopping Hamiltonian defined by Eq. (2), and this re-
inforces our claim that an understanding of this Hamiltonian is
the basic problem for topologically disordered systems.

VII. COMPOUND SEMICONDUCTORS

Amorphous III-V semiconductors add yet another dimension to
the problems posed by their Group IV counterparts. If their structures have odd numbered rings, then there must be pairs of nearest
neighbors of the same species. However we will here consider only
structures with even rings, so that every A atom is bonded to four
B atoms and vice versa. We introduce a diagonal term $\pm\Delta E|\phi_{ij}><\phi_{ij}|$
in the Hamiltonian (1) where the + sign is taken for states
associated with the A sites and the - sign for states associated
with the B sites. Also the term involving V_1 is generalized so
that V_1 becomes V_{1A}, V_{1B} for the A, B sites. The term which includes the parameter V_2 is unaltered. The eigenvalues for this
problem may be expressed in terms of the one band eigenvalues ε by
straightforward generalizations of the Green's function method (13)
or the method of §IV (see also J. Straley (14)). The eigenvalues
E for this problem are obtained from the solution of the quartic
equation.

$$[(E+\Delta E)(E-\Delta E-4V_{1A})-V_2^2][(E-\Delta E)(E+\Delta E-4V_{1B})-V_2^2] = V_{1A}V_{1B}V_2^2\epsilon^2/v^2 \quad (15)$$

The two structure independent delta functions occur at energies

$$\pm\sqrt{(\Delta E)^2 + V_2^2} \quad\quad\quad (16)$$

and the band gap is given by

$$\left|\sqrt{(\Delta E+2(V_{1A}-V_{1B}))^2+V_2^2}+\sqrt{(\Delta E)^2+V_2^2}\ -2|V_{1A}+V_{1B})|\right| \quad (17)$$

which in the case $V_{1A} = V_{1B} = V_1$ reduces to the expression

$$\left|2\sqrt{(\Delta E)^2 + V_2^2}\ -4|V_1|\right| \quad\quad\quad (18)$$

derived previously (13).

In Figs. 8a and 8b we show a comparison of the band structure by GaAs with the zincblende structure as calculated using pseudo-

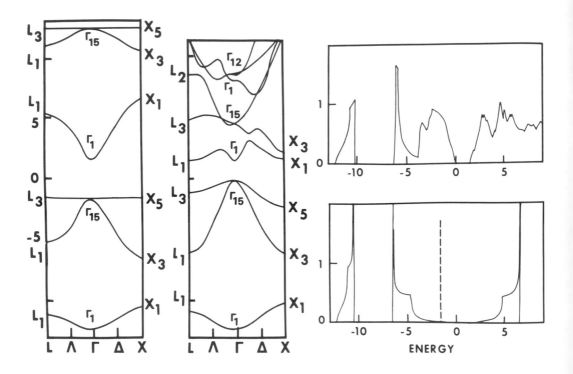

Fig. 8   (a) A comparison of the band structure for GaAs with the zincblende structure as calculated by Herman and Shay (26) and using the tight binding Hamiltonian of this section.   (b) The corresponding densities of states.   Energies are in eV.

potentials by Herman and Shay (26) and the present model with $V_{1A} = V_{1B} = -2.7$ eV, $V_2 = -6.1$ eV and $\Delta E = 3.2$ eV. These parameters have been chosen somewhat arbitrarily and should not be taken too seriously. It can be seen that the reflection symmetry previously noted (16) is still present--this would not be so if $V_{1A} \neq V_{1B}$. Also a gap develops in the middle of the valence band with a logarithmic infinity at both sides. It is intriguing to compare this with the pseudopotential density of states where there is a very sharp drop on either side of this gap.

VIII.   SOME RELEVANT EXPERIMENTS ON Si AND Ge

Our principal purpose has been to outline recent theoretical work but the experimental evidence should not be neglected entirely. The original motivation for this work was the assertion by W. Spicer and collaborators (27), on the basis of photoemission and optical spectroscopy, that there was a gap in the density of states, with essentially no states in it, in amorphous Ge. This was at variance with previous findings and it seems that the elimination of as many defects as possible, by suitable preparation conditions, is necessary. Given this proviso, the conclusion regarding the existence of the gap is now widely accepted, although the necessity to appraise the results of many different experiments on different samples leaves room for disagreement. There is very little agreement on the magnitude of the gap either in amorphous Ge or amorphous Si (28)--a highly suspicious circumstance which suggests that any search for detailed quantitative agreement between theory and experiment is still somewhat pointless.

As for the broad features of the density of states, the most direct evidence relating to this is to be found in soft x-ray emission and absorption experiments. The emission experiment (29), which gives a picture of the s-like and p-like parts of the density of states of the valence band (roughly speaking) lends some support to the suggestion, mentioned in §V, that the first dip in the valence band density of states should vanish on going from the crystalline to the amorphous state while the large peak at the top of the band (a delta function in our model) should survive. The absorption experiment of Rustgi and Brown (30) gives the s-like part of the conduction band density of states. Our results are somewhat academic for the conduction band, since the Hamiltonian is rather inadequate there. The experimental results suggest a remarkable absence of structure in the conduction band, save for a peak just above the gap. Such a peak would not be easily explicable from our point of view, while it would accord rather well with the Penn Model (31). However it may be a many-body enhancement effect rather than a genuine peak on the density of states (32).

IX. SILICA

This section is devoted to a brief sketch of the electronic structure of SiO_2, which in its vitreous form is perhaps the most familiar amorphous material. The tight binding method can be applied here with some confidence as the bands are expected to be rather narrow and well separated (33-36). There is a close analogy between SiO_2 and Si, since its various structures are fourfold coordinated with a silicon atom at each vertex and an oxygen atom at the center of each bond. We construct a Hamiltonian that contains a silicon term, an oxygen term, and a silicon-oxygen mixing term. The silicon term has matrix element V_1 between all four "sp^3" hydrized orbitals to give an s state at $4V_1$ and p states at zero. The oxygen term has a diagonal ΔE; we form two p states at right angles to the bond (sometimes referred to as lone pairs (33)), and form two states, which point towards the two neighboring silicons, from the s and remaining p state allowing an overlap V_1' between these two states. If no other terms were in the Hamiltonian, these two states could be diagonalized to give an s state at $\Delta E + V_1'$ and a p state at $\Delta E - V_1'$ (we also put the lone pair p states at $\Delta E - V_1'$; this is somewhat arbitrary but does not substantially effect the results). We now include the term V_2 that describes overlap between the silicon and oxygen states on the same bond. Using methods that are by now familiar, we obtain the eigenvalues of this problem E from the quartic equation

$$[E-4V_1][(E-\Delta E)(E^2-E\Delta E-V_2^2)-EV_1'^2]-V_2^2[E^2-E\Delta E-V_2^2] = V_1V_1'V_2^2\varepsilon/V \quad (19)$$

and delta functions that are made up from p bonding and antibonding states on the silicon at energies

$$\left.\begin{array}{c} \frac{1}{2}[(\Delta E+V_1') \pm \sqrt{(\Delta E+V_1')^2+4V_2^2}] \\[2mm] \frac{1}{2}[(\Delta E-V_1') \pm \sqrt{(\Delta E-V_1')^2+4V_2^2}] \end{array}\right\} \quad (20)$$

We choose reasonable values for the parameters $V_1 = -7/4$ eV, $V_2 = -7$ eV, $\Delta E = -14$ eV, $V_1' = -14$ eV and the resulting band structure and density of states is shown in Fig. 9 for the high form of the crystobalite structure (in which the silicon atoms form a diamond lattice). The number of states in the various bands is shown beside the density of states diagram. There is an accidental degeneracy in the conduction band due to the particular choice of parameters that results in no gap in the conduction band. The eightfold degenerate band at zero is formed by the lone pairs.

One can make the same kind of statements concerning bounds on the spectrum of this Hamiltonian, as we did for the simpler Hamiltonian appropriate to pure Si.

Finally we comment on the soft X ray absorption spectra. This

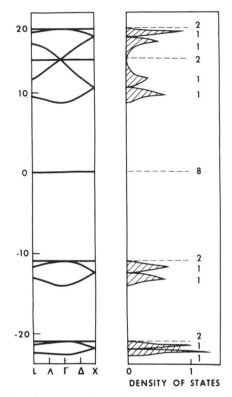

Fig. 9   The band structure for the high form of SiO$_2$ in the cristobalite structure using a tight binding Hamiltonian. Energies in eV.

measures the amplitude of the s,p like character of the wavefunction at Si in the L and K bands respectively. On the basis of Fig. 9 we would expect L and K spectra to be similar for the lone pairs band but peak at higher energies for the other two bands in the K spectra, which is very sensitive to the delta functions, whereas the L spectrum is not. This indeed happens in crystalline quartz (ref. 35, Fig. 5) and should be the same in other forms, both crystalline and amorphous, of SiO$_2$. The appropriate experiments do not appear to have been done.

X.   SI III AND GE III

In the introductory section, mention was made of the metastable crystalline phases Si III and Ge III. Since these are tetrahedrally bonded phases but have rather more complicated structures than diamond cubic, they have much in common with the amorphous

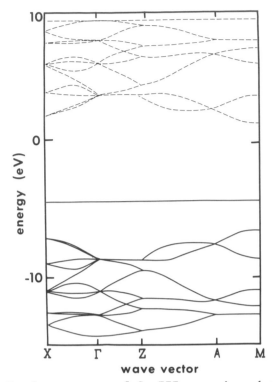

Fig. 10   Band structure of Ge III, as given by the simple
Hamiltonian (1).

phases of these elements.   They offer an ideal testing ground for
theories aimed at an understanding of the amorphous phase.   Theoretical and experimental study of Si III and Ge III would clearly
be most rewarding, but hardly anything has been done yet on either
front (despite the fact that these phases were discovered almost
a decade ago).   Band structures have recently been calculated (21)
for these crystal structures with the Hamiltonian given in Eq. (1).
The more interesting case is that of Ge III since this has five
fold rings.   The band structure is shown in Fig. 10.   The shift
of the bottom of the conduction band which was discussed in §V,
due to odd numbered rings, is found here.   While the exact magnitude of the shift (about 2 eV) is not meaningful, because the
Hamiltonian is rather inadequate in the conduction band, it illustrates the importance of the fivefold rings in determining the
band structure.   Furthermore, if the parameters V_1 and V_2 are
chosen to be appropriate to Sn (i.e. to give an "inverted" band
structure) this shift makes the band structure that of a semiconductor rather than a semimetal.   As yet, Sn has not been pre-

pared in an amorphous tetrahedrally bonded form.

XI. CONCLUSION

In the introduction it was stated that while the simpler crystalline semiconductors are today well understood, their amorphous
counterparts are not. This Jekyll-and-Hyde situation seems likely
to remain with us for some time, although, as we have suggested,
Si III and Ge III may help to bridge the gap in our understanding.
The study reviewed here is based on a frankly inadequate Hamiltonian, and even with such a simplified starting point, difficulties
abound. However, we believe the results are suggestive of various
things. They should contribute to our eventual understanding of
the electronic properties of these systems--particularly the existence of a gap. Since only topological disorder was considered,
the observed existence of this gap has certainly not been explained. How it is to be explained, when quantitative disorder is
incorporated, is still not understood. Phillips' suggestion (37),
that the minimisation of the free energy must play a key role in
producing the gap, presents a tantalising possible solution to the
dilemma. While an arbitrary tetrahedral structure might well have
no gap, because of quantitative disorder, in practice the solid
relaxes into a minimum of the free energy, and it is suggested
that, in the process, states are excluded from the gap. If there
is indeed such hidden method in the madness of the random network
structure, this further compounds the difficulty of the problem.

ACKNOWLEDGMENTS

This research was supported in part by NSF. Our recent work
has benefited greatly from the contributions of R. Alben and S.
Goldstein. We would also like to thank D. Polk for letting us have
some of the details of his model for amorphous Si and Ge.

REFERENCES

1. D. E. Polk, J. Non-crystall. Solids 5, 365 (1971).

2. M. L. Rudee and A. Howie, Phil. Mag. 25, 1001 (1972).

3. R. H. Wentorf, Jr. and J. S. Kasper, Science 139, 338 (1963).

4. J. S. Kasper and S. M. Richards, Acta Cryst. 17, 752 (1964).

5. F. Herman and J. P. Van Dyke, Phys. Rev. Lett. 21, 1575 (1968).

6.   D. Brust, Phys. Rev. Lett. <u>23</u>, 1232 (1969).

7.   B. Kramer, Phys. Stat. Sol. <u>41</u>, 649 (1970).

8.   T. C. McGill and J. Klima, Phys. Rev. <u>B4</u>, 1517 (1972).

9.   J. Keller, J. Phys. <u>C4</u>, 85 (1971).

10. M. Hulin and N. Pottier, Phys. Stat. Sol., <u>51</u>, 613 (1972).

11. For references to work on quantitative disorder, see the papers
     by Freed, and Eggarter and Kirkpatrick in this volume.

12. G. Leman and J. Friedel, J. Appl. Phys. <u>33</u>, Supp. 1, 281 (1962).

13. M. F. Thorpe and D. Weaire, Phys. Rev. <u>B4</u>, 3518 (1971).

14. J. Straley, to be published (1972).

15. M. Hulin, to be published (1972).

16. D. Weaire and M. F. Thorpe, Phys. Rev. <u>B4</u>, 2508 (1971).

17. The proof of the existence of the gap which was given in ref.
     (18) was rather different. In addition to the work of Hulin
     and Straley mentioned in the previous section, alternative
     proofs have been advanced by Heine (19) and Ehrenreich and
     Schwartz (20).

18. D. Weaire, Phys. Rev. Lett. <u>26</u>, 1541 (1971).

19. V. Heine, J. Phys. <u>C4</u>, L221 (1971).

20. H. Ehrenreich and L. Schwartz, to be published (1972).

21. R. Alben, S. Goldstein, M. F. Thorpe, and D. Weaire, to be
     published (1972).

22. F. Cyrot-Lackmann, J. Phys. <u>C5</u>, 300 (1972).

23. M. F. Thorpe, D. Weaire and R. Alben, Phys. Rev. to be
     published (1972).

24. For a direct proof, see M. F. Thorpe, J. Math. Phys. <u>13</u>, 294
     (1972).

25. I. B. Ortenburger, W. E. Rudge, and F. Herman, J. Non-Crystall.
     Solids <u>8-10</u>, 653 (1972).

26. Unpublished--quoted by R. C. Eden, Thesis, Stanford University (1967).

27. T. M. Donovan and W. E. Spicer, Phys. Rev. Lett. 21, 1572 (1968).

28. D. M. Kaplan, M. H. Brodsky, J. F. Ziegler, to be published (1972).

29. G. Wiech and E. Zöpf, Proceedings of the International Conference on Band Structure Spectroscopy of Metals and Alloys, Glasgow (1971), to be published.

30. F. C. Brown and Om. P. Rustgi, Phys. Rev. Letters 28, 497 (1972).

31. J. C. Phillips, Physica Status Solidi (b) 44, K1 (1971).

32. J. P. Van Dyke, Phys. Rev. B5, 4206 (1972).

33. M. A. Reilly, J. Chem. Phys. Solids 31, 1041 (1970).

34. W. Fischer in Advances in X-ray Analysis, edited by B. L. Henke, J. B. Newkirk and S. R. Mallett (Plenum, New York, 1970) Vol. 13, P. 159.

35. G. Wiech, Zeit. für Physik 207, 428 (1967).

36. T. H. Di Stefano and D. E. Eastman, Phys. Rev. Lett. 27, 1560 (1971).

37. J. C. Phillips, Comments in Solid State Physics 4, 9 (1971).

SOME POLYTYPES OF GERMANIUM AND A TIGHT BINDING MODEL FOR THE

ELECTRONIC STRUCTURE OF AMORPHOUS SOLIDS

Douglas Henderson and Irene B. Ortenburger

IBM Research Laboratory

San Jose, California 95114

There is now considerable evidence[1-3] that a useful, although idealized, model of amorphous silicon and germanium is the random tetrahedral network (RTN). In this model, each atom is bonded to four nearest-neighbors at very nearly the nearest-neighbor distance, R_o, of the crystal. The tetrahedral bonding is very nearly preserved in the RTN with the rms deviation in angle being about 10°-15° and a maximum deviation in angle of about twice this amount. Thus, on a local scale, the atomic arrangements in the RTN are much like those of the diamond lattice of crystalline Si and Ge. However, on a larger scale significant differences between the diamond lattice and the RTN become apparent. For example, the diamond lattice consists entirely of "staggered" arrangements of bonds on nearest-neighbors whereas the dihedral angle in the RTN is random. In addition, the RTN contains five-, six-, and seven-member rings whereas the diamond lattice contains only six-member rings.

The radial distribution function (RDF), $g(R)$, of amorphous Si and Ge has considerable similarity to that of their crystalline phases. The RDF of the amorphous solids has peaks at the nearest-neighbor distance, R_o, and the second-neighbor distance, $(8/3)^{1/2} R_o$, as do the crystals. Of course, the first-neighbor peak and especially the second-neighbor peak in the RDF's of the amorphous solids are broader than the corresponding peaks in the crystal RDF's. However, the main difference in the crystal and amorphous RDF's is the absence of a peak in the RDF of the amorphous solids at the third-neighbor distance, $(11/3)^{1/2} R_o$, of the crystals. The RTN accounts for these features. The distortions in the tetrahedral bonding cause the broadening of the first- and second-neighbor peaks and the random dihedral angle causes the third-neighbor

317

distances to be random and thus accounts for the absence of a peak
at the third-neighbor distance of the crystal. The RTN also
accounts for features of the RDF of the amorphous solids at larger
distances also.[2,3]

Polk[1] and Shevchik and Paul[3] have considered finite clusters
containing large numbers (500-1000) of atoms. On the other hand,
Henderson[2] has employed periodic boundary conditions. This permits
the use of much smaller number of atoms and eliminates surface
effects.

However, even the 61 and 64 atom systems considered by
Henderson require the diagonalization of large determinants.
Weaire and Williams[4] have recently suggested the use of the Si3 and
Ge3 structures as approximate realizations in nature of the RTN.
Germanium also exists in the Si3 structure and is called Ge4.
These structures have been relatively unknown by solid state
physicists although the counterpart of Ge3 in ice (ice 3) has been
known for some time. Both the Si3 (or Ge4) and Ge3 structures
contain distorted tetrahedral bonds. In addition, the Ge3 structure
has the further virtue of possessing five-member rings. Actually,
there is no factual basis for believing that five-member rings are
present in amorphous Si and Ge. However, a flat five-member ring
can be formed with only a 1° distortion from the tetrahedral angle.
Thus, it would be very surprising if such structures were not
present in amorphous Si and Ge. In addition, Weaire[5] has presented
other compelling arguments for the existence of five-member rings
in amorphous solids. In any case, the Si3 and Ge3 structures have
great merit since they have only 8 and 12 atoms in a unit cell and
thus are computationally tractable systems.

In Fig. 1 we have compared the RDF of the Ge3 and Ge4 (Si3)
structures with experimental RDF of Shevchik and Paul[3] for
amorphous Ge. Neither the Ge3 or Ge4 structure we considered are
exactly the experimental structures. We allowed the atoms to relax
slightly to a lower energy. The effects of this relaxation are
quite small.

The Ge4 (or Si3) RDF is in rough agreement with the experimental results. The first- and second-neighbor peaks are broadened
and the third-neighbor peak, although still present, is relatively
small. However, for larger distances the Ge4 RDF is unsatisfactory.
It consists of a series of delta functions and has peaks where none
should be present. On the other hand, the Ge3 RDF is fairly good.
The second-neighbor peak occurs at too small a value of R.   This
is its major fault. In addition, it has peaks at about 2.8 R_O and
3.4 R_O. The amorphous Ge RDF has no peaks at these distances.
Despite these deficiencies, the Ge3 structure looks quite promising--
especially since it contains only 12 atoms in a unit cell.

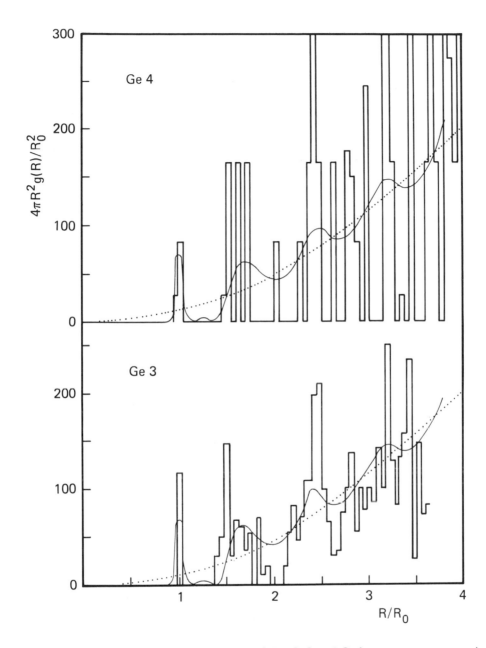

Figure 1. Radial distribution functions of the Ge3 and Ge4 structures compared with the experimental results of Shevchik and Paul[3] for amorphous Ge. The histograms give the Ge3 and Ge4 RDF's and solid curve gives the experimental results.

We have made tight-binding or LCAO calculations of the band structure of Ge3 and Ge4. In the tight-binding method, the energy levels are determined from the equation

$$|H_{ij} - E\delta_{ij}| = 0 , \qquad (1)$$

where

$$H_{\alpha'f';\alpha f} = \sum_{d} e^{i\underline{k}\cdot(\underline{d}+\underline{R}_f-\underline{R}_f')} \int \phi_\alpha^{*'} (\underline{r} - \underline{R}_f')H\phi_\alpha (\underline{r} - \underline{d} - \underline{R}_f)d\tau , \qquad (2)$$

and $\alpha = 1, \ldots, 4$ implies that an s, p_x, p_y, or p_z wave function is being used. We use the notation of Slater and Koster.[6] Thus, we write

$$E_{\alpha'\alpha}(x,y,z) = \int \phi_\alpha^{*'} (\underline{R})H\phi_\alpha (\underline{R} - x\underline{i}_1 - y\underline{i}_2 - z\underline{i}_3)d\tau , \qquad (3)$$

where $\underline{i}_1$, $\underline{i}_2$, $\underline{i}_3$ are unit vectors in the x, y, and z directions. Furthermore, we have assumed the two-center approximation. Thus,

$$E_{ss} = \xi(R) \qquad (4)$$

$$E_{sx} = \ell\eta(R) \qquad (5)$$

$$E_{xx} = \ell^2\lambda(R) + (1 - \ell^2)\mu(R) \qquad (6)$$

$$E_{xy} = \ell m\lambda(R) - \ell m\mu(R) , \qquad (7)$$

where ℓ, m, and n are the direction cosines of $\underline{R}$ and

$$\eta(0) = 0 \qquad (8)$$

$$\lambda(0) = \mu(0) \qquad (9)$$

The model of Weaire results from the further approximations:

$$\xi(0) = -3V_1 + V_o ,\tag{10}$$

$$\lambda(0) = V_1 + V_o ,\tag{11}$$

$$-\xi(R_o) = -\frac{1}{\sqrt{3}}\,\eta(R_o) = \frac{1}{3}\,\lambda(R_o) = \frac{1}{4}\,V_2 ,\tag{12}$$

and

$$\mu(R_o) = 0\tag{13}$$

We have parametrized the functions $\xi(R)$, $\eta(R)$, $\lambda(R)$, and $\mu(R)$ so as to reproduce, as well as possible, the density dependence of the band structure of crystalline Ge as calculated by Herman et al.[8] We have compared the results of our calculation for cubic Ge (i.e., in the diamond lattice) in Fig. 2. The agreement is fairly good. The density of states, calculated from our LCAO model is shown in Fig. 3. It is qualitatively correct.

We have calculated the density of states of Ge3 and Ge4. The Ge4 density of states is qualitatively incorrect. There is no gap and the peak at the upper edge of the valence band occurs at too low an energy. We have plotted our results for the Ge3 density of states in Fig. 3. Comparison with the density of states of cubic Ge shows the following effects. Firstly, the gap in Ge3 is slightly larger than in cubic Ge. Secondly, the peak in the upper edge of the valence band occurs at a higher energy in Ge3 than in cubic Ge. In addition, the band edges are sharper in Ge3. Finally, there is a mixing of the peaks in the lower part of the valence band and a smoothing of the conduction band in Ge3. These results are in qualitative agreement with the results of the pseudo-potential calculations of Joannopoulos and Cohen[9] and with the results of our preliminary pseudo-potential calculations.[10] In addition, they are in qualitative agreement with the experimental results of Donovan and Spicer.[11]

This indicates that Ge3 is a plausible and useful approximation to the structure of amorphous Ge. It accounts for the main features of the RDF, the density of states, and the optical properties[9,10] of amorphous Ge. Germanium 4 (or Si3) does not account for any of these properties of amorphous Ge. Since the main difference between the Ge3 and Ge4 is the presence of five member rings in the former structure we conclude that these rings are an important feature of the structure of amorphous Ge.

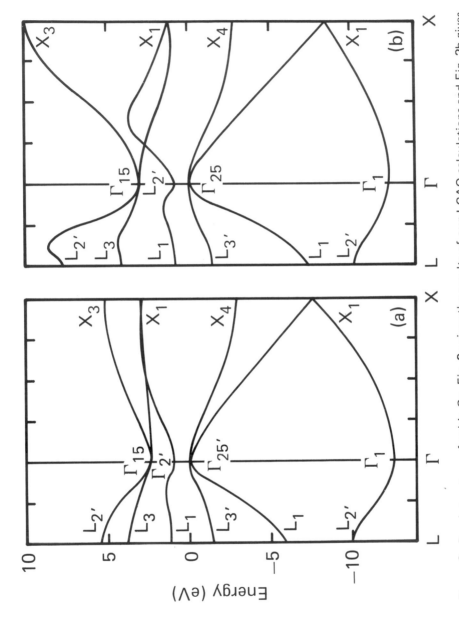

Figure 2. Band structure of cubic Ge. Fig. 2a gives the results of our LCAO calculations and Fig. 2b gives the results of Herman et al.[8]

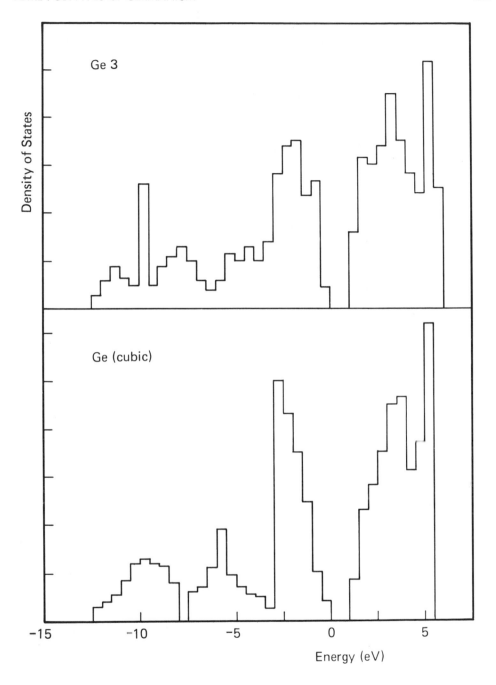

Figure 3. Density of states of cubic Ge and Ge3 calculated using our LCAO model.

The main deficiency of the Ge3 structure is the fact that it is about ten percent more dense than cubic Ge. Germanium 4 is also about ten percent more dense than cubic Ge. The RTN of Henderson[2] is also more dense than cubic Ge but not quite as dense as Ge3 and Ge4. The RTN's of Polk[1] and of Shevchik and Paul[3] do not say very much about the density because energy was not considered during the building of these structures. They appear to have densities close to those of the diamond structure. Conceivably, if the energy were considered and a minimum were sought, higher densities might be achieved. We are of the opinion that densities higher than those of the diamond structure are a feature of the RTN. This is what one might expect intuitively. The introduction of disorder should cause the relatively open structure of the diamond structure to collapse somewhat. For example, in contrast to the situation for many liquids, ice is less dense than water. If the RTN is more dense than the diamond structure then this is a problem since the density of amorphous Ge appears to be about the same as (or perhaps slightly less than) that of cubic Ge. Possibly, even the best-annealed laboratory samples of amorphous Ge are not sufficiently free of imperfections to permit accurate density measurements although one would be surprised to find that their densities are still ten percent less than that of the ideal amorphous Ge. However, at the very least, the RTN and the Ge3 structure are highly useful structures from which one may gain insight into the properties of amorphous solids.

Finally, we conclude that the LCAO model outlined here is of considerable use in investigating the band structures of systems, such as the RTN, which contain large numbers of atoms.

ACKNOWLEDGEMENTS

The authors are grateful to Drs. J. D. Joannopoulos and M. L. Cohen for a preprint of their paper. They are also grateful to Drs. F. Herman, W. E. Rudge and D. Weaire for many helpful discussions.

REFERENCES

1.   D. E. Polk, J. Non-Crystalline Solids 5, 365 (1971).

2.   D. Henderson, Computational Solid State Physics, (Edited by F. Herman, N. W. Dalton, and T. R. Koehler) Plenum Publishing Co., New York, 1972, p. 175; D. Henderson and F. Herman, J. Non-Crystalline Solids 8-10, 359 (1972).

3.   N. J. Shevchik and W. Paul, J. Non-Crystalline Solids (in press).

4.   D. Weaire and A. R. Williams, Phys. Stat. Sol. (b) 49, 619 (1972).

5.   D. Weaire, this volume.

6.   J. C. Slater and G. F. Koster, Phys. Rev. 94, 1498 (1954).

7.   D. Weaire, Phys. Rev. Letters 26, 1541 (1971).

8.   F. Herman, R. L. Kortum, C. D. Kuglin, and R. A. Short, in Quantum Theory of Atoms, Molecules, and the Solid State (Edited by P. O. Löwden) Academic Press Inc., New York, 1966, p. 381.

9.   J. D. Joannopoulos and M. L. Cohen, preprint.

10. I. B. Ortenburger and D. Henderson, Proceedings of the Eleventh International Conference on Semiconductors (in press).

11. T. M. Donovan and W. E. Spicer, Phys. Rev. Letters 21, 1572 (1968); T. M. Donovan, W. E. Spicer, J. M. Bennett, and E. J. Ashley, Phys. Rev. B2, 397 (1970); T. M. Donovan, E. J. Ashley, and W. E. Spicer, Phys. Letters 32A, 85 (1970).

COEXISTENCE OF LOCALIZED AND EXTENDED STATES?

Scott Kirkpatrick
IBM T. J. Watson Research Center

Thomas P. Eggarter
James Franck Institute, University of Chicago

The localized electronic states of a simple model of a substitutional alloy are studied in the strong scattering or separated
subband limit, using the tight-binding approximation. We observe
states localized in small isolated clusters (as studied in percolation theory), and localized states decaying exponentially at
large distances (as in the Anderson model), plus a class of small
localized states which have not been discussed before. These last
states occur at special energies in the middle of the band, yet are
not physically isolated from the rest of the material, or from the
extended states of the system.

The existence of localized states in the center of such a band
challenges the hypothesis that localized and extended states must
occur at different energies, separated by "mobility edges". However, numerical analysis suggests that the density of states goes
continuously to zero on either side of the energy at which the most
numerous small localized states occur, and that the states in these
"inner tails" are localized; i.e., localized states in this model
do not coexist with extended states, but instead repel them strongly.
Some implications of this result for disordered systems (electrons,
spins, lattice vibrations) with short-ranged interactions are
discussed.

INTRODUCTION

Computer calculations have played a rather different role in
the study of disordered systems than in the questions of molecular
and crystalline properties discussed in the earlier papers of this
conference. "Experiments" performed using computer simulations of

327

simple models of random materials have not only provided the opportunity to test approximate theories, but have also on frequent
occasion produced unexpected results, which have led to new physical insight. Two excellent examples of this are the early studies
of electrons in 1D alloys by Dean (1) and Mott and Twose (2). The
first revealed structure in the spectrum of the model as a sole
consequence of the disorder; the second gave direct evidence for
exponentially localized states of the sort predicted by Anderson (3).

We shall report below a series of computer experiments, performed on a very simple model of a substitutional alloy, in two or
three dimensions, which have produced several unanticipated results.
Some of the new features found can also be treated exactly. In
particular, we shall exhibit examples of a type of localized state
which has not, to our knowledge, been discussed before. A full
description of this work will appear elsewhere (4). In this paper
we shall review the results briefly, and discuss some possible
generalizations and consequences.

The model treated is a non-degenerate band in a substitutional
alloy, $A_x B_{1-x}$, described in the tight-binding approximation by the
Hamiltonian

$$H = \sum_{\text{sites } i} |i> E_i <i| + \sum_{\substack{i,j \\ \text{neighbors}}} |i> V <j| \qquad (1)$$

where $|i>$ represents an orbital on the i-th site of the lattice,
and the hopping matrix element, V, connecting only nearest neighbor orbitals, is assumed independent of the type of atoms connected.
The following conventions for the diagonal element, E_i, are used:

$$E_i \equiv 0 \quad \text{on A sites} \qquad (2)$$
$$\equiv \delta \quad \text{on B sites.}$$

It can be shown (5) that when $\delta > 2ZV$, where Z is the number of
nearest neighbors of a site, the spectrum of Eq. 1 divides into two
subbands, centered approximately about the energies 0 and δ. In
the "strong scattering limit", $\delta \to \infty$, the states near $E = 0$ are very
tightly confined to the A atoms, and the B atoms can be formally
removed from the problem (4,5). A restricted Hamiltonian of simpler
form than Eq. 1,

$$H_A = \sum_{\substack{i,j \\ \text{neighbors}}} P^A |i> V <j| P^A \qquad (3)$$

describes the A subband formed in this limit. (P^A denotes a projection operator that selects out A sites.)

RESULTS

We will first discuss the localization phenomena present in this model in the limit in which they are most prominent, where Eq. 3 is applicable. Nonetheless, all the numerical calculations to be reported were performed with the full Hamiltonian Eq. 1, and the consequences of varying the parameter δ from small to very large values are reported below.

In the strong-scattering limit, where tunnelling through the B sites is strictly forbidden to states in the A subband, adjacent B atoms may act as a boundary, isolating small clusters of A atoms from the rest of the material. The two most common cases of this are shown in Fig. 1.

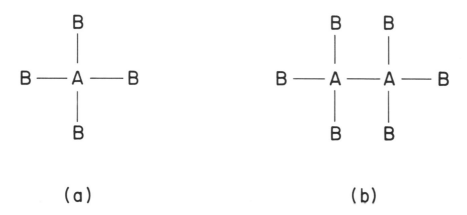

(a)   (b)

Fig. 1   The most common clusters of isolated A atoms.

At any concentration, x, a finite fraction of the A atoms will lie in such isolated clusters. Consequently the eigenstates of such similar clusters will give rise to delta function spikes in the density of states. In fact, if x is less than the "percolation concentration" (6), x_c, which is equal to .31 and .59, respectively, for the 3D simple cubic and 2D square lattices we consider, all A atoms are isolated, and the spectrum consists of a dense distribution of spikes. In general, the spikes at energies 0 and $\pm V$, to which the configurations in Fig. 1 contribute, are the most prominent.

When x exceeds x_c, some fraction of the A atoms will form a single large connected component, which is not bounded by B atoms, and extends across any sample (6). At sufficiently large concentrations, the spectrum of this component, forming a continuous band with tails (7) will dominate the density of states of the subband. By analogy with Anderson's arguments (3), as generalized

by Mott (8), one expects to find that the tails of the band consist
of exponentially localized states, separated from the extended
states by mobility edges (9).

A third type of localized state occurs in clusters like the
one sketched in Fig. 2A, with wave functions like that of Fig. 2B.

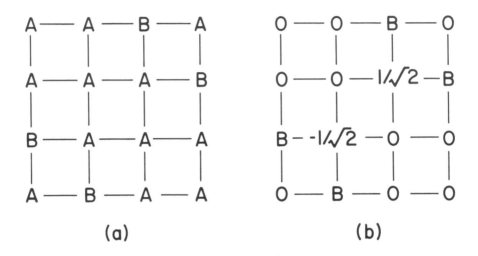

(a)   (b)

Fig. 2   A possible local configuration of A atoms (a)
which are not isolated, but still possess a localized
eigenstate, whose wavefunction is shown in (b).

The arrangement of atoms in Fig. 2A could be either part of a larger
A atom cluster, or taken from the unbounded component of a lattice
with $x > x_c$.   The eigenstate whose wavefunction is indicated in Fig. 2B
has energy 0, and is strictly zero on all but two sites.   Because
its energy is independent of the nature of the environment about
Fig. 2A, such a state will contribute to a spike in the density of
states.

The state shown in Fig. 2B is localized partly by a physical
boundary of B atoms, and partly by interference, because the effects
of hopping from the two central sites cancel on the immediate
neighbors, leaving no wavefunction to propagate further.   For
larger more complex localized states of this type, such as that
shown in Fig. 3A, the importance of interference is even more
apparent.   Localized states containing an arbitrarily large number
of A sites, but requiring only four B atoms in the boundary, can
be constructed (4) in 2D.

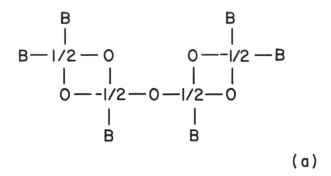

(a)

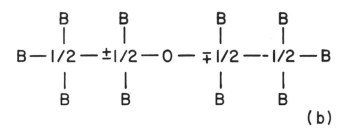

(b)

Fig. 3   Two examples of larger incompletely isolated localized states. In (b), taking the upper
signs gives an eigenfunction with energy +V, the
lower signs a state with E = -V.

Partially bounded localized states occur at energies other
than 0, although they (an example is given in Fig. 3B) are less
common. All examples shown here have been drawn from the 2D square
lattice. Equivalent states for the 3D simple cubic lattice may be
constructed from these by bounding the sites with B   atoms above
and below the plane shown on which the wave function is nonvanishing.

The prominence of the spikes resulting from these two types of
small localized states is evident in the densities of states shown
in Fig. 4, which were obtained numerically by a Sturm sequence algorithm (1) for large finite 2D Monte Carlo samples of several
concentrations.

Spikes at 0 and $\pm$V dominate Figs. 4A and 4B. In Fig. 4B,
x is still less than x_c^-, but the spikes arising from larger clusters create broad satellite bands about the two spikes at $\pm$V.

Densities of states obtained for 3D samples below the percolation concentration are similar to these. The striking persistence of the spike at 0 at concentrations above x_c was first remarked by Kohn and Olson (10), in a 3D calculation using a different calculational procedure. They noted that completely isolated configurations (of the type shown in Fig. 1) could not account for the strength of this spike.

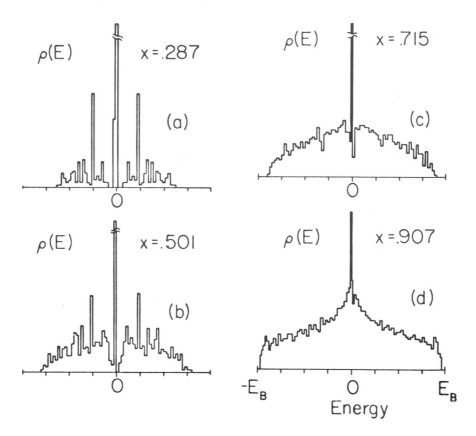

4. Density of states in the A subband of a 2D square lattice, calculated for a 30 by 50 site sample, at concentrations of (a) .287, (b) .501, (c) .715 and (d) .907.

We have carried out an enumeration of the configurations giving rise to both types of small localized states with energy 0, in both 2D and 3D, in order to see if they are a sufficient explanation of the observed spikes. Only the smaller states were counted (although one simple class of larger states was summed in the 2D enumerations), so we expect to obtain an underestimate

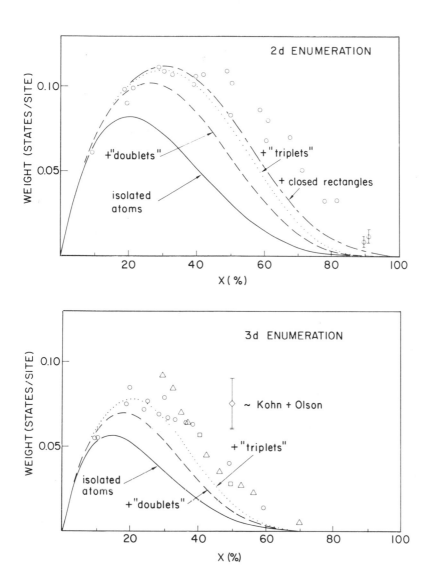

5.   Comparison of the enumerated spectral weights for localized states at E=0 with observed spikes in Monte Carlo samples.   The solid line denotes the contribution of isolated A atoms, the dashed line that of isolated atoms plus configurations in which the wave function is non-zero on two sites, and so on.   The data points represent observations on 30 by 50 site 2D samples (upper figure), and 3D samples (lower figure) with dimensions 7x7x10 (triangles), 8x8x20 (circles) and 10x10x30 (squares).

at high A concentrations. The results are contrasted in Fig. 5 with
observed spike weights on 2D and 3D Monte Carlo samples of various
sizes. Agreement between the enumeration and the data points plotted
in Fig. 5 is excellent at lower concentrations, and appears reasonably good at high concentrations.

We must, however, consider whether these localized states
within the band constitute a counterexample to the conventional
picture of an impurity band (8) in which mobility edges provide
a strict separation in energy between localized and extended
states. Cohen (11), in particular, has argued that localized and
extended states cannot occur at the same energy, since if they did,
they would mix, and all states would become extended. The localized
states occurring in isolated clusters do not challenge this assertion, since the boundaries of B atoms prevent any mixing. However,
the existence of incompletely isolated localized states at many
energies within the band seems to contradict this non-mixing
argument.

The difficulty is resolved, at least for the localized states
near $E=0$, in a surprising way. Numerical calculations strongly
indicate that there are no extended states at $E=0$. In the first
three cases of Fig. 4, the central spike sits in the center of a
dip in the density of states. The dip is broadest at low concentrations, narrows as x increases, and was not seen at the highest
concentration. Calculations for 3D samples, using a finer scale
near $E=0$, show this more clearly, and are displayed in Fig. 6.
At each of the concentrations shown, the spectrum of the finite
sample has a gap to either side of the central spike. As the
concentration is increased, the gap decreases until it is of the
order of a typical spacing between levels, and thus no longer
statistically significant. The gap remained significant, as is
seen in Fig. 6, up to concentrations much greater than the percolation concentration.

We conclude that for a wide range of concentrations the center
of the A subband is a "forbidden energy", about which the continuous
part of the density of states for an infinite sample goes smoothly
to zero from both sides. The gaps observed in Figs. 6A-C correspond
to regions in which the density of states in these inner tails is
too low for any states to occur in one finite sample.

It seems not unreasonable to suspect that $E=0$ remains forbidden
when any finite concentration of B atoms is present, or that similar
effects are present at the other energies for which partially isolated localized states exist. However, these conjectures cannot
be tested with the present numerical techniques, because the average
interval between eigenenergies limits the energy resolution with
which the continuum can be studied.

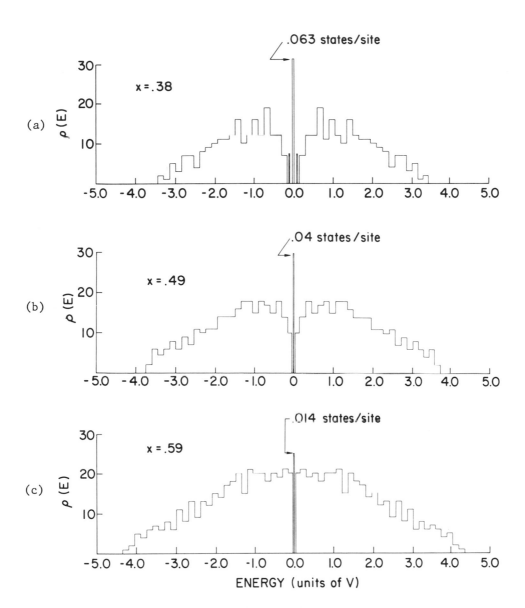

6. Density of states histograms calculated for 3D samples of 8x8x20 sites, at concentrations of (a) .38, (b) .49, and (c) .59. This range of concentrations shows the narrowing of the dip near the center of the band as the concentration increases above the percolation concentration (.31).

The structure near E=0 in the density of states is not a special feature of the limit $\delta \rightarrow \infty$, or of the Hamiltonian (3). Its persistence as δ is made smaller has been studied in a 2D example, as shown in Fig. 7. The spike seen in Fig. 7A broadens asymmetrically into a resonance in Figs. 7B and 7C, leaving the dip above E=0 unaffected. The total number of eigenstates below this dip remains constant until the dip disappears, at a value of δ small enough that the A and B subbands have merged into one another. In this case, shown in Fig. 7D, some 20 eigenenergies have shifted past E=0 to lower energies.

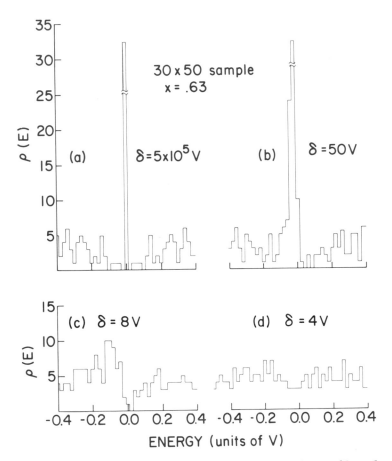

Fig. 7 Density of states for a 2D sample with x=.63, plotted on a fine scale to show the effects near the center of the band of permitting tunnelling through the "forbidden" B sites by decreasing the parameter δ, the energy separation between A and B orbital energies. The gap and dip persist until case (d), δ=4V, at which point the A and B subbands have merged.

Forbidden energies with similar associated phenomena have been shown to occur in 1D systems (12). Gubernatis and Taylor (13) have even shown that the analytic form of the spectrum in the inner tails near such an energy in a particular 1D model is the same as in the tails at the band edges, although the scale of energies is much finer. Efforts to date to generalize the theory to 2D and 3D disordered systems (14) have been restricted to the case $x < x_c$, and make the unphysical assumption that no clusters larger than a fixed finite size are present in the sample. Our results appear to be the first report of a forbidden energy for $x > x_c$, that is, in a system possessing extended states.

As a simple test of the spatial extent of the eigenstates, the sensitivity of specific eigenvalues to changes in the boundary conditions has been calculated for several finite samples. Roughly speaking, the smaller the region occupied by an eigenstate, the less sensitive is its energy to the boundary conditions (15). For cases with $x < x_c$, all eigenvalues studied are found to be completely insensitive, hence localized. In a 3D sample with x a few percent above x_c, the first extended states are found in two narrow regions of energy centered about E=+0.8V. In a sample whose concentration is close to that shown in Fig. 6A, all the states outside of +2.0V and all those inside +0.3V appear localized. Thus the "inner tails" surrounding the forbidden energy appear to show Anderson localization comparable to that anticipated for the tail at the band edges.

APPLICATIONS

The results of these model calculations have application in several areas. As its most immediate application, this work provides non-trivial exact results for a simple model, against which one may test approximate theories. Neither the well-known coherent potential approximation (16, 5), nor recent extensions of it to include scattering from pairs of sites (17) predicts the spikes or the structure observed here near E=0.

It appears reasonable to apply these results to several classes of systems, in which the simplifications of our model are approximately satisfied. The principal simplification is the restriction to short-ranged interactions. Although a poor approximation for the description of electronic energy bands in metallic alloys, nearest neighbor interactions do adequately describe interacting spins and lattice vibrations in many materials. Localized excitations with nodes in their wave functions are also found in these systems, e.g., vibrational excitations of light atoms in a heavy atom matrix, or magnetic excitations of mixtures of magnetic and non-magnetic materials. In the first case, an atom which remains at rest is a node of a localized vibration, while in the

second an impurity spin aligned in the direction of the net magnetization may act as an effective boundary.

We know of no relevant work on such magnetic systems, but
vibrational "impurity modes" of light mass atoms have been studied
extensively by Raman scattering in substitutional semiconductor
alloys (18) such as $Ge_{1-x}Si_x$. A sharp line is seen at the frequency characteristic of a single light atom vibrating while the
surrounding heavy atoms remain at rest. This line remains prominent when x exceeds 50%. We suggest that localized vibrations
involving several light mass atoms must contribute to this line to
account for its strength and persistence, and that the isolating
effects of nodes will permit these vibrations to occur in the unbounded light mass component of the material as well as in isolated
clusters.

The second observation made in this work, the non-coexistence
of localized and extended states, could easily hold in systems
more general than our simple models. In particular, topological
disorder which preserves short-ranged order, as is found in many
molecular and covalent glasses, should not affect our results.
Recent measurements of the specific heats of such glasses at low
temperatures (19) provide evidence for the existence of a large
number of low frequency vibrational modes. All explanations of
this anomalous specific heat to date (20) have pictured these
modes as highly localized.

If the non-coexistence of localized and extended states is a
general result, there should consequently be no extended states at
these low energies, or at least a very strong interaction between
the two. Numerical calculations of vibrational spectra in which
low frequency localized modes are present - see Dean's review
article (1) - do show a gap between these states and the next
lowest-lying modes. Experiments on glasses reveal strong damping
at thermal and ultrasonic frequencies (19).

Thus there is good reason to question whether long wavelength
vibrations of the usual sort are still present in these glasses.
Even though Anderson et. al. (20) have obtained qualitative agreement with experiments on energy transport in a simple calculation
including coupling between local and long wavelength modes to first
order only, this question clearly requires a more fundamental
theoretical understanding.

REFERENCES

1. P. Dean, Proc. Royal Soc. (London) A, 254, 507 (1960). Sharp structure in the density of states had previously been observed by R. Landauer and J. C. Helland, J. Chem. Phys. 22, 1655 (1954). For a review of Monte Carlo calculations on alloys, and description of numerical techniques, see P. Dean, Rev. Mod. Phys. 44, 127 (1972).

2. N. F. Mott and W. D. Twose, Adv. in Phys. 10, 107 (1961).

3. P. W. Anderson, Phys. Rev. 109, 1492 (1958).

4. S. Kirkpatrick and T. P. Eggarter, to appear in Phys. Rev. B15.

5. B. Velický, S. Kirkpatrick and H. Ehrenreich, Phys. Rev. 175, 747 (1968), and Appendix B of Phys. Rev. B1, 3250 (1970).

6. V. K. S. Shante and S. Kirkpatrick, Adv. in Phys. 20, 325 (1971).

7. S. Kirkpatrick, J. Non-Cryst. Solids 8, to appear as part of the proceedings, Int'l. Conf. on Liquid and Amorphous Semiconductors, Ann Arbor, Mich., 1971.

8. N. F. Mott, Adv. in Physics 16, 49 (1967).

9. E. N. Economou, S. Kirkpatrick, M. H. Cohen and T. P. Eggarter, Phys. Rev. Letters 25, 520 (1970).

10. W. Kohn and J. T. Olson, Proceedings of Conf. on Electronic Properties of Ordered and Disordered Solids, Menton, France, 1971 (to appear).

11. M. H. Cohen, Proceedings, Tenth Int'l. Conf. on the Physics of Semiconductors, Cambridge, Mass., (1970) p. 645.

12. H. Matsuda, Progr. Theoret. Phys. 31, 161 (1964); R. E. Borland, Proc. Phys. Soc. (London) 83, 1027 (1964).

13. J. E. Gubernatis and P. L. Taylor, J. Phys. C4, L94 (1971).

14. J. Hori and K. Wada, Progr. Theoret. Phys. Suppl. 45, 36 (1970), review this work.

15. J. T. Edwards and D. J. Thouless, J. Phys. C5, 807 (1972).

16. P. Soven, Phys. Rev. 156, 809 (1967).

17. B. G. Nickel, Phys. Rev. B4, 4354 (1971); L. Schwartz and E. Siggia, Phys. Rev. B5, 383 (1972); and private communications.

18. G. Lucovsky, M. H. Brodsky and E. Burstein, Phys. Rev. $\underline{B2}$, 3295 (1970).

19. R. C. Zeller and R. O. Pohl, Phys. Rev. $\underline{B4}$, 2029 (1971).

20. P. W. Anderson, B. I. Halperin and C. M. Varma, Phil. Mag. $\underline{25}$, 1 (1972); P. Fulde and H. Wagner, Phys. Rev. Letters $\underline{27}$, 1280 (1971); H. B. Rosenstock, J. Non-Cryst. Solids $\underline{7}$, 123 (1972).

CLUSTER SCATTERING IN AMORPHOUS SEMICONDUCTORS

AND LIQUID METALS

Jaime Keller

Facultad de Química, Universidad Nacional Autónoma

de México. México 20, D.F.

1. INTRODUCTION

Although the study of the electronic structure and derived properties of condensed matter has been one area of great development in the past decade, attention has been focused mainly on the band structure of crystalline solids, their fermi surfaces, etc. In recent years some understanding of the electronic properties of random alloys in periodic or almost periodic lattices has been obtained through the use of the coherent potential, the virtual crystal approximations, etc. The theoretical study of the electron structure of disordered crystals, amorphous materials and liquids, monotomic and molecular alloys rests mainly in semiempirical rules or ad hoc theoretical models. The study of non-crystalline materials through the use of more realistic Hamiltonians is now of basic importance not only to understand optical and electrical properties as such, but also to explain chemical behavior in solids and liquids, catalysis, clustering, etc. as well as cluster and molecular magnetism.

All these properties are related to the behavior of electrons in the region of positive energies inside the material. This is in contrast with those properties related mainly to electrons trapped in atomic or molecular wells. In this second case, almost bound states, the electrons have to tunnel from one well to another, the wave function being exponentially decaying outside a given well.

In the case of electrons with energies above muffin tin zero (MTZ) the electron densities and related properties depend strongly on the local geometry and on the boundary conditions imposed in a given region of the material by the rest of the system (see Fig. 1).

In other words, the mean free path is very much longer than atomic dimensions. It is then obvious that a method suitable for the systematic study of these properties in amorphous systems should at least consider the two main pieces of information about the geometric structure: the radial distribution function and the local short-range order (which includes contributions from higher order distribution functions). This is necessary in view of the poor theoretical and experimental knowledge of the actual geometry of the material.

In part two of this paper a "cluster method approach" to the electronic properties of amorphous materials will be developed in some detail using the formalism of multiple scattering theory (1, 2,3). In part three, some possible way of physical and mathematical approximation to the boundary conditions will be discussed, and in the last section some numerical results will be presented for amorphous semiconductors and liquid metals.

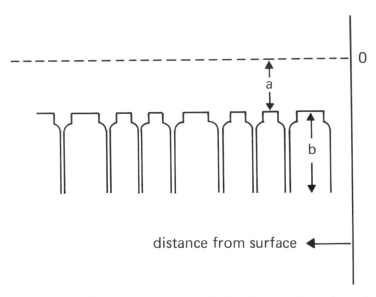

Figure 1. The one-electron potential in the condensed system divides into two distinct regions: (a) positive energy region, inside the material, properties strongly dependent on boundary conditions, and (b) negative energy region of "locally bound" states, properties dependent on boundary conditions through exponentially decaying factors.

2.   FORMULATION OF THE CLUSTER METHOD

The cluster method (4) assumes that the one-electron Schroedinger equation for a given set of atoms in a medium can be written as

$$(-\nabla^2 + \sum_{j}^{N} V_j(\underline{r}) + \sum_{i}^{N'} V_i(\underline{r}) + V^b - E)\psi = 0 \tag{1}$$

where V_j are the non-overlapping potentials of N atoms in the cluster, V_i are the potentials of N' "interstitions" used to represent large open non-overlapping regions of the interatomic space (as defined in Keller (5) or as discussed by Ziman (6)), and V^b is a real or fictitious potential which represents the effect of the rest of the medium on the cluster and defines the appropriate boundary conditions. A similar equation is written for each "cluster" in the system.

The clusters can be defined, for example, as chemical radicals, radical-ions or composite ions. In general they must form a representative part of the materials. Here is where the known or assumed or self-consistently determined short-range order should be introduced into the model.

The potentials V_i and V_j are to be determined in the form of a local potential using either a muffin-tin approximation (1,5) or a generalized non-overlapping form (7,8) using a local exchange of the Slater (9), Kohn-Sham (10), X_α (11), or $X_{\alpha\beta}$ (12) type. The explicit use of spin polarization is necessary in some applications like paramagnetic radicals or magnetic clusters of alloys. Potentials of a more general class can be used.

In all applications made of this model up to date, V^b has been determined in a similar way but an approximation has been made about the terms arising from it in the solution of the Schroedinger's equations, keeping in mind that it represents the effect of the rest of the medium on the cluster and defines the appropriate boundary conditions. This Hamiltonian will be used in connection with Eq. (9) below.

The usual expression for the density of states of an infinite system of scatterers can be written as

$$N(E) = -\frac{2}{\pi\Omega} \text{ Im Tr } \ln[H - E^+] \tag{2}$$

$$N(E) = N_o(E) - \frac{1}{\pi\Omega} \text{ Im Tr } \ln[S_{LL'}] \tag{3}$$

where $N_o(E)$ is the free electron part (at positive energies). This result was transformed by Lloyd (2,13) into the form:

$$N(E) = N_o(E) - \frac{2}{\pi\Omega} \text{Imln} ||D|| \tag{4}$$

where the determinant of the matrix D

$$||D|| = ||\delta_{LL'}\delta_{ij} - \Sigma_{L''}G^+_{LL''}(\underline{R}_i-\underline{R}_j) \ k_{L''L'}(\underline{R}_i)|| \tag{5}$$

is in mixed angular momentum-position representation, $K_{L''L'}(R_i)$ is the K-matrix of the scatterer at site R , and G^+ is a free-electron propagator. It can be factorized:

$$||D|| = \begin{vmatrix} \begin{bmatrix} A_a \end{bmatrix} & \begin{bmatrix} B'_{ab} \end{bmatrix} & \begin{bmatrix} B'_{ac} \end{bmatrix} & \cdots \\ \begin{bmatrix} B'_{ba} \end{bmatrix} & \begin{bmatrix} A_b \end{bmatrix} & \begin{bmatrix} B'_{bc} \end{bmatrix} & \cdots \\ \begin{bmatrix} B'_{ca} \end{bmatrix} & \begin{bmatrix} B'_{cb} \end{bmatrix} & \begin{bmatrix} A_c \end{bmatrix} & \cdots \\ \cdots & \cdots & \cdots & \cdots \\ \cdots & \cdots & \cdots & \cdots \end{vmatrix} \tag{6}$$

(in A_i only terms referring to cluster i appear, in B'_{ij} only the interaction terms between i and j, etc.), and the result expressed as:

$$||D|| = ||A|| \ x \ ||A^{-1}D|| = ||A|| \ x \ ||B|| \tag{7}$$

$$N(E) = N_o(E) + \Sigma_c N_c(E) + N_{interclusters}(E) \ . \tag{8}$$

The evaluation of the second term in Eq. (8) requires the sum for all possible clusters in such a way as to give the correct average when the density of states per atom or per unit volume is evaluated.

The third term in the R.H.S. of Eq. (8) can be approximated in several ways, but the idea of the cluster method is to write Eq. (8) in the form

$$N(E) = N_o(E) + \Sigma_\alpha P_\alpha \tilde{N}_\alpha(E) \tag{9}$$

where P_α is a probability factor for the occurrence of the cluster and $\tilde{N}_\alpha(E)$ is the Friedel sum for a cluster of type α in the presence of the fictitious potential $V_{boundary}$ of Eq. (1).

There are several different approaches one may take to evaluate $\tilde{N}_\alpha(E)$ in Eq. (9). A common starting point is the two potentials formalism beginning with the evaluation (or approximate evaluation)

of the Green's function for the propagation of a wave in the
presence of V^b alone.

$$G(V^b) = \frac{1}{E^+ - H^o - V^b} \tag{10}$$

"Outgoing" boundary conditions are used $(E+i\varepsilon)$. Nevertheless, this
allows the wave scattered by the cluster to be reflected back by
V^b. It is useful to consider the T-matrix

$$T(V^b) = V^b + G^+ V^b T(V^b) \tag{11}$$

when writing the wave function with scattering boundary conditions.
We also need the single site matrix $t(i,V^b)$ in the presence of the
background potential V^b

$$t(i,V^b) = \frac{V_i}{1-G(V^b)V_i} = t(i) + t(i)G^+ T(V^b)G^+ t(i) + \ldots \tag{12}$$

where multiple scattering between V_i and V^b is explicitly considered.
The integrated density of states and the Friedel sum can be calculated

$$N(E) = -\frac{2}{\pi\Omega} \text{ImTrln} [H^o + V^b + V_c - E^+] \tag{13}$$

where

$$V_c = \sum_i^N V_i(\underline{r} - \underline{r_i}) + \sum_j^{N'} V_j(\underline{r} - \underline{r_j}) \tag{14}$$

and Ω is the "volume" of the system

$$N(E) = N_o(E) + N_{vb}(E) + \frac{1}{\Omega}\mathcal{N}_c(V^b,E) \tag{15}$$

where the second term is the density of states due to V^b alone and
the Friedel sum for the cluster embedded in V^b has been defined to be

$$\mathcal{N}_c(V^b,E) = -\frac{2}{\pi} \text{ImTrln} [1 - G(V^b)V_c] \quad . \tag{16}$$

$G(V^b)$ is defined by Eq. (10). Equations (15) and (16) are exact.
The third term in Eq. (15) is vanishing small if the size of the
cluster is finite (and for practical purposes will not contain more
than a few dozen atoms) because it represents the effect of just a
small portion of the system. We are interested in the Friedel sum
per unit volume relative to the cluster

$$\tilde{N}_c(V^b,E) = \frac{1}{\Omega_{cluster}} \mathcal{N}_c(V^b,E) \tag{17}$$

which we will identify with that of Eq. (9).

In order to express this in the usual position-spin-angular momentum representation, we follow the same trick used by Lloyd of expanding everything out into a power series, reducing it in terms of individual scatterers and spin-angular momentum components (13). Let us use the composite spin-spherical harmonics functions $\mathcal{Y}_L(\theta,\phi,m_s)$ obtained by the composition of a spin ½ with the spherical harmonics of order (ℓ,m), as defined for example by Messiah (14). Now L stands for (ℓ,j,m_j), and

$$\mathcal{Y}_L(\theta,\phi,m_j) \quad \propto \quad Y_{\ell,m_j-\frac{1}{2}}(\theta,\phi)S_{\frac{1}{2}} \quad \text{or} \quad Y_{\ell,m_j+\frac{1}{2}}(\theta,\phi)S_{-\frac{1}{2}}$$

according to $m_s=\frac{1}{2}$ or $-\frac{1}{2}$, where $S_{\frac{1}{2}}$ and $S_{-\frac{1}{2}}$ are spin wave functions. For spherically symmetric local potentials (but spin or spin-orbit dependent), it is better to use combinations of these that decouple the radial wave functions equations.

If we define

$$\mathcal{G}_{LL'}(v^b,\underline{r}_i,\underline{r}_j) = \begin{cases} G^+_{LL'}(\underline{r}_i-\underline{r}_j)+\langle h^+_{L-i}\underline{r}_i|T(v^b)|h^+_{L',\underline{r}_j}\rangle \\ 0 \quad \text{if } \underline{r}_i = \underline{r}_j \end{cases} , \quad (18)$$

$$\langle h^+_{L-i}\underline{r}_i|T(v^b)|h^+_{L',\underline{r}_j}\rangle = \int d^3r \int d^3r'(-E)(-i)^{L_i}{}^{L'}h^+_L(K|\underline{r}-\underline{r}_i|)$$

$$\times\; Y^*_L(\underline{r}-\underline{r}_i)\langle\underline{r}|T(v^b)|\underline{r}'\rangle h^+_{L'},(K|\underline{r}'-\underline{r}_j|)$$

$$\times\; Y_{L'}(\underline{r}'-\underline{r}_j)S_M S_{M'} \quad . \quad (19)$$

The propagator $\mathcal{G}$ can be pictorially described as a free electron wave going directly from $\underline{r}_i$ to $\underline{r}_j$ plus another outgoing wave propagating outside the system and scattered back into $\underline{r}_j$ by the potential v^b.

The resummation of the single site contribution is

$$-\frac{2}{\pi}\,\text{ImTr}\Sigma_i \ln[1-G(v^b)v_i] \quad = \quad -\frac{2}{\pi}\,\text{ImTr}\Sigma_i \ln[1-G^o v_i]$$

$$\times\; [1+iK\Delta k(\underline{r}_i) - G^+ T(v^b)G^+ k(\underline{r}_i)] \quad (20)$$

and the multiple scattering contribution

$$-\frac{2}{\pi}\,\text{ImTr}\ln[\delta_{LL'}\delta_{ij}-\Sigma_{L''}\mathcal{G}_{LL''}(v^b,\underline{r}_i,\underline{r}_j)t_{L''L'}(v^b,\underline{r}_j)] \quad . \quad (21)$$

The sum of these two expressions and some lengthy algebra gives

$$\tilde{N}_c(V^b, E) = -\frac{2}{\pi\Omega_c} \, \text{ImTrln}[\delta_{LL'}\delta_{\underline{r}_i,\underline{r}_j}$$

$$- \Sigma_{L''}G^+_{L''L''}(V^b,\underline{r}_i,\underline{r}_j)k_{L''L''}(\underline{r}_k)] \qquad (22)$$

$G^+(V^b)$ is equal to $\mathcal{G}(V^b)$ without the restriction for $\underline{r}_i = \underline{r}_j$. The determinant is of the same form as that in Eq. (5), and reduces to it if $V^b = 0$.

Complementary part of the theory is the evaluation of the energy derivatives of the electron density

$$\frac{\partial\rho(\underline{r})}{\partial E} = -\frac{2}{\pi} \, \text{Im}\langle\underline{r}|G|\underline{r}\rangle \, ; \quad G^{-1} = [E^+ - H^o - V^b - V_c] \quad . \qquad (23)$$

The change in density of states evaluated with Eqs. (17) and (22) is normalized to the volume of the cluster but the Friedel sum is conceptually the modification to the free electron density in a sphere of very large volume produced by the potential V_c of the cluster considered. On the contrary, when evaluating $\partial\rho(\underline{r})/\partial E$ we have terms arising from the boundary potential V^b directly. We expect $\Sigma_m \int\rho(E,\underline{r})d^3r$ to coincide with $N_0(E)+N_c(V^b,E)$, but this identification will be discussed elsewhere in a future paper.

Consider the system's Green's function

$$G = G(V^b) + G(V^b) \, \Sigma_i \, t(V^b,i)G(B^V)$$

$$+ \, G(V^b) \, \Sigma_i \left(t(V^b,i)G(V^b) \, \Sigma_{j\neq i} t(V^b,j)G(V^b)\right) + \, \ldots \qquad (24)$$

and the energy derivative of the electron density

$$\rho'(E,\underline{r}) = -\frac{1}{\pi} \, \text{Im}\langle\underline{r}|G|\underline{r}\rangle \quad . \qquad (25)$$

Substituting Eq. (24) in (25) we obtain the first term of the expansion

$$-\frac{1}{\pi} \, \text{Im}\langle\underline{r}|G(V^b)|\underline{r}\rangle = \frac{K}{16\pi^3} \, \Sigma_L \, |\phi^o_L(V^b,\underline{r})|^2 \qquad (26)$$

where

$$\phi^o_L(\underline{r}) = 4\pi i^L j_L(K|\underline{r}-\underline{r}_o|) \, \mathcal{Y}_L(\widehat{\underline{r}-\underline{r}_o},m_s) + \langle\underline{r}|G^+T(V^b)|KL\rangle \, , \qquad (27)$$

The free electron density being obtained in Eq. (25) if V^b=0. The remaining of the contribution from this term for the actual value of V^b, when integrated over the volume of the cluster, must be equal to the Friedel oscillations outside Ω_c to restitute the electronic charge necessary to keep neutrality in the cluster (unless there is some charge transfer between the cluster and the embedding system).

The scattering matrices for a system of scatterers are quantities of related interest. A theoretical discussion and some numerical examples for clusters of carbon atoms is presented elsewhere (15).

3.   THE BOUNDARY CONDITIONS IN CONDENSED MATTER

For a molecule or any other cluster of atoms in a condensed system, the outside potential V^b consists of the ensemble of potential wells of all other atoms in the system (referred to MTZ). In an amorphous system as opposed to the case of a crystalline solid, we only have some statistical information for the position of those atoms. Far from the range of the short-range order considered in our basic cluster, only the two-body correlation function is known. We have to use a different technique to solve the multiple scattering equations to calculate $T(V^b)$ or to approximate this quantity in a realistic way. As it will be clear with the examples shown below, the answer we obtain is strongly dependent on the degree of approximation obtained. A simple consideration can illustrate this point. If we neglect V^b the electronic charge of the cluster will spread outside the volume Ω_c in which it is confined in the system. But if we overemphasize the backscattering, for example, by enclosing the cluster in some potential well, with or without a molecular coulomb tail, then the cluster behaves like a molecule; we lose all the bands and we are left with a discrete set of bound states.

Useful as that information can be, we need more precise knowledge of the electronic structure. The task of finding an adequate form to approximate $T(V^b)$ has been a central problem of our research.

From the many different ways of finding an approximation to $T(V^b)$ which can be handled in actual calculations, the following have been explored numerically:

a) $T(V^b) \rightarrow T(V = VMTZ)$.

It is the cluster that is considered to exist in "free space" where the zero of energy is the average interstitial potential. In this case there is no backward scattering at all. With these boundary conditions no energy gap is possible for a finite cluster because there is always the possibility of an electron wave arriving at and decaying inside the cluster. However, in large clusters $\rho(\underline{r},E)$ can

be very small far from the surface of the clusters. On the other
hand, the formation of bands and the splitting through further
interaction is clearly seen. (Compare Fig. 2.)

The first suggestion to use Lloyd's formalism to study amorphous semiconductors considering small clusters of atoms neglecting
the intercluster multiple scattering was made by Klima and McGill
(16). It is equivalent to this choice of boundary conditions.
They were able to show in a realistic calculation the role of short
range order in producing bands and a low density of states in the
energy gap region of a crystalline material. This showed a correct
and practical way of introducing short range order into a multiple
scattering approach.

b) $T(V^b) \rightarrow T(V_o)$

$V_o = V_{MTZ}$ inside the cluster

$V \rightarrow 0$ (atomic zero of energy) outside the cluster.

In this case the system behaves like a molecule, all states
being bound states for energies below the atomic zero. The molecular energy levels found are nevertheless related to some features
of the energy bands (see for example Watkins' paper in this volume
and Ref. (17)). Adopting this boundary condition, the formalism
is equivalent to the X_α multiple scattering method of Johnson,
Smith, Connolly, et al (18). (See also Johnson and Connolly, this
volume.)

c.) $T(V^b) \rightarrow T(V(E,\underline{k},m_s))$

where $V^b = 0$ if $\underline{r}$ is inside the domain of the cluster

$= V_R^b + iV_I^b$ if $\underline{r}$ is outside the domain of the cluster

The fictitious energy dependent potential $V(E,\underline{k})$ is obtained
from one of several possible approximations to the self energy of
an electron wave propagating outside the cluster (see for example
the discussion of Schwartz and Ehrenreich (19)). One possibility
is the use of a mean field approximation type

$$V(E,\underline{k}) = nT_E(\underline{k},\underline{k}) \tag{28}$$

where T_E is the transition matrix related to an atomic potential
or a cluster potential, n being the density Ω^{-1} atom or Ω^{-1} cluster.

Let us assume we were dealing with a cluster in an otherwise
periodic potential. Then $T_E(\underline{k},\underline{k})$ could be the transition matrix
of a unit cell or a group of unit cells, and the periodicity of the
host lattice could then be accounted by in a first approximation.

If on the contrary we have our cluster in an amorphous system, $T_E(\underline{k},\underline{k})$ would be different from region to region, long-range order being absent. But we can compute a spherically average $\tilde{T}_E(k,k)$ where the specific orientation of the array of scatterers has been eliminated on account of the randomness of the system.

A different approach being numerically investigated is a generalization of the Anderson and McMillan (20) procedure. They calculated a constant complex potential $V_c(E)$ in which $t(V^b,\underline{k},\underline{k})$ for a single atom vanishes.

Gyorffy and Keller have extended this idea, requiring

$$\tau^{\alpha\alpha} = t_\alpha(V^b) + \sum_{\beta\neq\alpha} t_\alpha(V^b)G^+(V^b)t_\beta(V^b)G^+(V^b)t_\alpha(V^b) + \ldots \quad (29)$$

to be vanishing small. t_α and G^+ are calculated when the system is embedded in $V_c(E)$ and the positions of the scatterers α are related to that of scatterers β only through the pair distribution function. This idea was originally proposed to improve the single site approximation method. $V_c(E)$ defined as mentioned above turns out to be a simple way of introducing the pair distribution function, which is the more important piece of information we have about the geometry of the medium. This procedure is related to CPA (21).

d) Finally, a very promising approach which has been investigated is the direct expansion of $T(V^b)$ in terms of single site scattering matrices and a resumation of the infinite series obtained in the form

$$T(V^b) = \sum_\alpha \sum_\beta \tau^{\alpha\beta} \quad ; \quad V^b = \sum_\gamma V_\gamma \quad (30)$$

where $\tau^{\alpha\beta}$ are the scattering operators relating an incoming wave arriving at α and being scattered from β after all possible multiple scattering between β,γ and α (in a diagramatic expansion this corresponds to the sum of all diagrams starting in β and ending in α). This approach can also be applied to a cluster in an otherwise periodic host. The sumations can then be made exactly.

Of course, we cannot in general solve Eq. (30), not even if we restrict ourselves to the use of the pair correlation function alone and the convergence of the series expansion is the subject of our present calculations. It must be stated that the sumations can only be performed by rather drastic approximations, but the results obtained so far are not significantly different from those obtained using method c) described above.

4. SOME RESULTS FOR AMORPHOUS SEMICONDUCTORS AND LIQUID METALS

We should mention some of the physicll quantities of interest related to the functions mentioned above. The reaction matrix of a cluster can be used to calculate the density of states, the introduction of the boundary condtions, according to Eq. (22), being straightforward.

From the T-matrix we can calculate band structures (specially with many atoms per unit cell, the cluster being the unit cell itself) and elastic scattering cross-sections.

From the Green's function we can evaluate, among other properties, the local density of density of states and map the electronic structure, not only as chemists do but also as a continuous function of energy. The latter scheme is particularly useful for understanding the formation of bands and the character of the energy bands. The computed electron density can be used as a basis for a self-consistent calculation.

One particular branch of computational solid state physics where a cluster approach provides a very valuable tool is in the calculation of optical properties where the transition is more or less localized in space. This is the case of soft x-ray spectra and related spectroscopies where a locally bound electron is excited to the continuum of conduction states or free electrons. Usual band structure calculations cannot account for the local relaxation of the bands because the atom or molecules where the transition takes place is an "impurity" with one (or more) electrons being removed. On the other hand, our cluster approach properly takes into account this factor. One of the atoms (or molecules) of the cluster having only a partial occupation of the core states, the peaks in the valence or conduction band shift "locally" to lower energies, and the full spectrum is further broadened in accordance with the observed soft x-ray spectra. Current research in this respect includes the calculations for crystalline C, Si, Fe, Cu, for amorphous carbon and silicon, liquid Fe and Cu and the FeGe alloy.

Application to Condensed Systems

The cluster method as defined by Eq. (9) has been used in the study of some models of liquid, transition metals, semi-metals, and amorphous semiconductors. It is currently being applied to the further study of these materials and also some alloys. The method is expected to be particularly useful for the investigation of impurities, vacancies and structural defects in condensed systems, including those organic and inorganic materials which are composed of

associations of more or less well defined molecular clusters.
Understanding of the electronic properties of those systems in
which there is clustering or chemical bonding in an otherwise uniform medium is an intrinsically difficult problem, and one which
is suitable for the application of this method.

Our study of the scattering properties of clusters of atoms
in semiconductors with tetrahedrally coordinated random network
structures (e.g., silicon and hypothetical amorphous diamond) includes calculations made for clusters of 1, 2, 5, 6, 8, 10, 17,
and 29 atoms in a series of different configurations. In a first
approximation the boundary conditions assumed that the clusters
were embedded in a potential equal to the interstitial potential
in the cluster (6).

For amorphous diamond there is a definite tendency to obtain
three distinct zones consisting of two bands separated by one region
of low density states. This result confirms the original work of
Klima and McGill. In all cases the energy region with low density
of states is at an energy range higher than that where an energy
gap is found in the perfect crystal using the same muffin-tin potential. Results for the large (really amorphous) clusters confirm
that the low density of states is not produced by the high symmetry
of the small clusters.

An analysis of the behavior of the density of states curve in
the energy gap region as a function of the size of the cluster
shows a definite tendency towards a perfect gap in a very large
cluster. The calculations also predict that this gap should be at
the same energy as the gap in the crystal.

The eventual cancellation of the free electron density of
states is a reasonable possibility. Similar results were also obtained for silicon and recently for germanium. Using the short
range order of the cluster properly introduces the higher order
correlation functions into the problem and this is obviously ad-
vantageous.

Unfortunately, this procedure of increasing the cluster size
cannot be carried much further without enormous computational expense, and with clusters of only a few dozen atoms, the intercluster
contribution from matrix B in Eq. (7) cannot be ignored. If we
throw away these terms we must introduce some other device to eliminate spurious effects from the boundary of the cluster.

To get some idea of the importance of the boundary conditions,
calculations have been made for different cases:

(i) The scattering properties of the environment of the cluster were neglected as in the previous calculations.

(ii) The cluster was subject to periodic boundary conditions.

(iii) The environment was represented by a dispersive medium outside the cluster or in the form discussed in d) above.

The main results were that in case (ii) the region of low density of states was transformed into a perfect gap region (see Ref. (6)), and in case (iii) into a region of very low density of states, below the accuracy of the numerical calculation.

Once the basic rules for the construction of the clusters have been given, i.e., a fixed coordination number and a restricted range of values for the nearest neighbor distance and for the angles between the bonds, the topologically disordered network so obtained represents the basic structure of an amorphous semiconductor. We can construct more realistic structures containing "defects" of two kinds: (i) topological defects in which the coordination number rule is not satisfied, and (ii) geometrical defects which are produced when any of the nearest neighbor distances or bond angles lie out-side their specified ranges.

If the cluster has geometrical defects the energy gap region is substantially altered. Decreasing the density of the material destroyed the gap by making the bands overlap.

A particular case of interest is obtained by changing some angles of the "bonds" from the tetrahedral configuration by about 15%. This corresponds to a highly disordered amorphous cluster in which the second to nearest neighbor distance of many atoms is drastically changed, and this new cluster gives rise to a larger number of states in the energy gap region. The use of periodic boundary conditions shows that the bands do not necessarily overlap; but in any case, the gap will be much smaller. (Compare Fig. 2.)

On the basis of our numerical calculations we conclude that with the correct combinations of scattering potentials in systems with appropriate short range order we can have sufficient conditions for the existence of an energy gap above muffin-tin zero even in the absence of long range order. In any material it is necessary to in-clude in the analysis clusters containing enough information about the local order and to use the appropriate boundary conditions. This can be seen clearly in two complementary examples, i.e., the calculation of the density of states of a semi-metal, graphite, and of the d-electron bands of the transition metals, iron, copper, and nickel.

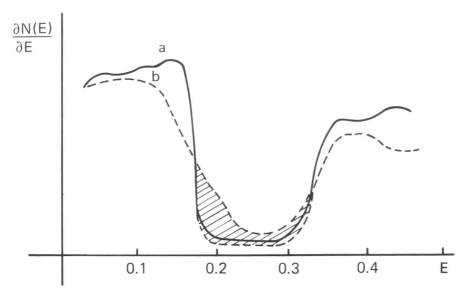

Fig. 2.   Calculated electronic density of states of amorphous germanium; (a) random network of tetrahedrally coordinated clusters, and (b) 20% of the clusters deviate from the fixed coordination number and nearest neighbors distance (and bond angles).   The shaded area is contributed from the "defective" clusters; electron transfer between those clusters presents a typical percolation problem.

In the first case, graphite (22), the boundary conditions are very important, as we have an electron density of states at the fermi level of the order of the free electron value.   In the second case, the height of the density of states in the d-band is more than an order of magnitude larger than the free electron value, and meaningful results are obtained even for very small clusters (23). Our present work in the d-electron alloys confirms these results.

A very attractive result of our analysis of the scattering properties of clusters of atoms has been the clear indication that the lines joining the atoms, represented usually as chemical bonds, are also the channels along which a wave of electrons scatter.   The channeling of electrons in the direction of the bonds, which is clearly seen at energies 10 or 15 eV above the muffin-tin zero (where scattering is less isotropic), provides supporting evidence for the picture of the bonds as regions of high electron density. We feel this concept will be important for the full understanding of the subtleties of the chemical bond.   Indeed, our calculations (15) show that the bonds themselves behave as strong electron scatterers.

5. CONCLUSION

We have discussed how some quantities derived from cluster scattering can be used to analyze the electronic properties in condensed matter. The method is practical and requires less effort than usual ab initio band structure techniques.

We want to point out again that many experimental measurements on the condensed system deal with phenomena occurring in the region of energies where we have a continuous spectrum, and here the techniques discussed above are useful. If we study a condensed system (as opposed to the case of isolated molecules where the same range of energies corresponds to discrete states), quantum chemistry needs practical techniques to use in these cases, complementary to conventional quantum chemistry.

I would like to thank my colleagues in the U.N.A.M. as well as all my former colleagues in the Theoretical Physics group of the University of Bristol for their encouragement and help, and especially Professor J. M. Ziman who suggested this type of approach (24).

REFERENCES

1. J. Korringa, Physica 13, 392 (1947); and W. Kohn and N. Rostoker, Phys. Rev. 94, 1111 (1954).

2. P. Lloyd, Proc. Phys. Soc. 90, 207 (1967).

3. J. L. Beeby, Proc. Roy. Soc. (London) A279, 82 (1964).

4. J. Keller, J. Phys. C: Solid St. Phys. 4, 3143 (1971); J. Keller, J. Phys. Suppl., Menton Conference (1971).

5. J. Keller, J. Phys. C: Solid St. Phys. 4, 1, L85 (1971).

6. J. M. Ziman, J. Phys. Suppl., Menton Conference (1971).

7. R. Evans and J. Keller, J. Phys. C: Solid St. Phys. 4, 3155 (1971).

8. A. R. Williams and J. W. Morgan, to be published.

9. J. C. Slater, Phys. Rev. 81, 385; and 82, 538 (1951).

10. W. Kohn and L. J. Sham, Phys. Rev. 140, A1133 (1965).

11. J. C. Slater and K. H. Johnson, Phys. Rev. B5, 844 (1972).

12. F. Herman, J. P. Van Dyke and I. B. Ortenburger, Phys. Rev. Letters 22, 807 (1969).

13. P. Lloyd, Multiple Scattering in Liquid Metals, unpublished (1969).

14. A. Messiah, Quantum Mechanics (Amsterdam), (1961).

15. J. Keller and P. V. Smith, J. Phys. C. Solid St. Phys. 5, 1109 (1972).

16. T. C. McGill and J. Klima, J. Phys. C. 3, L163 (1970).

17. R. P. Messmer and G. D. Watkins, Phys. Rev. Letters 25, 656 (1970); R. P. Messmer, General Electric Report No. 71-C-150 (1971); and present volume.

18. K. H. Johnson, J. Chem. Phys. 45, 3085 (1966); K. H. Johnson and F. C. Smith, Phys. Rev. B5, 831 (1972) (see paper in this volume for a description of their work).

19. L. Schwartz and H. Ehrenreich, Ann. Phys. 64, 100 (1971).

20. P. W. Anderson and W. L. McMillan, Proc. Int. School Physics, Enrico Fermi, ed. W. Marshall (New York: Academic Press, 1967).

21. P. Soven, Phys. Rev. 156, 809 (1967).

22. T. Spiridonov and J. Keller, to be published.

23. J. Keller and R. Jones, J. Phys. F. Metal Phys. L33, 36 (1971).

24. J. Keller and J. M. Ziman, paper at Fourth Int. Conf. on Amorphous and Liquid Semiconductors, Ann Arbor, Michigan (August 1971), published in J. Non-Crystalline Solids 8/9, 111 (1972).

PROPERTIES OF LOCALIZED STATES IN DISORDERED MATERIALS*

Karl F. Freed[+]

The Department of Chemistry and James Franck Institute

The University of Chicago, Chicago, Illinois

ABSTRACT

The transition between localized and extended electronic states in disordered materials is considered by the use of two different models. The mathematical analogy between this transition and phase transitions and critical points in fluids and magnets is stressed and exploited. The localization probability and range in the neighborhood of the transition are shown to have critical exponent type behavior with the exponents 13/6 and -2/3, respectively. A preliminary discussion of electron mobility in disordered materials is also presented.

I. INTRODUCTION

The electronic structure of disordered materials has been studied for many years by both chemists and physicists. The systems studied exhibit a wide variety of qualitatively different types of behavior. In metal-liquid ammonia solutions, for instance, at low metallic concentrations, the solution has a characteristic blue color due to solvated electrons which are trapped in localized states. At higher concentrations the solutions become bronze colored and exhibit metallic properties. Metal-molten salt mixtures likewise have a metal-nonmetal transition as a function of metal

---

* Research supported by NSF Grant GP-28135, ACS PRF Grant 5767-AC6, and ARPA facilities for materials research at The University of Chicago

+ Alfred P. Sloan Foundation Fellow

concentration. Other systems which have extensively been studied involve excess electrons in various liquids such as liquid argon and helium. Excess electrons in liquid argon behave as if they were almost free-electron like, but at low argon densities it appears that the electrons are in localized states. In liquid helium, on the other hand, because of the strong repulsive interaction between the excess electron and the closed shell helium atoms, the electron is localized in a bubble with a diameter of the order of 30 Å. At lower helium densities it appears that the electrons are in extended states. Liquid metals such as mercury have been explained by a nearly free electron theory, while elemental amorphous solids like silicon and germanium are semiconductors. Chalcogenide glasses such as $Te_{0.5} As_{0.3} Si_{0.1} Ge_{0.1}$ are amorphous semiconductors which have been studied partially because of their switching behavior.

The above cited examples of disordered materials whose electronic structure has been studied exhibit a number of different kinds of disorder. There is spatial disorder in amorphous solids and liquid metals which is due to the loss of the three dimensional periodicity of corresponding crystalline materials. Metal-molten salt solutions, random alloys, and chalcogenide glasses have compositional disorder. Lastly, there can also be temporal disorder due to the thermal motion of the atoms or ions of the system.

There are a number of fundamental questions concerning electronic structure in disordered materials which these examples evoke. Given the lack of a rigid lattice in liquids there can be a rapid rearrangement of the local liquid structure in the presence of the electron. Much effort has gone into the elucidation of this local liquid structure in, for instance, metal-ammonia solutions. However, even when this local liquid structure is completely understood, the fundamental problems of the description of the nature of the electronic wavefunctions and transport processes still remain. Herein, we shall confine attention to only the latter problems (1,2).

Some of the examples cited above exhibit a metal-nonmetal transition. Although this transition is of considerable interest, it can be considered in purely crystalline systems. Thus, it is unnecessary to introduce the added complication of disorder in the discussion of this interesting transition.

One of the general features of the electronic structure of disordered materials such as those noted above involves the occurrence of localized and extended electronic states. The nature of these localized and extended states is central to the understanding of the properties of amorphous semiconductors and is the prime focus of this lecture.

Before considering the theoretical description of electronic structure in disordered materials, it is instructive to reflect upon the corresponding theories in crystalline materials. In perfect crystals the perfect periodicity enables the use of the Bloch-Floquet theorem to reduce the problem to one involving the solution within a unit cell. This use of simple group theory immediately predicts the existence of continuous bands of electronic states which are extended throughout the crystal. These bands are easily shown to be separated by band gaps of forbidden energies. The existence of bands of extended states which are separated by gaps provides a general qualitative understanding of the properties of metals, insulators, and semiconductors. The science of solid state physics has developed to provide the detailed quantitative understanding of the electronic structure of crystalline materials.

In discussing the electronic structure of disordered materials we should first inquire whether there exists corresponding universal features of the electronic structure of disordered systems. However, because of the disorder, these materials completely lack symmetry, thereby removing the periodicity which enabled the qualitative understanding of the crystalline materials. It would therefore appear that it is necessary to start from the beginning when confronting electronic structure in disordered systems.

If there are universal features of the electronic structure of disordered materials, it should be necessary to use only the simplest models to uncover these general properties, provided these models are physically qualitatively realistic. As an example, we may recall the simplicity, yet generality, of the Ising model in describing the nature of critical phenomena. It will become apparent that because of the general complexity of disordered materials, the use of simple models is currently mandatory.

In the next section a brief review is given of some of the general features of electronic structure in disordered materials that are expected on the basis of the very limited experimental and theoretical work to date. Section III considers a cluster theory approach to localized electronic states in disordered materials, while the final sections discuss a self-consistent field path integral approach to the problem.

II. MODELS OF ELECTRONIC STRUCTURE IN DISORDERED SYSTEMS

In contrast to crystalline materials, disordered materials do not have a unique structure by definition. Thus it is only meaningful to consider the ensemble average properties of such a system. When the disorder is small, it represents a small perturbation on the electronic motion. In this case the electron is primarily affected by the average effects of the disorder. It can therefore be expected to occupy an extended state which has some mean free path or coherence length which results from the presence of the disorder. This is the situation encountered in liquid argon and in liquid metals such as mercury. In the latter case a nearly free electron theory is successful despite the fact that the mean free path is of the order of six angstroms. On the other hand, when the disorder becomes great, large potential fluctuations can trap electrons in localized states (3). For instance, in random bindary alloys various clusters of atoms can easily be chosen in many instances which are expected to give rise to these localized states.

Mott-Cohen, Fritzsche, and Ovshinsky (CFO) have presented a model to explain the electronic structure of amorphous semiconductors (2,4). In the absence of broken bonds, the Mott-CFO model assumes that there are bands of electronic states with extended states in the band center and localized states in the band tails. The energies E_c where there is a transition between localized and extended states have been termed mobility edges since the mobility is taken to vanish for electrons occupying localized states at $0^{\circ}K$. An equivalent definition of a localized state is one with a wavefunction which is decaying, e.g., exponentially, about some given center. When the disorder becomes sufficiently large, all the states become localized. This is termed the Anderson transition (5). A reasonable goal of the theory of electronic structure of disordered materials is to verify, correct, and expand upon this Mott-CFO model in order to make it quantitative.

As noted above, the states in the center of the band correspond to extended ones where the electron experiences occasional scattering from the potential fluctuations, while those in the band tails are localized due to local fluctuations in composition, intermolecular spacing, etc., which can lead to states with energies near the band edges. The most difficult problem arises in the description of the electronic states in the region of the mobility edges where there is a transition between localized and extended states. This is the question to which the remainder of the lecture is devoted.

In order to get a qualitative feel for the difficulties involved, it is instructive to first consider a classical description. Picture an electron with energy E in a disordered system which, in general, is characterized by potential wells of varying depth and potential barriers between these wells of varying heights. Upon averaging over all the configurations, i.e., values of the random potentials, we wish to ask whether the motion of this classical electron is bounded or not. This is basically the problem considered by percolation theory (6,7). If E is low enough, the classical electron cannot percolate off to infinity; it is in a localized state. For E suffi-

ciently large, the electron percolates in an extended state. When
E is near the critical energy for percolation, in order to see if
the electron is to be localized or extended it must look throughout
all of space to see whether on the average there are roadblocks
everywhere which impede its unbounded motion. Thus, it is the infinite range potential energy fluctuations which ultimately determine the nature of the electronic states in the region of the mobility edges. This makes the mobility edge for this classical electron
analogous to critical points or points of phase transitions in
fluids and magnets (8). For this reason it is clear that only the
simplest models can and need be considered.

The above use of percolation theory considers only classical
electrons, but the real quantum mechanical situation is complicated
by the presence of tunneling. Thus, we could imagine that localized states found for classical electrons might be distributed
throughout space such that tunneling may occur between them. These
classical localized states would then become extended states because of tunneling. As tunneling is most favorable between states
of similar energies, the probability of finding a site to which it
is favorable to tunnel, of course, increases as the distance from the
original site increases. However, as this distance increases the
probability for tunneling decreases exponentially. When the decrease in the latter is greater than the increase in the former,
the electron on the average cannot find enough favorable sites
for tunneling, so it remains in a localized state. Even in the
quantum mechanical problem, it will be useful to consider the analogy between the mobility edge and critical points and phase transitions in fluids and magnets.

III. CLUSTER THEORY OF ELECTRON LOCALIZATION

Consider a one-electron, one-band model in which the electron
moves on a lattice where the diagonal site energies V_α (at site α)
are random variables, but the matrix elements between sites, the
hopping integrals, are not. Many realistic systems would also require the latter to be random, but there is no need to include
these additional mathematical complications herein. Chemists will
recognize this model as Huckel theory for a random lattice.

The Green's function $\mathcal{G}(E)$ for an electron with energy E with
a given unaveraged configuration of the lattice and a time independent Hamiltonian $H_o + V$ is defined by the Schroedinger equation

$$(E - H_o - V)\mathcal{G}(E) = 1 ,$$ (1)

where V is the random potential, H_o is fixed, and 1 is the unit
operator. The ensemble averaged Green's function is then

$$G_o(E) = \langle g(E) \rangle \quad . \tag{2}$$

Localized states can be sought by focusing attention on those clusters n of fixed potential which may provide localization (9). Let $\langle \rangle^{\underline{n}}$ imply an average only over the potentials on clusters $\underline{n}$, and $\langle \rangle_{\underline{n}}$ an average over all sites outside of $\underline{n}$. Thus the Green's function for a material in which the cluster $\underline{n}$ is specified and the rest is averaged may be written as

$$G_{\underline{n}} = \langle g \rangle_{\underline{n}} \equiv \langle g \rangle^{\underline{N-n}} \quad . \tag{3}$$

Averaging Eq. (1) over all but the cluster $\underline{n}$ leads to the equation of motion

$$(E - H_o - V_{\underline{n}})G_{\underline{n}} - \sum_{\alpha \notin \underline{n}} \langle V_\alpha G_{\underline{n}\alpha} \rangle^\alpha = 1 \quad , \tag{4}$$

where $V_{\underline{n}}$ is the random potential on cluster $\underline{n}$. Equation (4) is a member of a hierarchy in which the n-cluster function $G_{\underline{n}}$ is coupled to the $(n+1)$-cluster function $G_{\underline{n}\alpha}$. The exact form of $G_{\underline{n}}$ can easily be deduced from Eq. (4) to have the general form

$$G_{\underline{n}} = (E - H_o - V_{\underline{n}} - \Sigma_{\underline{n}})^{-1} \quad , \tag{5}$$

where $\Sigma_{\underline{n}}$ is the complex energy-dependent self-energy which provides the net effects of the disorder due to lattice sites outside of $\underline{n}$. An approximation to $G_{\underline{n}}$, call it $\Gamma_{\underline{n}}$, would involve the replacement of $\Sigma_{\underline{n}}$ by some approximation $s_{\underline{n}}$ which is formally written as a super-position of potentials $s_{\underline{n}}^{(\alpha)}$ which are centered at the lattice sites $\alpha \notin \underline{n}$. Adding and subtracting $s_{\underline{n}}G_{\underline{n}}$ on the right in Eq. (4) leads to the formal solution of the hierarchy

$$G_{\underline{n}} = \Gamma_{\underline{n}} + \Gamma_{\underline{n}} \sum_{\alpha \notin \underline{n}} \langle (V_\alpha - s_{\underline{n}}^{(\alpha)}) G_{\underline{n}\alpha} \rangle^\alpha \quad , \tag{6}$$

which can be iterated to obtain an exact formal solution of the hierarchy for G_o in terms of the dressed propagators $\Gamma_{\underline{n}}$ which are still at our disposal (9). Some approaches take all the $s_{\underline{n}}^{(\alpha)}$ to be the self-energies $\Sigma_o^{(\alpha)}$ of G_o and then attempt to obtain these self-energies self-consistently. The formal hierarchy obtained from Eq. (6), in which the terms can be classified as involving scattering off of all possible clusters of one, two, ... sites, contains the extra degrees of freedom of the choice of $s_{\underline{n}}$, $n \neq 0$, and is therefore more flexible. In order to assure that all the terms representing scattering off of clusters of n sites vanish, it is sufficient to require that the average scattering off of any site $\alpha \notin \underline{n}$ vanish; this is the CPn approximation which determines the $s_{\underline{n}}$ self-consistently. The $s_{\underline{m}}$ for $m \leq n$ can likewise be chosen so that the average scattering off all clusters of size m≤n vanishes, but this choice is not unique and therefore leads to a variety of different possible approximations. In any such case the result is that $G = \Gamma_0 +$ terms involving the scattering off of clusters involving at least n+1 sites.

This full procedure is somewhat involved and is more than necessary in order to investigate the properties of localized electronic states (9). In obtaining the CP$\underline{n}$ approximation Γ_n to G_n, attention can be focused upon those clusters $\underline{n}$ which are expected to lead to localized states. In general, information concerning the cluster $\underline{n}$ cannot persist far from $\underline{n}$, so the self energy s_n must become s_0 far from $\underline{n}$. The CP0 approximation is just the coherent potential approximation of Soven, Taylor, and Yonezawa and Matsubara (10). Thus, the solution for Γ_n on each iteration of the self-consistency cycle just corresponds to a Koster-Slater treatment of the impurity cluster $\underline{n}$ with its surrounding perturbed averaged medium in the CP0 band. This CP0 band corresponds to one with complex energy-dependent potentials s_0. It is therefore clear that localized states can be found in this manner; however, they will be of two general types (9). In the first type, the wavefunction is completely localized within the cluster $\underline{n}$ and therefore would be unaffected by the approximations inherent in CP$\underline{n}$ or in choosing a larger cluster size. In the second type, the localized wavefunction extends into the surrounding averaged medium, so the tails of the wavefunction may be sensitive to the approximations invoked. In order to check this, it is necessary to consider the behavior of these type two localized states as the cluster size is increased. Some of these will ultimately become type one states, while others will continue to be type two. It is therefore only in the limit of infinite sized clusters that we can ultimately decide the fate of states in the neighborhood of the mobility edge, in conformity with the expectations on the basis of percolation theory.

Given a type one localized state for a cluster $\underline{n}$, this finite cluster has nonzero measure, so this localized state must contribute with nontrivial weight to the averaged Green's function G_0. The most general procedure would involve introducing Γ_n into the hierarchy obtained from Eq. (6) and thereby providing a translationally invariant approximation to G_0 which contains the localized state with its suitable weight. However, it is simpler to introduce auxiliary averaging procedures to maintain the translational invariance which results from the fact that the cluster $\underline{n}$ could be anywhere in the lattice. In any event, the nature of the localized states in the band tails can be obtained from a self-consistent generalized Koster-Slater procedure, but a new theory is required to describe the states in the region of the mobility edge. Such a theory, employing a different model, is presented in the subsequent sections.

IV.   RANDOM SCATTERER MODEL: PATH INTEGRALS

Again only a one-electron model is considered; in this case
the electron is moving in the presence of randomly distributed
scattering centers.   This model has been employed to consider electrons in heavily doped semiconductors when correlations between
impurities may be neglected (11).   It also represents a very crude
model of an electron in a liquid where the liquid structure and
thermal motion is ignored.   In any event, this provides a mathematical model which contains the qualitative effects of randomness
and is therefore sufficient for a discussion of the region near
the mobility edge.   Actually, this model is considered only in a
special limit, the limit in which the scatterers are dense but the
electron-scatterer interaction $v(\underline{r}-\underline{R})$ is weak.   This dense, weak,
random scatterer limit is equivalent to the model of an electron
in a Gaussian random potential (1,8,12,13).   It will be convenient
to consider the model from these two equivalent points of view.

More explicitly, the one-electron Hamiltonian is

$$H = -\frac{\hbar^2}{2m} \nabla_{\underline{r}}^2 + \sum_{j=1}^{N} v(\underline{r}-\underline{R}_j) \quad , \tag{7}$$

where $\rho=N/\Omega$ is the scatterer density, and the assumption of random
scatterers implies that the ensemble averaging procedure to be
taken is

$$< \quad >_{Av} = \prod_{j=1}^{N} [\int d\underline{R}_j/\Omega] \quad . \tag{8}$$

On the average the electron is in a localized state if it can-
not "percolate" off to infinity.   Let $c_0(t|\{\underline{R}_j\})$ be the probability
amplitude that an electron return to its starting point after a time
t when the scatterers are at the points $\{\underline{R}_j\}$, then the quantity

$$P_0 = \lim_{t\to\infty} <|c_0(t|\{\underline{R}_j\})|^2>_{Av} \tag{9}$$

gives the ensemble averaged probability that the electron can return to its starting point after an infinite amount of time.   If
$P_0=0$, the electron cannot return, so it is in an extended state
(the volume $\Omega\to\infty$).   If $P_0\neq0$, some of the electronic states must be
localized.   It is, of course, more revealing to examine the energy
dependence of this localization criterion.   It can be shown that a
quantity $p_0(E)$ can be introduced such that (1,5,14)

$$P_0 = \int_{-\infty}^{\infty} dE p_0(E) \quad , \tag{10}$$

and $p_o(E)$ is the ensemble averaged probability that electrons with energy E return to their starting point after an infinite time. If $p_o(E)=0$, states at energy E are extended, and if not, they are localized and $p_o(E)$ is then some measure of the density of localized states at energy E at any point in space and of the spatial volume occupied by these localized states.

Rather than basing a consideration of electron localization upon p_o or $p_o(E)$ in Eqs. (9) or (10), a number of workers have erroneously presumed that electron localization can be determined from the average probability amplitude $<c_o(t|\{R_j\})>_{Av}$. Consider a single impurity on a lattice where the impurity energy has a probability distribution such that for all energies allowed by this distribution it has a bound state. The ensemble averaged probability that an electron on this impurity be in a localized state is consequently unity; however, $<c_o(t|\{R_j\})>_{Av}$ vanishes as $t \to \infty$ and is therefore not a true localization criterion. (A simple example is one for which the probability density is uniform for $E_o-\delta \leq E \leq E_o+\delta$ and is zero otherwise.)

Although the average probability amplitudes cannot be used as a localization criterion, it is still convenient to employ them in order to introduce the requisite concepts. $p_o(E)$ will be treated afterwards. For each configuration of the random scatterers, in principle, it is always possible to determine the eigenfunctions $\psi_j(\underline{r}|\{R_k\})$ and energies $E(\{R_j\})$. Thus the ensemble averaged probability amplitude for going from $\underline{o}$ to $\underline{r}$ in time t is

$$<c_{or}(t|\{R_k\})>_{Av} \equiv G(\underline{ro};t)$$

$$= \sum_j <\psi_j(\underline{r}|\{R_k\})\psi_j^*(\underline{o}|\{R_k\})\exp[-iE_j(\{R_k\})t/\hbar])>_{Av} \quad , \quad (11)$$

and is just the off-diagonal matrix element of the average time-dependent Green's function. We can consider $G(\underline{ro};t)$ to be the probability amplitude for some "effective" or "average" particle or "quasi-particle" and search for the effective Schroedinger equation obeyed by $G(\underline{ro};t)$. Such an equation of motion may be obtained directly from path integral methods (13,15), but its form can be deduced on the basis of purely physical arguments (13). In Eq. (11) an average has been performed over the positions of all of the scatterers. Therefore, the "average electron" must be moving in free space and $G(\underline{rr'};t)$ is only a function of $\underline{r}-\underline{r'}$. Since the average electron is not a free electron, there must be some interactions. With the scatterers removed by the averaging, the only thing left in space is the average electron, so the average electron must interact with itself in some manner in order to simulate the net effects of the ensemble average over the positions of all the scatterers. It is convenient first to consider the average electron classically and then make the transcription to quantum mechanics. If $\underline{r}(t)$ is the classical trajectory, the most general form of a translationally invariant self interaction is symbolically

$$V_{eff} = V[\underline{r}(\tau)-\underline{r}(\tau')] + \text{three time terms} + \ldots \quad , \tag{12}$$

where the average electron interacts with itself at different times. A similar situation arises in the consideration of the interaction of an electron with the radiation field. If an average is performed over all the variables of the radiation field, an effective electron self interaction results which, naturally enough, provides the electron's self-energy (16).

Qualitatively the effects of randomness in the present problem are contained in the leading two time terms in Eq. (12) which are the only two that survive in the convenient high density, weak interaction limit $v \to 0$, $\rho \to \infty$, $v^2\rho \to$ finite. It is amusing to pause to compare this limit to the weak coupling, long time limit, $v \to 0$, $t \to \infty$, $v^2 t \to$ finite, in classical transport theory where in this limit all problems are, in principle, solvable. In the present case our limit has turned the impossible problem posed by Eq. (12) into one that is just intractable as it is now one that is mathematically isomorphic to the description of self-avoiding random walks, the polymer excluded volume problem (1,8,13,15,17).

For the classical average electron in the weak coupling, high density limit Hamilton's principal function would be of the form (13)

$$A = \int_0^t \frac{m}{2} [\underline{\dot{r}}(\tau)]^2 d\tau - \int_0^t d\tau \int_0^t d\tau' \, V[\underline{r}(\tau)-\underline{r}(\tau')] \quad . \tag{13}$$

It is clear from Eq. (13) that no simple conserved or nearly conserved Hamiltonian may be obtained from Eq. (13), so the Lagrangian, path integral (16), formulation of quantum mechanics must be used. The details are omitted, but it should again be noted that the problem can alternatively be expressed as an electron in a Gaussian random field $\phi(\underline{r})$ (Refs. 1,8,13,15,17).

We now have an averaged, self-interacting electron which moves in translationally invariant free space, and we are looking for localized states. But localized states are not translationally invariant, so they involve a breaking of this translational symmetry in the same manner that the occurrence of a spontaneous magnetization in one direction breaks the rotational symmetry of space. In an approximate theory, the symmetry breaking can be directly introduced in order to find the localized states. In the cluster theory symmetry is broken when a particular cluster $\underline{n}$ is chosen (9). In the present model the definition of probability amplitudes, Green's functions, etc. can be used to introduce this symmetry breaking

naturally (13). Since $G(ro;t)$ is the probability amplitude that if
the electron is started (created) at o, it can be found (annihil-
ated) at r after a time t. The acts of creation and annihilation
of the electron at o and r make these points special in an otherwise translational invariant space; it breaks symmetry automatically.

The problem at hand is a mathematically complicated one, but
the analogy with the self-avoiding polymer problem can be helpful
in providing approximate solutions. The physically simplest non-
trivial approximation to the polymer problem involves the use of
a self-consistent field approach (15,17). First one end of the
polymer chain is tied down (thereby breaking the symmetry of space).
The solution to the problem would provide the polymer density $\rho(r)$
everywhere in space. Given the polymer density and the polymer-
polymer long-range interaction (along the chain) $v(r-r')$, the excluded volume field can be calculated as usual by

$$V(r) = \int dr' \, v(r-r') \, \rho(r') \quad . \tag{14}$$

The self-consistent field (SCF) procedure then evaluates the polymer
density by considering diffusion in the self-consistent field $V(r)$
of Eq. (14). The difficulty of solving the SCF equations arises
from the fact that $\rho(r)$ must contain fractional powers of r, etc.,
and it would therefore be quite unlikely if the solution could be
found unless these "critical exponents" and the general analytic
structure of $\rho(r)$ were known (they presently are not). By analogy
it should then be expected that there be a SCF theory for the electron problem. Halperin and Lax and Zittartz and Langer employed
such a SCF for the density of states of the deep traps in this
model (12). Their approach must be generalized to treat the localization probability $p_o(E)$ and to analyze the nature of the electronic states near the mobility edge (13).

There is one feature of the SCF which can immediately be deduced, its symmetry. The SCF for $G(ro;t)$ has $D_{\infty h}$ symmetry, cylin-
drical symmetry about the $o-r$ axis and a plane of symmetry bisecting
this axis. Thus the SCF procedure requires the solution of $c_{or}(t)$
for an arbitrary nonseparable field of $D_{\infty h}$ symmetry as an explicit
functional of that field. Then the field is determined by the SCF
equations. Such a problem is currently intractable unless $r=o$,
whereupon the SCF has spherical symmetry.

The SCF for $G(oo;t)$ is not of great interest, but its Fourier
transform Im $G(oo;E)$ is related to the density of states at o. The
SCF for Im $G(oo;E)$ may be obtained, but it differs from that for
the localization probability $p_o(E)$ by some uninteresting factors of
two in a fundamental scaling length (13). From the fact that the
SCF for $p_o(E)$ is spherically symmetric it can be automatically de-
duced that $|\phi_{SCF}(R)|$ is real and is less singular than $R^{-3/2}$ as
$R\to0$ and vanishes faster than R^{-1} as $R\to\infty$. In order to obtain the

solution for the SCF, it is necessary to have an explicit solution
for the eigenstates and energies for an arbitrary spherically symmetric potential as an explicit functional of the field. By analogy with the polymer problem, simple perturbation theory is useless,
so WKB techniques are chosen in preference to the diagram summations
of Zittartz and Langer since the criteria for the validity of the
WKB method are well understood.

The final result is as follows: If E=0 is chosen as the zero
of energy, the SCF is energy dependent and behaves for E<0 as in
Ref. (13)

$$\phi_{SCF} \sim - C(-E)^{1/9} R^{-4/3} \quad , \quad R \to 0 \tag{15a}$$

$$\sim + |D(E)| \exp(-\kappa R)/R^{-2} \quad , \quad R \to \infty \quad , \tag{15b}$$

where κ is $(-2mE/\hbar^2)^{\frac{1}{2}}$.

The localization probability is found to vary as in Ref. (13),

$$p_0(E) \propto (-E)^{13/6} \quad , \quad E \leq 0 \tag{16a}$$

$$= 0 \quad , \quad E \geq 0 \quad , \tag{16b}$$

and consequently the mobility edge occurs at the average energy
E=0 which is eminently reasonable for a mean field theory. Equation
(16a) implies that the localization probability tends to zero as the
mobility edge is approached from below. Although the density of localized states is increasing as $E \to 0^-$, the size of these states grows
as Eq. (13)

$$<R> \propto (-E)^{-2/3} \tag{17}$$

and off-sets the increase in the density of states. The resulting
Eq. (17) corresponds to a spherical average of the states near the
mobility edge. From percolation theory it would be expected that
the localized states below the mobility edge have a finite network
type structure (1). It should be noted, however, that White and
Anderson have shown that the large size (cf. Eq. (17)) of the localized states near the mobility edge contribute to the enhancement of
the diamagnetic susceptibility of amorphous semiconductors (e.g.,
As_2S_3 and As_2Se_3) relative to the corresponding crystals (13).

Some of the general features of the SCF can be understood from
purely physical considerations. The SCF for $p_0(E)$ is real. If the
SCF were constant throughout space, there would be no localized
states, and consequently it is not constant, but is spherically
symmetric as noted above. If we consider all of the possible configurations of the scatterers which can lead to localized states of

energy E, the point $\underline{o}$ will fall at various places inside these localized states. The scatterers will provide a potential that keeps
the electron within the confines of the localized state. When an
average is performed over all such scatterer positions, the net result must be an effective potential which localizes the electron
near the point $\underline{o}$. If the electron is moved a distance R from the
origin $\underline{o}$, the ensemble averaged probability that it has been moved
outside the confines of the localized states must increase as R
increases. Thus, the potential must be deepest at the origin o.
Given a SCF with a deep well for R=0 and the fact that the Gaussian
random potential is on the average zero everywhere in space, it is
quite natural for the SCF to become positive for large R in order
to "conserve probability." This also reflects the high probability
that for large R the electron has been moved outside the confines
of the overwhelming majority of the localized states at energy E.

As noted above, the SCF's for Im $G(\underline{oo};E)$ and $p_o(E)$ have identical analytic structures. Thus, in this case inferrences made concerning electron localization solely from Im $G(\underline{oo};E)$ and its SCF
would be correct although there is no rigorous reason known for
this to be so. There is therefore the possibility that information
concerning electron localization in disordered materials can be deduced from the analytic structure of $G(\underline{oo};E)$ even though rigor currently requires an analysis of at least $p_o(E)$ which is an average
of a product of two Green's functions.

V.   ELECTRON MOBILITIES

It is of prime importance to be able to also evaluate electronic mobilities in disordered systems. Although a full solution to
this problem cannot yet be presented for the random scatterer model,
some conclusions can be drawn. If the mobility corresponded to dif-
fusive type motion, the limit of $<|c_{or}(t|\{R_j\})|^2>_{Av}$ as $t\to\infty$ and $R^2 \propto t$
would enable the description of this mobility. Of course, near the
mobility edge t would probably scale as a different power of R.
Thus, let us consider the SCF describing this quantity. Since
$c_{or}(t|\{R_j\})$ breaks symmetry at two points, $\underline{o}$ and $\underline{r}$, this SCF must
have the full $D_{\infty h}$ symmetry. Even if semiclassical procedures were
invoked, this would necessitate the formal solution of the classical
equations of motion for the motion of a particle in a plane under
the influence of an arbitrary non-separable potential of $D_{\infty h}$ symmetry. This, of course, is one of the classic problems of celestial
mechanics. Some progress, however, can be made for the case of the
low lying levels, even though these states are expected to be localized and yield a vanishing mobility.

From the restriction to potentials of $D_{\infty h}$ symmetry it is pos-
sible to conclude that the SCF could tend to $-\infty$ as $|R|$ or $|r| \to 0$

and must asymptotically become 0 as $|\underline{R}-\underline{r}|$ or $|\underline{R}|\to\infty$. Using physical arguments similar to those which explain why the spherically symmetric SCF corresponds to a deep well about $\underline{o}$, it is also possible to deduce that the $D_{\infty h}$ SCF must also have deep wells about $\underline{o}$ and $\underline{r}$, and for the low lying states there should be a positive potential barrier in the neighborhood of $\underline{r}/2$. Such a result is in fact obtained if the planar problem is approximately solved along the one-dimensional "reaction coordinate" corresponding to the path of minimum potential energy which connects $\underline{o}$ with $\underline{r}$ (13). This result implies that the mobility of electrons in the lowest lying states must vanish as expected. (A mobility edge should appear near E=0, but the precise details in this region are not yet amenable to solution.) For the lowest lying states the SCF has pairs of degenerate gerade and ungerade states for r large enough. These degenerate g and u states can be combined to yield two separate degenerate levels on each of the potential wells surrounding $\underline{o}$ and $\underline{r}$. These resultant single well states (13) correspond to the model used by Mott to show that the a.c. conductivity $\sigma(\omega)$ vanishes as $\omega^2\ln\omega$ as the frequency $\omega\to 0$ (Ref. 2). It is also the ad hoc model of Lukes which was employed to describe hopping between two Zittartz-Langer type potential wells (19), and it also justifies the neglect of other wells (12) when determining the properties of the low lying states.

VI.   CONCLUSIONS

The presence of localized and extended electronic states is discussed as fundamental to the understanding of the generic properties of the electronic structure of disordered materials. By the use of two different quantum mechanical models, it was shown how a description of the transition between localized and extended states is reminiscent of the theory of phase transitions and critical points in fluids and magnets. As specific results, for the model of an electron in a system of dense, weak, random scatterers, it is shown that the localization probability, the ensemble averaged probability that electrons with energy E return to their starting position after an infinite time, vanishes as $(-E)^{13/6}$ as E tends to the mobility edge (at E=0) and the range of the localized states varies as $(-E)^{-2/3}$ near the mobility edge. These critical index behaviors support the conjecture that the mobility varies as $(E-E_c)^\alpha$, $\alpha>0$, above the mobility edge at E_c. However, more work is necessary to elucidate this point.

REFERENCES

1. For a review and further references see E. N. Economou, M. H.
    Cohen, K. F. Freed, and E. S. Kirkpatrick in Amorphous and
    Liquid Semiconductors, J. Tauc, ed., (Plenum, New York) to be
    published.

2. N. F. Mott and E. A. Davis, Electronic Processes in NonCrystal-
    line Materials (Clarendon Press, Oxford, 1971).

3. I. M. Lifshitz, Adv. Phys. 13, 403 (1964).

4. M. H. Cohen, H. Fritzsche, and S. R. Ovshinsky, Phys. Rev.
    Letters 22, 1065 (1969).

5. P. W. Anderson, Phys. Rev. 109, 1492 (1958).

6. J. M. Ziman, J. Phys. C. 1, 1532 (1968).

7. V.K.S. Shante and S. Kirkpatrick, Adv. Phys. 20, 325 (1971).

8. S. F. Edwards, J. Phys. C. 3, L30 (1970); J. Non-Cryst. Solids
    4, 417 (1970).

9. K. F. Freed and M. H. Cohen, Phys. Rev. B3, 3400 (1971).

10. F. Yonezawa and T. Matsubara, Progr. Theoret. Phys. (Kyoto)
    35, 357, 759 (1966); P. Soven, Phys. Rev. 156, 809 (1967);
    D. W. Taylor, ibid. 156, 1017 (1967); P. L. Leath, ibid. 171,
    725 (1968); B. Veclický, S. Kirkpatrick and H. Ehrenreich,
    ibid. 175, 747 (1968); P. Soven, ibid. 178, 1136 (1969); and
    S. Kirkpatrick, B. Velický and H. Ehrenreich, ibid. B1, 3250
    (1970).

11. S. F. Edwards and Y. B. Gulyaev, Proc. Phys. Soc. (London) 83,
    495 (1964).

12. B. Halperin and M. Lax, Phys. Rev. 148, 722 (1966); J. Zittartz
    and J. S. Langer, ibid. 148, 741 (1966).

13. K. F. Freed, J. Phys. C. 4, L331 (1971); Phys. Rev. B5, 4802
    (1972).

14. E. N. Economou and M. H. Cohen, Phys. Rev. B5, 2931 (1972).

15. K. F. Freed, Advan. Chem. Phys. 22, 1 (1972).

16. R. P. Feynman and A. R. Hibbs, Quantum Mechanics and Path
    Integrals (McGraw-Hill, New York, 1965).

17. S. F. Edwards, Proc. Phys. Soc. (London) 85, 613 (1965);
     Natl. Bur. Std. (U.S.) Misc. Publ. 273, 225 (1966).

18. R. M. White and P. W. Anderson, Phil. Mag. 25, 737 (1972).

19. T. Lukes (private communication).

7. Banquet Speech

ARE THERE CULTS OF THEORETICIANS?

George S. Hammond, Chairman

Division of Chemistry and Chemical Engineering

California Institute of Technology

At other times I have discussed the fact that international science as a whole has much of the character of a world-wide sub-culture and that the style and even many of the goals of scientific work are dictated by the mores of that culture. I think that my mandate in this symposium is consideration of the applicability of this principle to the fields of molecular and solid-state theory. To lay it on the line as an opener, I will state that I believe that theoretical science is filled with small, clubish cults. However, I hasten to add that this particular symposium is a great example of fine strategy for evading the restrictions imposed by small cults. Rarely, if ever, has a symposium brought together, as speakers, theoreticians with such disparate styles as Lionel Salem, Wilse Robinson, Paul Flory, Klaus Ruedenberg, and Karl Freed. The fact that spin waves in solids, transition states, inorganic reactions and statistical physics of macromolecules are all under discussion in one meeting is a remarkable accomplishment.

Having given much deserved plaudits to the organizers and participants in the symposium, I will return to the serious subject. There are cults in theoretical science, even in a subfield as narrow as molecular quantum mechanics. For example, Wilse Robinson once told me, "There are two kinds of theoretical guys, people who believe in wavefunctions and people who believe in Hamiltonians." Similarly, Roald Hoffman, who has made a sensational contribution to nonquantitative theory, feels that it is mandatory for him to keep his credentials alive in the area of theoretical calculations. In fact I have often heard Hoffman referred to as "really an organic theoretician." Similarly, I have heard my colleague, Bill Goddard, described as, "a sort of physicists' theoretician." Yet the interests of Goddard and Hoffman, in terms of the properties of molecules, are remarkably similar. Although I am hideously ignorant of the details, I have the strong impression that for a long time theories of electrical conductivity of solids were largely restricted to bands and continue with little regard for the short-range interactions that give rise to the phenomena. There has probably been a tendency for the solid state theoreticians to ignore short-range interactions and for the short-range experts to concentrate on molecules and ignore solids. The change in this situation is truly heartening.

I have another concern which affects the affairs of my own scientific club, the systematic modeling of reaction processes. In my opinion, most workers in this field have struggled for years with the handicap of very weakly interacting approaches to theory. Physical organic chemists have gathered enormous volumes of data concerning variations in the relative rates of reactions in solution, including much highly ingenious experimental dissection of complex reaction mechanisms. The results have usually been analyzed in terms of semiempirical theory based upon concepts generated by the Eyring-Polanyi theory of rate processes and a sort of mystical electron-pushing model generated many years ago by Ingold and elaborated by many others. It is my own opinion that Eyring theory, as it is usually used, is not very effective except as a source of vocabulary for discussion of reactivity relationships.

The situation is not necessarily undesirable but there have been some undesirable side effects. For many years theoreticians have attempted to improve the situation, often with disappointingly unproductive results. Usually people have simply used the methods of semiempirical molecular quantum mechanics to calculate molecular parameters which are solely dictated by the semiempirical notions of physical organic theory, for example, electron densities, bond orders, free valence indices, etc. I am oversimplifying, but I believe that the effort has largely been abortive because it has been an attempt to make dynamic processes a simple corollary of structural chemistry. The process of conversion of one chemical species to another should, in principle, be expressible in terms of initial and final wave functions and dynamic operators which generate finite matrix elements giving the probability amplitude for the molecular transition.

It is my feeling that an enormous amount of effort has been put into generation of molecular wave functions of varying quality. On the other hand, the operators are left to intuition and hope. At the same time another small group of theoreticians, doing scattering theory and trajectory calculations, have been trying to deal directly with chemical reactions as inelastic scattering processes. Unfortunately, I have heard members of the molecular quantum mechanician's fraternity speak of scattering theory as silly little exercises in ballistics.

Provincialism in theoretical science, as anywhere else, is harmful for two reasons. First, people are inhibited from testing their ideas in new areas. Second, there are many times when important new perception goes unnoticed because there is stylism in perception as well as reception. People tend to hear from a man what they expect to hear from him, and they usually expect to hear whatever is the usual line of the members of his club. For example, the Woodward-Hoffmann Rules for prediction of preferred courses of reactions really were a brilliant contribution to the slowly evolving new theory of chemical dynamics and the alternative formulations of the same principles in alternate forms by other people have also led to growth in insight. However, Fukui had for many years been developing the so-called "frontier electron theory" which elucidates most of the same ideas. I believe that Fukui's ideas were ignored by many people, certainly including myself, because we expected a particular brand of theory from someone associated with semiempirical molecular orbital theory.

Because numerical calculations are a powerful test of the utility of theory, numerology has become a fetish of the cult of theoretical science. I am particularly sensitive about this matter because it touches on my own tenuous position in science. When I am classified, I am called an experimentalist. This bothers my conscience a good deal because it has been many years since I have done an experiment with my own hands. I am associated with a research group, some of whom do very nice experiments. I think about the experiments and daydream about the phenomenology of chemical reactions. Yet I can not join any of the various theoretical clubs because I usually express my ideas in words rather than in mathematical statements. Furthermore, when I do put something in mathematical form, I always crib the equations from someone else. When I am honest with myself, I realize that I have no credentials for membership in a theoretical club and that to claim membership in the experimental fraternity is arrant hypocrisy. Yet, I still feel that I am a chemical scientist.

Having taken time to bleed a little for my own plight, I will return to more significant matters. The mathematical ritual of theoreticians can be both powerful and emasculating. Mathematics is a tool, which is often used as a status symbol. There can be a beautiful compactness in a mathematical statement, but premature forcing of a physical concept into mathematical form

can also destroy a new thought or convert mediocre theory into
hopelessly bad theory.

Interestingly, the worship of mathematics increases as
scientists move away from the fields where mathematics is most
immediately valuable. Murray Gell-Mann used mathematical statements in describing the "eightfold way," but few people have
complained that he also uses verbal exposition extensively and
does very little numerical evaluation. In chemical and solid-
state theory a much higher premium is placed on computation; and
my friends in economics tell me that the name of the game is
casting theory into forms which allow one to generate second
derivatives. The heretics in economics tell me that the drive
for second derivatives creates more economic theory than does the
whole economic experience of mankind. Remember, these are the
heretics, and I imagine that they give me a distorted view of
microeconomic theory. However, the mere existence of heretical
views attests to the existence of an orthodoxy!

Allow me a final point. I have spoken of the existence of
cultism in theoretical science and mentioned some of the ways in
which cultish behavior may inhibit growth of the field. However,
I have the feeling that science could not exist at all without its
cults. The intensive short-range interaction between people in
the same field, a sense of unity and mutual appreciation of non-
spectacular progress, and the self-discipline imposed by a demanding subculture all provide guidance and security in maintaining
our day-to-day activities. There are many times when a rigid
style allows us to get into a day's work efficiently, without
unnecessary spinning of the wheels while we choose among myriad
conceivable approaches to the day's work. I believe that we
profit from the existence of our little cultures as long as we do
not allow them total mastery over our minds, and as long as we do
not make them the tools for divisive struggle among scientists who
should interact constructively. Once again I congratulate Dr.
Herman and his coorganizers for creating a symposium consistent
with these principles.

8. Symposium Summary

SUMMARY AND CONCLUDING REMARKS

R. K. Nesbet

IBM Research Laboratory

San Jose, California 95114

This conference has covered a great deal of ground, bringing together people from both theoretical chemistry and solid-state physics. The main theme of common interest is the calculation of electronic wave functions for complex systems. Another important theme has been model building in general, the subject of several interesting papers. In many cases this becomes a study in statistical mechanics, but the parameters must be obtained from electronic wave functions. My remarks will be concerned primarily with the choice of methods for computing electronic wave functions.

These methods can be classified according to the two dimensional diagram shown as Table 1. One of the coordinates is the degree of sophistication. The sophistication coordinate runs from naive, through practical, to sophisticated. Naive methods are used the very first time that any attempt is made to apply theory; such methods are discovered in retrospect not to work out very well. After the naive methods are found to be too simple, practical methods evolve that really do work and give results. Sophisticated methods are often practical methods in the course of evolution, sometimes bright ideas that need detailed applications to test cases to try their merit, and sometimes methods of such complexity that they tend to be used only by specialists.

At right angles to this, particular methods for wave functions of molecules, hopefully large molecules, and for localized states in solids, can be put into roughly three categories, which have already been talked about at this conference. The first category, semiempirical methods, is most dependent on parametrization. In

general, the parameters are matrix elements either of a model
Hamiltonian, or of parts of the true electronic Hamiltonian.

The Hückel theory is the typical naive matrix semiempirical
method. People in chemistry who work in this field consider it to
be an oversimplification.

Table 1. Computational methods for electronic structure

	Semiempirical	One-electron models	Ab initio
Naive	Hückel	Free electron	Minimum basis LCAO
Practical	Extended Hückel, modified for charge distribution Pariser-Parr-Pople CNDO(INDO) LEDO	Band theory, OPW,APW,KKR Scattered wave Xα Pseudopotentials	Split shell & polarization Direct integration SCF & empirical correlation
Sophisticated	Integrals modified by correlation Inclusion of polarizabilities, dielectric constant	Correlation "potential" Green's function, response function	Selective CI Bethe-Goldstone hierarchy Large orbital basis, pushed to convergence

The second broad category of methods indicated in Table 1 is
one-electron models. A typical, practical, everyday usable one-
electron model is the energy-band theory of solids. It comes in
various forms, with different ranges of application. The APW and
KKR methods work well for metals, but depend on the use of muffin-
tin potentials. The OPW method is less efficient, but can get by
without the muffin-tin approximation, and therefore is more suitable for things that are actually held together by chemical bonds,
such as semiconductors. At the present time, a great deal of
effort is being devoted to improving and generalizing these band
structure methods. Recent versions of the KKR and APW methods no

longer are limited to muffin-tin potentials, and recent versions of
the OPW method are applicable to transition metals as well as
semiconductors, and are considerably more efficient computationally
than earlier versions.

The final category of methods is <u>ab initio</u>, referring to a
method that starts from the basic structure of quantum theory and
cranks away on it, eventually to achieve convergent quantitative
results. Presumably, the purpose is not in all cases to come in
after an experiment has been done and compute the same number, but
ultimately to deal with a situation so complicated that experimental
numbers don't exist. Or possibly, where experimental numbers are
so plentiful that there is no hope of interpreting them, to provide
a definitive calculation that helps sort out the data.

As typical sophisticated <u>ab initio</u> methods I list selective
configuration interaction, or the particular procedure that uses a
hierarchy of Bethe-Goldstone calculations. This gives us an
ultimately convergent procedure for computing any property of any
bound state of any atom or molecule, if only we could afford to do
the computations. For large molecules, this must be put off some
time into the future.

The free-electron theory is an example of naive one-electron
models. "Naive" here refers to the physics of the model, not to
the formal mathematics needed to work with it. It has been applied
in various forms to molecules and found to be comparable to Hückel
theory. For solids, the free-electron theory applies to too
simplified a special situation. It doesn't apply directly to real
solids, but has a certain qualitative validity for simple metals
that is best rationalized in terms of pseudopotentials.

In <u>ab initio</u> methods, the minimum basis calculations were
obviously the first to be done. They did not work very well for
diatomic molecules, where better calculations have been done, and
there is no reason to expect large molecules to be better behaved
than diatomics. In a minimum basis LCAO calculation, molecular
orbitals are made up as linear combinations of atomic orbitals,
one for each inner shell or valence orbital in the free atom. What
comes out is not very useful with regard to the structure and
properties of molecules, because the distortion of atomic valence
orbitals due to charge displacement and the nonspherical environment
of each atom is ignored.

The point of impact on real interests of chemistry and physics
comes with the practical methods. An excellent lecture by Zerner
summarized the practical semiempirical methods. The simplest and
most widely used is extended Hückel theory, not just Hückel theory
including inner shells, but including a bit of the physics of

electronic interactions. Electrons repel each other, so that
matrix elements must be modified when the electronic charge density
piles up at one end of a molecule. This begins to look like the
self-consistent field, very important in a molecule or solid because
electrons can move about in a molecule, and can move within the unit
cell of a solid. To represent the resulting change in interaction
energy you must have Coulomb integrals, two-electron integrals, and
you can't just diagonalize a matrix. The non-uniform Coulomb
interaction depends on the orbitals and occupation numbers in a
self-consistent way.

The two-electron integrals are parametrized directly in various
forms of Pariser-Parr-Pople theory, so that movements of charge
throughout a molecule can be followed in detail in a matrix Hartree-
Fock formalism. The popular acronyms here are CNDO, INDO, etc.,
for various ways of neglecting overlap charge densities in the two-
electron integrals.

Another class of methods, approximate non-empirical methods,
is typified by LEDO (limited expansion of diatomic overlap). Here
the full structure of an ab initio calculation is used, but two-
electron integrals are in part estimated and in part taken as
empirical parameters. The simple estimate of zero for overlap
charge densities is replaced by formulas drawn from various expansion
procedures, where the quantitative validity of the approximation
can be judged from accurate computations.

A practical ab initio method that is both systematic and
efficient was expounded by Pople. A limited orbital basis set is
used in an ab initio matrix Hartree-Fock calculation. If the
orbital basis is extended to completeness, this moves into the area
of sophisticated computations, which McLean talked about. This is
where we are getting really quantitative results. But if we set up
a working approximation, the physical essence of ab initio calculations at the Hartree-Fock level is that electrons are aware of the
existence of other electrons, and of the nuclear framework of a
molecule. The electrons can move from one atom to another. This
is already possible in naive, minimum basis LCAO, but if you put
an atom next to another atom it's no longer isolated; the atom is
in a very strongly polarized and polarizing environment. If you
move two atoms together the valence orbitals, which are most easily
polarized, are going to change their shape.

This can take place in two ways, both extremely important in
molecules. When we first began to do ab initio calculations
seriously we found that this had to be taken into account in order
to get reasonable results. But if done to a very modest extent,
the results are very good. The first step is double-zeta or split
valence shell orbitals, using two basis functions for each atomic

valence orbital. This is one of the features of Pople's method.
It allows for radial distortion when an atom is in its molecular
environment, where its net charge is somewhere on the scale of
negative ion, neutral atom, or positive ion. The outer valence
orbitals must expand or contract according to this net charge,
amounting to radial distortion. In addition, an atom in a molecule
is in a nonspherical environment. This must change the nature of
the angular dependence of the valence orbitals, requiring an
admixture of spherical harmonics not found in a free atom. Because
of this a molecular basis set must include what are called polarization functions. In particular, p-orbitals are added to hydrogen
and d-orbitals to the valence basis sets of first-row atoms, to
describe non-spherical distortion of the electronic charge cloud.
In ab initio calculations, very good results are achieved with only
a few physically obvious polarization functions. The method
described by Pople puts these elements together systematically,
using Gaussian orbitals as the underlying basis so that all
integrals needed for general polyatomic molecules can be evaluated.

     Pople's method stays within the framework of the Hartree-Fock
approximation, as does the practical method of direct integration
of multidimensional integrals currently being exploited by Ellis
and Freeman at Northwestern, but not mentioned at this conference.
Because these methods are limited to Hartree-Fock, they cannot deal
with the details of electronic correlations. The crucial thing
about electronic correlation is that whenever you make or break a
bond, when you change the number of pairs of electrons or drastically change their structure, correlation energy is gained or lost.
However, the charge density distribution of a molecule is very well
determined in the Hartree-Fock approximation for ordinary stable
bonded molecules. Hence molecular conformations, that is, bond
lengths and bond angles, and other properties primarily determined
by the charge density, are obtained to useful practical accuracy
in the Hartree-Fock approximation. So Pople is using the Hartree-
Fock approximation with orbital basis sets that contain all that's
necesary, but no more, to compute bond angles and bond lengths.
But he leaves out correlation effects.

     An important unresolved question is whether a practical method
can do better than this. If we wish to contribute to a practical
theory of excitations, for photochemistry, or spectroscopy in
general, then the theory has to cope with significant changes of
correlation energy. But it turns out that you can parametrize the
correlation energy in some straightforward and simple ways that
are based on our knowledge of correlation energy in the valence
shells of atoms. We can now do very accurate direct calculations
of correlation energy corrections for atoms. We also know
empirically pretty well what these numbers are. The combination
of a molecular Hartree-Fock calculation with an empirical treatment

of correlation energy, largely in terms of electron pair energies,
turns out to be a viable practical way to deal with molecular
excitations. Quite a bit of such work has been done, particularly
for diatomic molecules.

These ideas might lead to a viable method for solid-state
theory. A good energy band calculation could be corrected in terms
of the correlation energy of localized electron pairs, and then
should be good enough to treat the energy differences involved in
excitations. Wannier functions or the Lennard-Jones kind of
localized orbitals described by Ruedenberg for organic molecules
may be the best line to pursue for parametrizing correlation
energies in solids.

A new practical method has recently emerged, based on experience in solid-state theory with the one-electron model. This is
the scattered wave $X\alpha$ method, presented here by Johnson. This
method applies the KKR formalism to molecules, with ideas that tend
to go with this and the APW method for energy bands, including the
muffin-tin potential and a statistical treatment of electronic
exchange. The statistical exchange is crucial to the formalism, but
somewhat hard to swallow for anyone familiar with the detailed
treatment of antisymmetric wave functions in atomic and molecular
theory. But very recently work done here by Herman and Ortenberger
has shown that the arbitrariness of the parameter α, which multi-
plies the statistical exchange potential depending on electron
density, can be eliminated by including an additional parameter β,
multiplying a potential depending on the density gradient. The
resulting parameters α and β appear, from a study of atoms
ranging through the periodic table, to be something like universal
constants. So the scattered wave $X\alpha\beta$ method, which someone will
try in the near future, may be free of any arbitrary parameters in
the exchange potential.

The scattered-wave $X\alpha$ method, as a one-electron model, is
in the context of the Hartree-Fock approximation. It should not
be expected to get results that depend on correlation energies.
Nevertheless, because of the statistical model, the method can
follow continuous changes of occupation numbers from one state to
another. If the one-electron energies are considered to be derivatives of total energy with respect to occupation numbers, then
energy differences are obtained by the trapezoidal rule in terms
of the transition state idea introduced by Slater. The transition
state is defined by occupation numbers half way between the two
states whose energy difference is required. Very good results
using the transition state idea have been reported here by Johnson,
especially for the inner shell ionization energies measured in ESCA
experiments. These results show that the transition state method

accounts very well for relaxation energies, but correlation energy differences are not explicitly included in the formalism.

One very important feature of the scattered wave $X\alpha$ method is that the orbital wave functions within a muffin-tin cell are directly integrated, not obtained as an expansion in basis orbitals. In this context, it is not much more difficult to use Dirac's equation than to use Schrödinger's. This fact has been exploited in solid-state theory, where relativistic band calculations are standard practice. But there has been almost no work in molecular theory, either semiempirical or ab initio, with relativistic systems. This means that the scattered wave $X\alpha$ method opens up a whole new range of heavy-atom molecules for theoretical calculations. It can be expected to become a very useful technique.

Looking to the future, what might be looked for in a sophisticated semiempirical method? Here I'm putting in some of my own thoughts, although they're not necessarily original. In fact, one important idea, that electronic correlation can be parametrized by modifying the two-electron Coulomb interaction integrals, was expressed in the first publications of the Pariser-Parr theory. This goes back to Brueckner's theory of nuclear structure where the two-nucleon interaction is modified by correlation to get rid of the effect of a hard-core interaction. So in a sophisticated semiempirical theory, the two-electron integrals should be modified by electronic correlation. In work not discussed here, Freed has considered such a systematic rationalization of semiempirical theory in a formal way. If the modified integrals can be fitted from experimental data, then the theory inherently goes beyond Hartree-Fock.

Unfortunately, there are so many different integrals that unique values are difficult to establish. The theory has too many free parameters. What is required is a formal theory of the parameters that will allow computed data on atoms and diatomic molecules to be combined with empirical data to produce parametrized integrals suitable for specific applications.

For larger systems, polarizabilities should be included in the empirical input to a semiempirical scheme. This has been done by Little, in work not discussed here, dividing Coulomb integrals by a dielectric constant. In formal theory this is a correlation effect, but the asymptotic form can be represented by a dielectric constant. To my knowledge this was first done by Kuhn, working with the free-electron theory of molecules more than a decade ago.

The same thing should be true in one-electron models. Static potentials should be modified to include polarizability effects or

dielectric constants.   In solid-state theory this is not a new idea.
One would also like to include effective one-electron potentials
that somehow account for correlation effects.   It is not at all
clear that this can be done within the context of the one-electron
model, because the formal theory in terms of coupled Green's functions leads to integral operators rather than just to potential
functions.   However, if the exchange integral operator can be
replaced by a function of the electronic density, perhaps a correlation operator can be approximated in the same way.

As a final point, Taylor and collaborators, in work not
formally on the program, have presented some very interesting
results using the idea of a one-particle Green's function coupled
to an RPA response function for the helium atom.   Electron-helium
scattering is described in quantitative detail, together with
excitation energies and the frequency-dependent polarizability of
the atom.   All of this comes from one calculation.   This Green's
function-response function formalism maps all dynamical properties
of the atom onto a one-electron model.   This formalism might provide the ultimate sophisticated generalization of one-electron
models.

This concludes my summary of methods that have been talked
about, that should have been talked about, that are important now,
or may be important in the future.

LIST OF CONTRIBUTORS

Anderson, P. W., Bell Laboratories and Cavendish Laboratory, Cambridge University

Bagus, P. S., IBM Research, San Jose

Barker, J. A., IBM Research, San Jose

Batra, I. P., IBM Research, San Jose

Brower, K. L., Sandia Laboratories

Connolly, J. W. D., University of Florida

Corbett, J. W., State University of New York at Albany

Eggarter, T. P., University of Chicago

Erdős, P., Florida State University

Eschenfelder, A. H., IBM Research, San Jose

Freed, K. F., University of Chicago

Frisch, H. L., State University of New York at Albany

Gilbert, T. L., Argonne National Laboratory

Hammond, G. S., California Institute of Technology; presently at University of California at Santa Cruz

Harris, F. E., University of Utah

Harrison, W. A. Stanford University

Henderson, D., IBM Research, San Jose

Herman, F., IBM Research, San Jose

389

Herndon, R. C., Nova University

Jepsen, D. W., IBM Research, Yorktown Heights

Johnson, K. H., Massachusetts Institute of Technology

Keller, J., Universidad Nacional Autónoma de México

Kirkpatrick, S., IBM Research, Yorktown Heights

Kohn, W., University of California at San Diego

Little, W. A., Stanford University

Liu, B., IBM Research, San Jose

Marcus, P. M., IBM Research, Yorktown Heights

McLean, A. D., IBM Research, San Jose

Messmer, R. P., General Electric Research and
    Development Center

Mueller-Westerhoff, U. T., IBM Research, San Jose

Mulliken, R. S., University of Chicago

Nesbet, R. K., IBM Research, San Jose

Norman, J. G. Jr., Massachusetts Institute of
    Technology and University of Washington
    (permanent address)

Ortenburger, I. B., IBM Research, San Jose

Peak, D., State University of New York at Albany

Pople, J. A., Carnegie-Mellon University

Robinson, G. W., California Institute of Technology
    (permanent address) and University of Canterbury,
    Christchurch, New Zealand

Ruedenberg, K., Iowa State University

Salem, L., Harvard University and Université de
    Paris-Sud, Orsay (permanent address)

Segal, G. A., University of Southern California

Seki, H., IBM Research, San Jose

St. Peters, M., State University of New York at Albany

Taylor, H. S., University of Southern California

Thorpe, M. F., Yale University

Vook, F. L., Sandia Laboratories

Watkins, G. D., General Electric Research and
      Development Center

Weaire, D., Yale University

Weeks, J. D., Cavendish Laboratory, Cambridge University

Weissbluth, M., Stanford University

Yoshimine, M., IBM Research, San Jose

Zerner, M. C., University of Guelph, Ontario, Canada